In the last few years, important changes have taken place in the field of x-ray and radium physics. Most the material in this new, completely revised and up-dated edition has been rewritten to simplify and clarify further the important physical concepts, although the fundamentals have undergone relatively little revision. Many of the illustrations have also been revised or completely redrawn.

New sections have been added, covering:

—*Solid state rectification*

—*Newer types of mobile apparatus, such as battery-operated, capacitor-discharge, and field-emission units*

—*Special tubes for mammography, using molybdenum targets and filters*

—*Xeroradiography*

—*Saturable reactor for control of filament current*

Despite the many changes, the format has been kept essentially intact. However, the following areas have been expanded:

Improvements in x-ray tube design

Radionuclides

Irradiation therapy

Radiographic quality with inclusion of some of the newer concepts

Tomography in greater detail, with more complete discussion of principles and comparison of various systems

Nuclear medicine, including many of the newer radionuclides

Health physics in greater depth, with more details concerning radiation protection.

The Fundamentals of
X-ray and Radium Physics

(Fifth Edition)

The Fundamentals

of X-ray and

Radium Physics

By JOSEPH SELMAN, M.D.

Clinical Assistant Professor of Radiology
University of Texas, Southwestern Medical School
Director, School of Radiologic Technology
Tyler Junior College
Mother Frances Hospital
Medical Center Hospital
Attending Radiologist
Mother Frances Hospital
Attending Radiologist
Medical Center Hospital
Consultant in Radiology, East Texas Chest Hospital
Tyler, Texas

CHARLES C THOMAS · PUBLISHER
Springfield · Illinois · U.S.A.

Published and Distributed Throughout the World by
CHARLES C THOMAS • PUBLISHER
Bannerstone House
301-327 East Lawrence Avenue, Springfield, Illinois, U.S.A.

First Edition, First Printing, 1954, 3000 copies
First Edition, Second Printing, 1955, 3000 copies
Second Edition, First Printing, 1957, 4000 copies
Second Edition, Second Printing, 1959, 4000 copies
Second Edition, Third Printing, 1960, 3000 copies
Third Edition, First Printing, 1961, 4000 copies
Third Edition, Second Printing, 1962, 4000 copies
Third Edition, Third Printing, 1963, 4000 copies
Third Edition, Fourth Printing, 1964, 4000 copies
Fourth Edition, First Printing, 1965, 10,000 copies
Fourth Edition, Second Printing, 1967, 5000 copies
Fourth Edition, Third Printing, 1967, 5000 copies
Fourth Edition, Fourth Printing, 1968, 7000 copies
Fourth Edition, Fifth Printing, 1969, 5000 copies
Fourth Edition, Sixth Printing, 1969, 5000 copies
Fourth Edition, Seventh Printing, 1970, 5000 copies
Fourth Edition, Eighth Printing, 1970, 5000 copies
Fourth Edition, Ninth Printing, 1971, 7500 copies
Fourth Edition, Tenth Printing, 1972, 2000 copies
Fifth Edition, First Printing, 1972, 10,000 copies

With THOMAS BOOKS *careful attention is given to all details of
manufacturing and design. It is the Publisher's desire to present books
that are satisfactory as to their physical qualities and artistic possibilities
and appropriate for their particular use.* THOMAS BOOKS *will be true
to those laws of quality that assure a good name and good will.*

Printed in the United States of America
P-4

Dedicated to my wife

PREFACE TO FIFTH EDITION

In the seven years since publication of the Fourth Edition, important changes have taken place in the design of x-ray equipment, although fundamental concepts have undergone relatively little revision. Accordingly, new material has been introduced covering such subjects as solid state rectifiers; newer mobile apparatus, including battery powered and capacitor-discharge units; field emission tubes; special tubes for mammography, using molybdenum targets and filters; xeroradiography; and saturable reactor for control of filament current.

Many of the illustrations have been revised or completely redrawn, and corresponding changes have been made in the descriptive legends.

Despite the many changes, the format has been kept essentially intact. However, certain areas have been expanded; for example, improvements in x-ray tube design; tomography; radiographic quality; radionuclides and nuclear medicine; irradiation therapy; and health physics.

Throughout the book, many sections have been rewritten on the basis of the author's further teaching experience, with a view toward improving the explanation of fundamental principles and the description of devices and equipment. Wherever possible, the aim has been in the direction of greater simplification. The material has been generally updated on the basis of the most reliable data available at the time of revision.

Credit is again due to the previously mentioned sources. Additionally, we would like to thank the Field Emission Corporation, the Machlett Laboratories, the Dunlee Corporation, the Xerox Corporation, and the CGR Medical Corporation for their

vii

kindness in supplying important material relative to their products.

The author again wishes to express his gratitude to Charles C Thomas, Publisher, for providing the opportunity of preparing this new edition.

J.S.

PREFACE TO FOURTH EDITION

In order to keep abreast of new developments in the field of radiologic physics, especially those concerning the radiologic technologist and the radiology resident, a number of significant changes have been introduced into this, the Fourth Edition. Only the more important ones will be mentioned here.

Again, the format will be retained intact, all changes being made within its framework. Outdated material, such as mechanical rectification, has been deleted.

The concept of electrical field and the relationship between static electricity, electric discharge, and current electricity have been given greater emphasis. The factors in series and parallel circuits, often difficult for the student technologist, have been discussed more fully with the aid of numerical examples.

X-ray quantity and quality have been described in greater depth, and more attention given to half value layer, including a summary of the methods of measuring it. The determination of tumor exposure and absorbed dose is explained more completely and is based on the latest method of converting the data in published depth dose tables so that they become applicable to any particular therapy unit. In accordance with recent changes in terminology proposed by the International Commission on Radiologic Units and Measurements (ICRU), the Capital R now designates the exposure in roentgens. The exposure rate is indicated as R/min or R/sec.

Due to lack of clarity, generally, concerning saturation current as it applies to the modern diagnostic x-ray tube, this subject is covered more fully than before.

In view of the wide use of automatic processing of x-ray films, mainly as the result of improvement in quality and reduction in cost, more space has been devoted to the principles and construction of this type of equipment. In addition, the basic theory of photographic image formation is presented in greater detail for those students who seek a deeper understanding of this subject.

The chapters on radiographic quality and its control have been rearranged and largely rewritten. Radiographic grids are discussed in greater depth, the latest methods of designating grid quality presented, and newer terminology used throughout in accordance with the recommendations of the ICRU.

Under special procedures, bright fluoroscopy (image intensification) have been emphasized because of its wide acceptance by radiologists. Stereoscopic viewing of radiographs has been presented in a much more fundamental manner than before.

A completely new chapter on radioactive isotopes has been added. It is not meant to replace standard textbooks devoted exclusively to this subject, but rather to provide a well-rounded background for the medical use of radioisotopes for the general radiologic technologist. The chapter on health physics has been expanded to include the more commonly used radioisotopes.

It must be pointed out, that insofar as possible, the latest terminology is used throughout this book, conforming to the recommendations of the ICRU, and the International Union of Pure and Applied Chemistry and its twin in Physics. For example, the form "x ray" refers to the noun, whereas "x-ray" designates the adjective; thus, x rays, but x-ray physics. The form ^{131}I instead of I^{131} exemplifies the designation of radioisotopes. Attempts at international standardization of scientific terms are highly commendable in this age of automatic retrieval of information by means of computers.

The author again acknowledges with thanks the cooperation of a number of commercial suppliers in furnishing important data about their products. In addition to those mentioned in earlier editions are: Picker X-Ray Corporation; Liebel-Flarsheim Company; United States Radium Corporation; General Aniline and Film Corporation; and Smit-Roentgen Company.

Grateful recognition is due Charles C Thomas, Publisher, for interest in the preparation of the Fourth Edition, and for the excellence in publication of previous editions.

J.S.

PREFACE TO THIRD EDITION

ONLY FOUR YEARS have passed since the publication of the Second Edition, but a number of changes, some major and some minor, are considered to be of sufficient importance to warrant a fairly thorough revision. This is dictated in part by changing concepts, and in part by the author's further experience with students of widely different degrees of ability and educational background.

The format and figures have been kept essentially intact, although some figures have been revised and several new ones added. Every effort has been made to retain the simplicity of the first two editions. A decimal system of numbering the illustrations has been introduced, a practice which is gaining acceptance in scientific books.

One of the major changes in the present edition is in the chapters on the electric current, electromagnetism, and electric generators and motors. Whereas the conventional direction was emphasized previously, this has now been relegated to history, and all the directional rules have been altered to conform to the flow of electrons. This step was taken only after considerable deliberation, but was finally adopted in compliance with the vast majority of present-day physics textbooks. For the same reason, the term *condenser* has been changed to *capacitor*.

The theory of magnetism has been brought up to date, with the introduction of the concept of atomic magnets and magnetic domains. Otherwise, the same simple approach to magnetic phenomena has been retained.

There have been no radical innovations in the field of x-ray equipment generally, but certain theoretical aspects of this subject have been rewritten in an effort to improve their comprehensibility.

The sections on radiation dosage have been completely revised and expanded, although the student should be encouraged to read more advanced material on this subject, especially during his second year. The modern concept of absorbed dose and its unit, the rad, have been described in some detail. Filtration and radiation quality have received increased attention.

In the sections on radiography, special attention has been devoted to intensifying screens. Included are revised concepts of the relative importance of the effect of crystal size and active layer thickness on image sharpness; the intrinsic and extrinsic factors affecting screen speed; and comparative data on medium and high speed screens.

Because of the increasing interest in automatic processing of films, a section has been included to summarize present information in this important field.

The chapter on radiographic quality has been carefully reworked. The more accurate term *definition* has been substituted for *detail* to indicate image sharpness. The factors influencing definition have been reclassified and clarified. In the chapter on devices for improving radioactive quality, emphasis has been placed on modern collimators in preference to conventional cones, especially with regard to patient protection. The newer types of Bucky mechanisms have also been described.

In view of the recent downward revision of the maximum permissible dose, the chapter on radiation protection has been reoriented accordingly. Emphasis has also been placed on radiation monitoring methods for radiologic personnel. Furthermore, the current agitation regarding the exposure of the population to ionizing radiation has led to a more detailed treatment of the various methods that can be utilized to reduce patient dosage in radiography and fluoroscopy.

Finally, answers to the sample problems at the end of certain chapters have been included in the Appendix as an aid to the student in checking the correctness of his solutions.

As before, the author expresses his sincere appreciation to Mr. Charles C Thomas, Mr. Payne Thomas, and their competent staff for their interest in the preparation of the Third Edition.

J.S.

PREFACE TO SECOND EDITION

Although great advances are continuously being made in Physics, the basic concepts required for the instruction of student x-ray technicians and Radiology residents have changed very little since the publication of the First Edition. Nevertheless, new data which are considered to be of sufficient interest and importance have been incorporated in the Second Edition.

The original format has been preserved almost intact because it has proved successful in actual classroom instruction in numerous schools of x-ray technology. Simplicity is retained throughout, and abundant line diagrams are employed again to clarify the explanation of basic principles and the description of equipment.

The introductory chapter on Mathematics now includes the notation of large and small numbers as powers of 10. Terminology applied to the factors in an electric current and the heating effects of currents has been revised and simplified.

Brems radiation has received more emphasis in the discussion of x-ray production. The descriptive material relating to the interaction of radiation and matter has been enlarged to include pair production.

Fluoroscopic image intensification is introduced and described in detail, since this device and its future modifications promise to revolutionize fluoroscopy in the next few years.

More space has been allotted to the inverse square law, a simplified formula being presented for use in protection and therapy problems. Considerable revision of the chapter on radiation protection was necessitated by recent changes in the recommendations of official bodies. Emphasis is placed also on the roentgen dosage received by patients during various diagnostic procedures. Obsolete data have been eliminated.

Finally, the Bibliography has been expanded to include additional choice references for the more venturesome student.

In addition to the acknowledgments already made in the First Edition, the author wishes to thank the Westinghouse Electric Corporation and the North American Philips Company for valuable data on image intensification.

Again, the author appreciates the interest shown by the publisher, Mr. Charles C Thomas, in making possible the Second Edition.

J.S.

PREFACE TO FIRST EDITION

I<small>T IS OBVIOUS</small> to anyone who has had experience in teaching radiologic physics to student x-ray technicians that the majority of students have had poor preparation for this course of study. Those students who may have been exposed to physics and mathematics in high school often retain so little knowledge of these subjects that they must learn anew even the simplest principles. The student x-ray technician with a background of college training in the sciences is indeed a rarity.

Many teachers devoted to the training of student x-ray technicians have long recognized the dearth of textbooks in physics designed for technicians. A desirable textbook of this type must lead the student, as painlessly as possible, from the most elementary considerations all the way to radiologic physics, a task which is by no means simple. The subject matter must be presented in great detail in order that the student may derive a more complete understanding of it. At the same time, the text must be as nonmathematical as possible and couched in language that is readily comprehensible.

On the basis of lecture notes and other data used in the instruction of student technicians over a period of years, the author has written the present book with the primary purpose of simplifying for the student x-ray technician the subject of radiologic physics in all its theoretical and practical aspects as it confronts the technician. With this aim in view, basic principles of physics and chemistry are emphasized and presented in greatest possible detail. At the same time, the simplest available terminology is employed in order to avoid a language barrier to the comprehension of these important principles. In accordance with the familiar pedagogic rule of repetition, the more significant physical principles are repeated, presented from various standpoints, and correlated wherever possible throughout the text.

Recognizing the value of visual aids to education, the author has made free use of numerous line drawings to facilitate the study of the text. Tables have been included only to emphasize certain points rather than as a source of reference for specific data. Throughout the book, fundamental principles are stressed, and if these are grasped by the student, he is then in a better position to understand the specific data furnished by the manufacturer for the equipment or supplies in use in his radiology department.

Mathematics of the simplest type is employed only where deemed essential to the understanding of fundamental principles. A summary of elementary mathematics introduces the book so that the student's memory can be more easily refreshed. It embodies all of the mathematical principles that are to appear in later chapters, so that if the first chapter is mastered the student should encounter no difficulty with the algebra and geometry employed in subsequent chapters.

There has been included a somewhat detailed consideration of the intimate structure of matter, and some space has been devoted to the quantum theory. This may be criticized by those teachers who feel that too much emphasis is already being placed on physics in the students' overburdened curriculum. However, it is felt, in view of the great strides that have been made in atomic physics in the last few years and the growing importance of radioactive isotopes, that the student should have at least a minimum concept of atomic structure and atomic energy.

The material has been so organized that certain sections may be omitted at the instructor's discretion without seriously interfering with the continuity of the text. On the other hand, these more advanced considerations will appeal to the student technician fortunate enough to have a better-than-average scientific background.

Although it has been pointed out repeatedly that this book has been conceived primarily for the student x-ray technician, it is felt that it will also be of distinct advantage to the resident in radiology who is just entering upon his period of training and needs an introductory textbook to prepare him for more advanced reading as his knowledge of the subject develops.

With some exceptions, the author makes no claim to originality, but it is impossible to mention by name the thousands of radiologists, physicists, and technicians who have developed the science of radiology to its present high position. Grateful acknowledgment is hereby accorded to the following who so kindly took time from their busy curriculum to review certain sections of the text and offered invaluable suggestions and criticisms: T. W. Bonner, Ph.D., Professor of Physics at the Rice Institute; Otto Glasser, Ph.D., Professor of Biophysics at the Cleveland Clinic Foundation; and Mr. W. S. Cornwell and his associates at the Eastman Kodak Company. The illustrations were prepared from the author's sketches by Mr. Howard Marlin, Tyler Texas. The manuscript and typing was ably executed by Miss Betty Presley of Tyler, Texas.

The author appreciates the kindness of the Machlett Laboratories, Inc., and of the General Electric X-ray Corporation for furnishing the basic drawings of their x-ray and valve tubes; and of the Eastman Kodak Company, E. I. duPont de Nemours, Inc., and Ansco for the data on x-ray films, film processing, and screens. Finally, the author is grateful to the publisher, Mr. Charles C Thomas, for his encouragement, patience, cooperation, and sound advice.

JOSEPH SELMAN, M.D.

Tyler, Texas

CONTENTS

The Fundamentals of
X-ray and Radium Physics

CHAPTER 1 SIMPLIFIED MATHEMATICS

ALL OF THE PHYSICAL SCIENCES have in common a firm basis in mathematics. This is no less true of radiologic physics which is an important branch of the physical sciences. Clearly, then, in approaching a course in radiologic physics you, as a student technologist, should find your path smoothed by an adequate background in the appropriate areas of mathematics.

We shall assume here that you have had at least the required high school exposure to mathematics, although this may vary widely from place to place. However, realizing that much of this material may have become hazy with time, we shall review the simple but necessary aspects of arithmetic, algebra, and plane geometry. Such a review should be beneficial in at least two ways. First, it should make it easier to understand the basic principles and concepts of radiologic physics. Second, it should aid in the solution of such everyday problems as conversion of radiographic technics, interpretation of tube rating charts, determination of radiographic magnification, and many others that may arise from time to time.

The discussion will be subdivided as follows: (1) arithmetic, (2) algebra, (3) ratio and proportion, (4) geometry, (5) graphs and charts, and (6) large and small numbers. Only fundamental principles will be included.

ARITHMETIC

Arithmetic is calculation or problem solving by means of numerals. We shall assume that you are familiar with addition, subtraction, multiplication, and division and shall therefore omit these operations.

Fractions

A fraction is defined as a part of a whole number. For instance, ½, ⅓, ⅖ are fractions. The fraction ⅗ may be read "three divided by five," and represented also as $3 \div 5$. Thus, a fraction denotes the division of one number by another.

In any fraction, the numeral above the line is the ***numerator*** and that below the line the ***denominator.*** In the fraction ⅗, 3 is the numerator and 5 is the denominator.

If the numerator is smaller than the denominator, as ⅗, we have a ***proper fraction.*** If the numerator is larger than the denominator, as 5/3, we have an ***improper fraction,*** because $5 \div 3 = 1\frac{2}{3}$, which is really an integer plus a fraction.

In ***adding*** fractions, all of which have the ***same*** denominator, we add all the numerators first and then place the sum over the denominator:

$$\frac{2}{7} + \frac{3}{7} + \frac{6}{7} + \frac{5}{7} = \frac{2 + 3 + 6 + 5}{7} = \frac{16}{7}$$

$$\frac{16}{7} = 2\frac{2}{7}$$

Subtraction of fractions having identical denominators follows the same rule:

$$\frac{6}{7} - \frac{4}{7} = \frac{6 - 4}{7} = \frac{2}{7}$$

If a series of fractions is to be added or subtracted, and the denominators are ***different,*** then the ***least common denominator*** must be found. This is the smallest numeral which is exactly divisible by all the denominators. Thus,

$$\frac{1}{2} + \frac{2}{3} - \frac{3}{4} = ? \tag{1}$$

The smallest numeral which is divided exactly by each denominator is 12. Place 12 in the denominator of a new fraction:

$$\frac{\qquad}{12} \tag{2}$$

Divide the denominator of each of the fractions in (1) into 12, and then multiply the answer by the numerator of that fraction;

each result is then placed in the numerator of the new fraction (2):

$$\frac{6 + 8 - 9}{12} = \frac{5}{12}$$

When numbers are *multiplied, we take their product.* In multiplying fractions, we take the product of the numerators and place it over the product of the denominators,

$$\frac{4}{5} \times \frac{3}{10} = \frac{4 \times 3}{5 \times 10} = \frac{12}{50}$$

The resulting fraction can be reduced by dividing the numerator and the denominator by the *same* number, in this case, 2:

$$\frac{12}{50} \div \frac{2}{2} = \frac{6}{25}$$

which cannot be further simplified.

Note that when the numerator and the denominator are both multiplied or divided by the same number, the value of the fraction does not change. For instance,

$$\frac{3}{5} \times \frac{2}{2} = \frac{6}{10} \text{ is the same as}$$

$$\frac{3}{5} \times 1 = \frac{3}{5}$$

When two fractions are to be divided, as $\frac{4}{5} \div \frac{3}{7}$, the fraction that is to be divided is the *dividend,* and the fraction that does the dividing is called the *divisor.* In this case, $\frac{4}{5}$ is the dividend and $\frac{3}{7}$ the divisor. The rule is to invert the divisor (called "taking the reciprocal") and multiply the dividend by it:

$$\frac{4}{5} \div \frac{3}{7}$$

$$\frac{4}{5} \times \frac{7}{3} = \frac{28}{15} = 1\frac{13}{15}$$

Per Cent

A special type of fraction, *per cent,* is represented by the sign % to indicate that the number standing with it is to be divided by 100. Thus, $95\% = \frac{95}{100}$. We do not use percentages directly in mathematical computations, but first convert them to fractions or decimals. For instance,

$$150 \times 40\% \text{ is changed to}$$
$$150 \times {}^{40}\!/_{100} \text{ or } 150 \times {}^{2}\!/_{5}$$
$$\text{or } 150 \times 0.40.$$

Decimal Fractions

Our common method of representing numbers as multiples of ten is embodied in the *decimal system*. A *decimal fraction* is a proper fraction whose denominator is 10, or 10 raised to some power such as 100, 1000, 10,000, etc. The denominator is symbolized by a dot in a certain position. For example, the decimal $0.2 = {}^{2}\!/_{10}$; $0.02 = {}^{2}\!/_{100}$; $0.002 = {}^{2}\!/_{1000}$, etc. Decimals can be multiplied or divided, but care must be taken to place the decimal point in the proper position:

$$
\begin{array}{r}
2.24 \\
\times\, 1.25 \\
\hline
1120 \\
448 \\
224 \\
\hline
2.8000
\end{array}
$$

Note that we add the total number of digits to the right of the decimal points in the numbers being multipled, which in this case turns out to be 4. We then point off 4 places from the right in the answer to determine the correct position of the decimal point. In the physical sciences, decimal fractions are used almost exclusively in preference to conventional fractions.

SIGNIFICANT FIGURES

The precision of any type of measurement is limited by the precision of the measuring instrument. For example, if a scale is calibrated in grams as shown in Figure 1.1, we can estimate to the nearest tenth of a gram. Thus, the scale in Figure 1.1 reads 8.4 grams. The last figure, 0.4, is estimated and is the *last significant figure*—that is, it is the last figure that has meaning. Obviously, no greater precision is possible with this particular in-

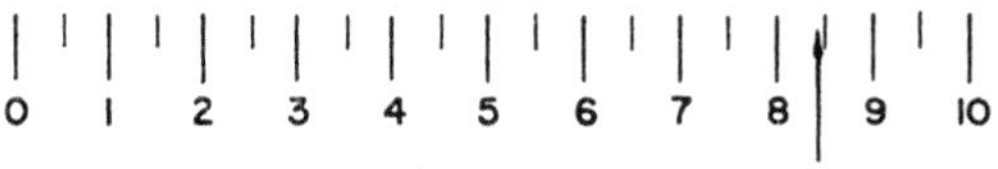

Figure 1.1. With this calibrated scale we can estimate to the nearest tenth. Thus, the position of the pointer indicates 8.4 units, the 0.4 being the last significant figure.

strument. To improve precision, the scale would have to show a greater number of subdivisions.

Now let us see how significant figures are used in various operations. For example, in addition:

item 1	98.26 grams
item 2	1.350 g
item 3	260.1 g

359.710 g

Notice that there are three figures after the decimal point in the answer. But in item (3) there is only one figure—a 1—after the decimal point; beyond this, the figures are unknown. Consequently, the digits after the 7 in the answer have no meaning, since the answer can be no more precise than the least precise item being added. In this case, the answer should be properly stated as 359.7. Thus, in addition and subtraction the answer can have no more significant figures after the decimal point than the item with the *least number of significant figures after its decimal point.*

A different situation exists in multiplication and division. Here, the total number of significant figures in the answer is equal to that in the items having the *least total number of significant figures.* For example, in

25.23 cm
× 1.21 cm

2523
5046
2523

30.5283 cm²

1.21 has fewer significant figures—three in all. Therefore the answer should have three significant figures and be read as 30.5 (dropping the 0.0283).

In general, to *round off* significant figures, we observe the following rule: if the figure following the last significant figure is equal to or greater than 5, the last significant figure is increased by 1; if less than 5, it is unchanged. The rule is applied in the following examples:

$$45.157 \text{ is rounded to } 45.16$$
$$45.155 \text{ is rounded to } 45.16$$
$$45.153 \text{ is rounded to } 45.15$$

where the answer is to be expressed in four significant figures.

ALGEBRA

The word *algebra,* derived from the Arabic language, connotes that branch of mathematics which deals with the relationship of quantities by the use of letters or symbols. Algebra may be regarded as mathematical shorthand. Ordinarily, the symbols used in algebra are the letters of the alphabet.

Operations. Mathematical operations with letter symbols are the same as with numerals, since both are symbolic representations of numbers which, in themselves, are abstract concepts. In algebra, the fundamental operations are addition, subtraction, multiplication, and division. There are fractions, equations, and proportions just as in arithmetic. *Algebra provides a method of finding an unknown quantity when the relationship of certain known quantities is specified.*

Algebraic operations are indicated by the same symbols as arithmetic:

$+$ (plus) add
$-$ (minus) subtract
$\times$ (times) multiply
$\div$ (divided by) divide
$=$ equals

To indicate addition in algebra, we may use the general expression

$$x + y \tag{1}$$

The symbols x and y, called *variables*, may represent any number or quantity we choose. Thus, if $x = 4$ and $y = 7$, then

$$4 + 7 = 11$$

Similarly, to indicate subtraction in algebra, we may use the general expression

$$x - y$$

If $x = 9$ and $y = 5$, then

$$9 - 5 = 4$$

Notice that the symbols in algebra apply to whole numbers, fractions, zero, and negative numbers, among others. Negative numbers are those whose value is less than zero and are designated as ^-x. In algebraic terms, we may add a positive and a negative number as follows:

$$x + {}^-y$$

If $x = 8$ and $^-y = {}^-3$, then

$$8 + {}^-3$$

is the same as

$$8 - 3 = 5$$

The $+$ sign is omitted in the designation of positive numbers, being reserved to indicate the operation of addition.

On the other hand, we may subtract a negative quantity from a positive one:

$$x - {}^-y$$

If $x = 4$ and $^-y = {}^-6$, then

$$4 - {}^-6$$

is the same as

$$4 + 6 = 10$$

Multiplication in algebra follows the same rules as in arithmetic. However, in the multiplication of letter symbols the $\times$ sign is omitted, $x \times y$ being written as xy. If $x = 3$ and $y = 5$, then substituting in the expression xy,

$$3 \times 5 = 15$$

Division in algebra is customarily expressed as a fraction. Thus, $x \div y$ is written as x/y. If $x = 3$ and $y = 5$, then

$$3 \div 5 = \tfrac{3}{5}$$

In general, when two negative quantities are multiplied, the answer is positive: $^-x \times {}^-y = xy$. When two negative quantities are divided, the answer is positive: $^-x/{}^-y = x/y$. When a positive and a negative quantity are multiplied or divided, the answer is negative; thus, $x \times {}^-y = -xy$ and $x \div {}^-y = -x/y$.

In solving an algebraic expression consisting of a collection of *terms* we must perform the indicated *multiplication and division first,* and then carry out he indicated addition and subtraction. An example will clarify this:

$$ab + c/d = f$$

Let $a = 2$, $b = 3$, $c = 4$, $d = 8$, and $f = 5$. Substituting in the preceding expression,

$$2 \times 3 + \frac{4}{8} - 5 = ?$$

Performing *multiplication and division first,*

$$6 + \frac{4}{8} - 5 = ?$$

Then, performing addition and subtraction,

$$6\frac{4}{8} - 5 = 1\frac{4}{8} = 1\frac{1}{2}$$

A set of parentheses inclosing a group of terms indicates that all of the terms inside the parentheses are to be multiplied by the term outside the parentheses. Thus, in the expression

$$6\,(8 - 4 + 3 \times 2) =$$
$$6\,(8 - 4 + 6) =$$
$$6 \times 10 = 60$$

Equations. The simpler algebraic equations can be solved without difficulty if fundamental rules are applied. You can easily verify these rules by substituting numerals. In the equation

$$a + b = c + d \tag{2}$$

$a + b$ is called the *left side,* and $c + d$ the *right side.* Each letter is called a *term*. If any quantity is added to one side of the equation, the same quantity must be added to the other side in order for this to remain an equation. Similarly, if any quantity is subtracted from one side, the same quantity must be subtracted from the other side. The concept of the equation is simplified if one pictures it as a see-saw as in Figure 1.2. If persons of equal

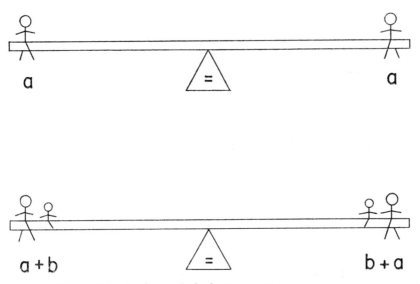

Figure 1.2. Analogy of algebraic equation to a see-saw.

weight are placed at each end, the board will remain horizontal—the equation is balanced. If a second person is now added to one end of the see-saw, a person of similar weight must be added to the other end in order to keep the board level.

Return again to the simple equation:

$$a + b = c + d$$

in which each term is a variable. If any three of the variables are known, the fourth can be found. Suppose, a is 3, b is 4, c is 1, and d is unknown. Substituting these values in the equation,

$$3 + 4 = 1 + d$$

How can d be found? Simply rearrange the equation so that d is alone, that is, **the only term in its side.** In this case, subtract 1 from both sides of the equation. However, a mathematical short cut can be used: *a term may be transposed from one side of an equation to the other side, provided it is given the opposite sign.* Following this rule, 1 becomes a minus 1 when moved to the left side. Thus,

$$3 + 4 - 1 = d$$
$$6 = d$$
$$\text{or,} \quad d = 6$$

Usually, in algebraic calculations, the terms are rearranged before their numerical values are substituted. In the equation

$$a + b = c + d$$

To find d, transpose c to the left side and change its sign.

$$a + b - c = d$$
$$d = a + b - c$$

(Reversing both sides of an equation does not alter its equality.)

In algebraic equations in which terms are multiplied or divided, analogous rules apply. For example, equation

$$x = y/z$$

may be solved for y by multiplying both sides by z,

$$xz = yz/z$$
$$xz = y$$
$$\text{or,} \quad y = xz$$

The same result may be obtained by moving z from the denominator of the right side to the numerator of the left side. Thus, we have the short-cut rule: *if the denominator of one side of an equation is moved, it multiplies the numerator of the other side and, conversely, if the numerator of one side is moved, it multiplies the denominator of the other side.*

Suppose that in the equation $x = y/z$, x and y are known; then z is solved as follows: move z into the numerator of the opposite side as a multiplier,

$$xz = y$$

and move x into the denominator of the right side as a multiplier, and

$$z = y/x \qquad (3)$$

The above rule can be readily tested. Suppose that y is 12, and x is 3. Substituting in equation (3),

$$z = 12/3$$
$$z = 4$$
$$4 = 12/3 \qquad (4)$$

If we wish to move 3, we must place it in the numerator of the left side of equation (4),

$$4 \times 3 = 12$$

Note that numerical equation (4) balances.

Now, referring again to equation (4), suppose we wish to move 12. We must place it in the denominator of the left side:

$$4/12 = 1/3$$

Again, it is evident that the equation balances.

RATIO AND PROPORTION

A *ratio* is a fixed relationship between two quantities, simply indicating how many times larger or smaller one quantity is relative to another. It has essentially the same meaning as a fraction. One symbol that expresses a ratio is the colon (:). Thus, $a{:}b$ is read "*a* is to *b*." Or, 1:2 is read "1 is to 2." In modern mathematics ratios are usually represented as fractions:

$$a{:}b \text{ is the same as } a/b$$
$$1{:}2 \text{ is the same as } \tfrac{1}{2}$$

As noted above, the fraction $\tfrac{1}{2}$ indicates that the numerator is $\tfrac{1}{2}$ as large as the denominator. Similarly, $\tfrac{2}{3}$ indicates that the numerator is $\tfrac{2}{3}$ as large as the denominator.

The meaning of ratio is important for the technologist, because it underlies the concept of *proportion,* defined as *an expression showing that two given ratios are equal.* Thus, we may have an algebraic proportion,

$$a/b = c/d \tag{5}$$

which is read "*a* is to *b* as *c* is to *d*." The same idea can also be presented numerically. For example,

$$3/6 = 4/8$$

If any three terms of a proportion are known, the fourth may easily be determined. Suppose in proportion (5) *a* is 2, *b* is 4, *d* is 8, and *c* is to be found. Then,

$$2/4 = c/8$$

Moving 8 to the numerator of the left side (crossmultiplying),

$$\frac{2 \times 8}{4} = c$$

$$c = 16/4 = 4$$

There are two general types of proportions that are of interest in radiography.

1. **Direct proportion.** Here, one quantity maintains the same ratio to another quantity as the latter changes. For example, the statement "*a* is proportional to *b*" means that if *b* is doubled, *a* is automatically doubled; if *b* is tripled, *a* is tripled, etc. In order to represent this mathematically, let us assume that the quantity a_1 exists when b_1 exists; and that if b_1 is changed to b_2, then a_1 becomes a_2. Thus,

$$a_1/b_1 = a_2/b_2$$

The numbers below the letters are called "subscripts" and have no significance except to label a_2 as being different from a_1. Such a direct proportion is solved by the method described above.

2. **Inverse proportion.** In this type, one quantity varies in the opposite direction from another quantity as the latter changes. Thus, the statement "*a* is inversely proportional to *b*" means that

if b is doubled a is halved, if b is tripled, a is divided by 3, etc. Such a proportion is set up as follows:

$$\frac{a_1}{1/b_1} = \frac{a_2}{1/b_2}$$

Crossmultiplying,

$$a_1/b_2 = a_2/b_1 \qquad (6)$$

A numerical example should help clarify this. Suppose that when a_1 is 2, b_1 is 4; if a is inversely proportional to b, what will a_1 become if b_1 is changed to 8? In this case, a_2 is the unknown, and b_2 is 8. Substituting in equation (6),

$$2/8 = a_2/4$$
$$a_2 = 8/8 = 1$$

Thus, a is halved when b is doubled.

PLANE GEOMETRY

Plane geometry is that branch of mathematics which deals with figures lying entirely in one plane; that is, on a flat surface. Some of the elementary rules of plane geometry will now be listed.

1. A *straight line* is the shortest distance between two points. It has only one dimension—length.

2. A *rectangle* is a plane figure composed of four straight lines meeting at right angles. Its opposite sides are equal. The sum of the lengths of the four sides is called the *perimeter.* The area of a rectangle is the product of two adjacent sides. Thus, in Figure 1.3 the perimeter is the sum $a + b + a + b = 2a + 2b$. The area is $a \times b$ or ab.

3. A *square* is a special rectangle in which all four sides are equal. Therefore, the area of a square equals the square of one side. This is shown in Figure 1.4.

4. A *triangle* is a figure made up of three straight lines connecting three points that are not in the same line. The perimeter of a triangle is the sum of the three sides. The area of a triangle

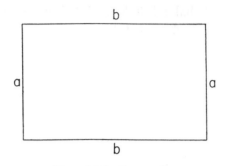

Figure 1.3. Rectangle.
Perimeter $= 2a + 2b$
Area $= ab$

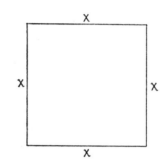

Figure 1.4. Square.
Perimeter $= 4x$
Area $= x \times x = x^2$

is ½ the base times the altitude (the altitude is the perpendicular distance from the apex to the base). This is shown in Figure 1.5.

5. A *circle* is a closed curved line which is everywhere at an equal distance from one point called the center, lying in the same plane as the center. A straight line from the center to any point on the circle is the *radius*. A straight line passing through the center and meeting the circle at two points is the *diameter*, which is obviously equal to twice the radius. The length of the circle is its *circumference*, obtained by multiplying the diameter by a constant, π (pi). Pi always equals 3.14 (approximately). The area enclosed by a circle equals π times the square of the radius. These relationships can be more easily grasped by referring to Figure 1.6.

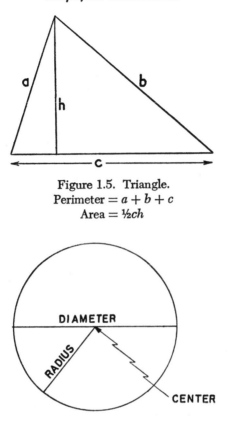

Figure 1.5. Triangle.
Perimeter $= a + b + c$
Area $= \frac{1}{2}ch$

Figure 1.6. Circle.
Circumference $= \pi \times$ diameter $= \pi d$
Area $= \pi \times$ radius^2 $= \pi r^2$

Similar Triangles

Of great importance in radiology, and closely related to ratio and proportion, is the proportionality of *similar triangles*. These are triangles which have the same shape, although they differ in size. *In similar triangles the corresponding sides are directly proportional and the corresponding angles are equal.* Thus, similar triangles may be regarded as the geometric expression of direct proportion.

Figure 1.7 shows two similar triangles. The corresponding

sides (those opposite the equal corresponding angles) are proportional, so that

$$a/A = b/B, \text{ or}$$
$$a/A = c/C, \text{ or}$$
$$b/B = c/C$$

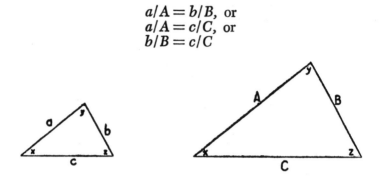

Figure 1.7. Similar triangles. The corresponding angles are *equal;* that is, $x = x$, $y = y$, $z = z$. The corresponding sides are proportional; thus, $a/A = b/B = c/C$.

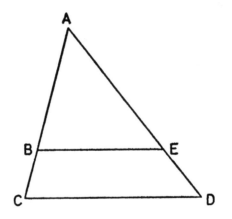

Figure 1.8. Similar triangles. *BE* is parallel to *CD*. Triangle *ABE* is similar to triangle *ACD*. Therefore,
$$AB/AC = AE/AD = BE/CD.$$

If a line is drawn across any triangle parallel to its base, there will result two similar triangles, one partly superimposed on the other. This is shown in Figure 1.8. Line *BE* has been drawn parallel to the base *CD*. Then triangle *ABE* is similar to triangle *ACD* and their corresponding sides are proportional. This may be simplified by separating the two triangles as in Figure 1.7.

A thorough comprehension of similar triangles is essential to the understanding of photographic and radiographic projection. Let us assume that an object is placed in a beam of light originating at a point source (see Figure 1.9). The shadow or image of

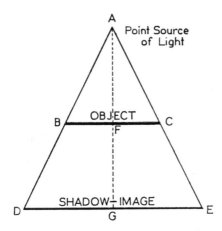

Figure 1.9. Projection of the shadow image of an object by a point source of light. $DE/BC = AG/AF$

the object is allowed to fall on a surface which is parallel to the long axis of the object. *Note that the image is larger than the object.* If the distances of the object and of the image from the source of light are known, and if the size of the object is known, then the size of the image can be determined from the similarity of triangles *ABC* and *ADE* by setting up the proportion,

$$\frac{\text{image size}}{\text{object size}} = \frac{\text{image distance}}{\text{object distance}}$$

Substituting the known values in the equation, we can predict the size of the image by solving the equation. Note that if any three of the values are known, the fourth can be readily obtained from the same question. This same principle applies in radiography, as will be shown in Chapter 18.

GRAPHS AND CHARTS

For practical purposes, graphs and charts may be regarded as diagrams representing the relationship of two quantities, one of which depends on the other. The dependent factor is called the *dependent variable.* The factor which changes independently is called the *independent variable.*

When data are accumulated, showing how a dependent variable changes with a change in the independent variable, they can be compiled in the form of a table, or they can be represented graphically. Tables are often less convenient to read than graphs, and, besides, do not give intermediate values as do graphs.

The construction and interpretation of a graph may be exemplified by the following observation: the optimum developing time for x-ray film increases as the temperature of the developing solution decreases (not inversely proportional, however). By actual test with a certain developer, the data in Table 1.1 are obtained. To chart these on graph paper, first plot the independent variable along the horizontal axis of the graph. Then plot the developing time, or dependent variable on the vertical axis, as shown in Figure 1.10. In mathematics, the dependent variable is said to be a *function* of the independent variable. To graph the tabulated data, take the first temperature listed in the table and locate it on the horizontal axis. Trace vertically upward from this point to the horizontal line corresponding to the correct developing time, and mark the intersection with an x. Repeat this for all the values in the table, and then draw a line that best fits the x's. This constitutes a *line graph* or *curve.*

Having constructed a graph, how do we read it? Suppose the

TABLE 1.1

TIME-TEMPERATURE DEVELOPMENT DATA

Temperature in F	Developing Time in Min
60	5¼
62	4½
64	3¾
65	3½
68	3
70	2¾
72	2½
75	2

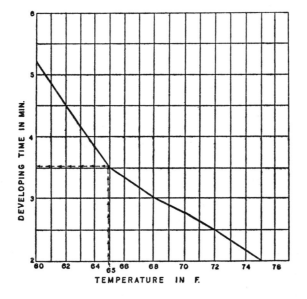

Figure 1.10. Method of using a time-temperature development curve. Select the correct temperature on the horizontal axis, in this case 65 F. Follow vertically upward, as shown by the arrows, to the curve. Then follow horizontally to the time axis where the correct developing time in this case is seen to be 3½ min.

temperature is found to be 65 F, and we wish to determine the correct developing time. Locate 65 F on the horizontal axis and trace vertically upward to the intersection with the curve; from this point of intersection trace horizontally to the vertical axis on the left, where the correct developing time can be read. This is shown in Figure 1.10, where the tracing lines are formed by broken lines with arrows. The horizontal tracing line meets the vertical axis at approximately 3½ minutes, which is the correct developing time at this temperature. Thus, a properly constructed graph gives the developing times for temperature readings that did not appear in the original table, a process called *interpolation.*

LARGE AND SMALL NUMBERS

When extremely large numbers are required in physics, ordinary notation is so cumbersome that scientists use the *exponen-*

tial system in powers of 10. For example, the number 10^3 means that 10 is multiplied by itself three times: $10 \times 10 \times 10$. In the quantity 10^3, 10 is the *base* and 3 the *exponent,* and is read as "ten to the third power." In general,

$$10^1 = 10$$
$$10^2 = 100$$
$$10^3 = 1000, \text{ etc}$$

Note that the exponent of 10 indicates the number of zeros required to express the number in ordinary notation.

Let us now apply this principle. The velocity of light in air is 30,000,000,000 cm per sec. Place a decimal point after 3; there are 10 zeroes to the right of the decimal point. Therefore, this number may be simply expressed as 3.0×10^{10} cm per sec, read as "three point oh times ten to the tenth." A more complicated number such as 4,310,000,000,000,000 may be simplified in the same way by placing a decimal point after the first digit, 4, and counting the number of *places* to the right, in this case 15. The corresponding number in the powers-of-ten system is 4.31×10^{15}.

Suppose we are given 6.3×10^6. How is this converted to *ordinary notation?* Count 6 places to the right of the decimal point, filling out the places with zeros. Since the first place is already occupied by the digit 3, 5 zeros must follow. Therefore, the number is 6,300,000. In general, any number followed by 10^6 is expressed in *millions.*

The same system can be applied to *extremely small* numbers, by the use of negative exponents. This is based on

$$10^{-1} = \tfrac{1}{10} = 0.1$$
$$10^{-2} = \tfrac{1}{100} = 0.01$$
$$10^{-3} = \tfrac{1}{1000} = 0.001, \text{ etc}$$

From this you can see that a negative exponent indicates the number of decimal places to the *left* of a digit, *including the digit.* Thus, 0.004 would be expressed as 4×10^{-3}. To obtain this result, count the number of places to the right of the decimal point, including the first digit,

$$\overset{\cdots}{0.004}$$

The counted number of places (three in this instance) then becomes the negative exponent of 10; thus, 4×10^{-3}. A more complicated number such as 0.00000000000372 would be simply expressed as 3.72×10^{-12}, read as "three point seven two times ten to the minus twelfth."

The simplified system of notation facilitates the multiplication and division of large and small numbers. To multiply such numbers, we multiply the digits and make the exponent of 10 in the answer equal to the algebraic sum of the exponents of 10 in the multipliers. Thus,

$$4 \times 10^3 \times 2 \times 10^2 = 4 \times 2 \times 10^{(3+2)} = 8 \times 10^5$$
$$3 \times 10^6 \times 2 \times 10^{-4} = 3 \times 2 \times 10^{(6-4)} = 6 \times 10^2$$

To divide numbers in this system, we divide the digits and derive the exponent of 10 in the answer by subtracting algebraically the exponent in the divisor from the exponent in the dividend:

$$\frac{6 \times 10^5}{2 \times 10^3} = \frac{6 \times 10^{(5-3)}}{2} = 3 \times 10^2$$

$$\frac{9 \times 10^7}{3 \times 10^{-2}} = \frac{9 \times 10^{(7+2)}}{3} = 3 \times 10^9$$

Note in the above examples that when a power number is moved from the denominator of a fraction to the numerator, or *vice versa*, one simply changes the sign of the exponent. Thus,

$$\frac{8}{3 \times 10^{-5}} = \frac{8 \times 10^5}{3}$$

QUESTIONS AND PROBLEMS

1. Reduce the following fractions:

    (a) $\frac{4}{8}$ (b) $\frac{9}{12}$ (c) $\frac{10}{15}$ (d) $\frac{4}{5}$

2. Solve the following problems:

    (a) $\frac{3}{5} + \frac{2}{5} + \frac{4}{5} =$
    (b) $\frac{2}{3} + \frac{3}{4} + \frac{3}{7} =$
    (c) $\frac{1}{2} + \frac{2}{3} + \frac{4}{5} =$

3. Mr. Jones has 100 bushels of potatoes and sells 75 per cent of this crop. How many bushels does he have left?

4. Divide as indicated:

   (a) $\frac{4}{5} \div \frac{3}{5}$ (b) $\frac{4}{9} \div \frac{7}{16}$ (c) $\frac{8}{9} \div \frac{3}{5}$

5. If $a = b/c$, what is b in terms of a and c? What is c in terms of a and b?

6. Solve the following equations for the unknown term:

   (a) $x + 4 = 7 - 3 + 8$
   (b) $6 + 3 - 1 = 4 + 1 - a$
   (c) $x/3 = 7/9$
   (d) $4/y = \dfrac{6 + 3}{10}$

   (e) $\dfrac{x + 4}{8} = 7 + 2 - 3$

7. Solve the following proportions for the unknown quantity:

   (a) $a/7 = 2/21$ (c) $4/6 = 7/y$
   (b) $3/9 = x/15$ (d) $3/5 = 9/x$

8. The diameter of an x-ray beam is directly proportional to the distance from the tube target. If the diameter of the beam at a 20-in. distance is 10 cm, what will the diameter be at a 40-in. distance?

9. The exposure of a radiograph is directly proportional to the time of exposure. What will happen to the exposure if the time is tripled?

10. What is the area of a circle having a diameter of 4 in.?

11. What is the area of a rectangular plot of ground measuring 20 ft on one side and 30 ft on the adjacent side?

12. Using the time-temperature curve in Figure 1.10, determine the proper developing time when the temperature is 64 F.

13. Convert 3.6×10^6 to the ordinary number system.

14. Change 424,000 to the exponential system.

15. Solve $320,000 \times 1,200,000 \div 240,000$ using the exponential system.

CHAPTER 2   PHYSICS AND THE UNITS OF MEASUREMENT

Physics is an exact science. Therefore, in studying physics, we must learn its language as precisely as possible so that each word will have a definite meaning. Not only does this simplify the learning process but it also makes it easier for us to organize our ideas and communicate them to others.

In order to appreciate the position of physics within the framework of science, we may use the term *natural science* to include the study of objects in general. Figure 2.1 shows the subdivision of natural science into its major categories. Here you can readily

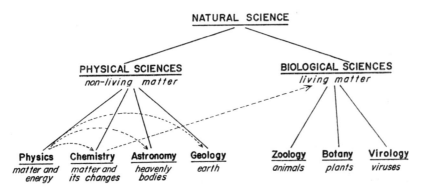

Figure 2.1. Subdivisions of natural science. Notice that physics directly or indirectly plays a part in all branches of science (dashed lines).

see that while physics is one of the physical sciences, it actually underlies all of them. In fact, it is ultimately an important contributor even to the biologic sciences.

What, then, is *physics?* It may be defined as that branch of science which deals with matter and energy, and their relation to each other. It includes mechanics, heat, light, sound, electricity and magnetism, and the fundamental structure and properties of matter. For our purpose, we shall be interested in those aspects of physics pertaining to the origin, nature, and behavior of x rays and related types of radiation; that is, *radiologic physics.*

Thus far, we have used the word science rather freely without defining it. *Science is organized and classified knowledge.* Although we may not always be aware of them, natural occurrences continually surround us. But these do not constitute a science until the scientific method has been applied to them. The *scientific method* comprises the systematic collection of facts, the study of their interrelationship, and the drawing of valid conclusions from these data. Underlying the scientific method is the assumption that nature is orderly and, for the most part, predictable, so that we can discover the rules by which she behaves.

Not only does the scientific method seek to describe and classify natural occurrences, but it also attempts to correlate them by deriving certain principles known as *laws.* Scientific laws are based on human experience, being derived by observation of natural phenomena, by laboratory experiments, or by the application of mathematical reasoning. Such laws state as clearly and simply as possible, according to our present state of knowledge, that certain events will always follow in the same order, or that certain facts will always be related in the same way. For example, we know that all bodies near the earth are attracted to it; this is a simple statement of the law of gravity.

Scientists are not satisfied with merely discovering the laws of nature. They also attempt to correlate various laws in order to determine their conformity to a *general pattern.* Thus, wherever possible, natural laws are so tied together that their operation can be expressed as a general concept, often on a mathematical basis. Such a broad, unified concept of the laws underlying certain natural phenomena is called a *theory.* Not only do theories provide a better insight into nature but they also suggest new lines of scientific research, since it is common experience that the solution of one problem in science may give rise to a host of new problems.

STANDARD UNITS

An exact science (as contrasted, for example, with some phases of biological science) must include not only the observation and classification of natural occurrences, but also their *measurement*. In other words, physics as an exact science deals with both the "how" and the "how much" of physical phenomena. In order for these measurements to have the same meaning everywhere, they must be expressed in certain standard units. A *unit* is a quantity adopted as a standard of measurement by which other quantities of the same kind can be measured, such as inch, ounce, or second. *Standard units* are the basic units in any particular system of measurement; they are used mathematically to describe natural phenomena, derive laws, and predict future events within the realm of nature. For example, astronomers can foretell precisely the time of an eclipse of the sun or moon.

Units should have a size compatible with that which is being measured. Thus, we measure the length of a baby in inches, but we use a larger unit, the foot, to measure the height of a building; and a still larger unit, the mile, to measure the distance between cities. We shall return later to the subject of conversion of units of different sizes and types.

The standard units employed in physics are usually divided into two general types, the simpler *fundamental units* dealing with *length, mass,* and *time;* and the more complicated *derived units,* obtained by various combinations of the fundamental units.

Fundamental Units

The fundamental units are arbitrarily selected and named, but are so standardized that a given unit has the same meaning everywhere. We now have two widely used systems of measurement, the *English* and the *metric.*

Since the metric system is used internationally in the exact sciences, it will be emphasized here, but the more familiar equivalents in the English system units will also be indicated. The metric system is also known as the MKS (meter-kilogram-second), or the CGS (centimeter-gram-second) system.

1. **Length.** The unit of length in the metric system is the

meter, originally defined as the distance between two scratches on a bar of platinum, kept at Sèvres, France, at the International Bureau of Weights and Measures. (It was redefined in 1960 on a much more precise basis.) Roughly equal to a yard, the meter can be conveniently subdivided into smaller units such as a centimeter (0.01 or 10^{-2} meter) and millimeter (0.001 or 10^{-3} meter); or it can be treated as a multiple such as a kilometer (1,000 or 10^3 meter). In x-ray physics, we even have occasion to use very tiny fractions of these units; thus, the angstrom is $\frac{1}{100,000,000}$ or 10^{-8} centimeter. The following common equivalents should prove useful:

1 meter = 100 cm (centimeters) = 1000 mm (millimeters)
1 Å (angstrom) = $\frac{1}{100,000,000}$ cm = 10^{-8} cm
1 km (kilometer) = 1,000 m (meters) = about 1,000 yards
1 in. = 2.54 cm

2. **Mass.** Designating the quantity of matter in a body, its mass is determined by weighing, a procedure which measures the attraction the earth has for it at some location; the more massive the body, the greater the gravitational force and, therefore, the greater its weight. The unit of mass is the *kilogram*, which is the weight of a standard piece of platinum-iridium (International Kilogram Prototype) kept at the International Bureau of Standards. All other kilograms are more or less exact copies of this standard unit. A more convenient unit is the *gram*, which is $\frac{1}{1000}$ of a kilogram.

1 kg (kilogram) = 1000 or 10^3 g (grams)
1 kg = 2.2 lb
28 g = 1 oz

(The above discussion of mass and weight is not a complete presentation, but has necessarily been simplified for our purposes.)

3. **Time.** This is a measure of the duration of events. We are all aware of the occurrence of events and the motion of objects, but to measure duration with relation to our senses alone is inaccurate. The standard unit of time is the *second*, defined as $\frac{1}{86,400}$ of a mean solar day. In other words, the second is a

definite fraction of the average time it takes for the earth to make one rotation on its axis.

DERIVED UNITS

There are numerous derived units, but only those having a practical bearing on x-ray physics will be mentioned.

1. **Area** is the measure of a given surface, and depends on length. Thus, a square or rectangle has an area equal to the product of two sides. The area of a circle equals the radius squared times π. Figures 1.3, 1.4, 1.5, and 1.6 in Chapter 1 explain area in detail. In the metric system, area is represented by square meters for larger surfaces and square centimeters for smaller ones. Square centimeters are abbreviated either as sq cm or cm^2.

2. **Volume** is a measure of the capacity of a container, and is also derived from length. The volume of a cube equals the product of three sides. In metric units, volume may be expressed in cubic centimeters (cc) or milliliters (ml). One liter equals 1,000 ml or about 1,000 cc—approximately one quart.

3. **Density** is the mass per unit volume of a substance, and may be expressed in g per ml.

4. **Specific gravity** has *no units*. It is the ratio of the density of any material to the density of water. The density of water is 1.

5. **Velocity** is speed in a given direction, and can be expressed in cm per sec, or km per sec, or some other convenient units. Speed is commonly expressed in miles per hr (English system).

6. **Temperature** is a measure of the average energy of motion of the molecules of matter. Two systems are commonly employed today. The centigrade system is used mainly by scientists, whereas the Fahrenheit system is the one generally used in the United States. The following data exemplify the differences between these systems:

CENTIGRADE   0 C = freezing point of water
                           100 C = boiling point of water

FAHRENHEIT   32 F = freezing point of water
                           212 F = boiling point of water

To change values from one system to the other, one may refer to tables, or the following formulas may be used:

$$C = \tfrac{5}{9} (F - 32)$$
$$F = \tfrac{9}{5} C + 32$$

QUESTIONS AND PROBLEMS

1. What is meant by science; law; theory?
2. Why are standard units necessary?
3. Define natural science; physics; radiologic physics.
4. Which branch of natural science underlies all the others?
5. How were standard units in the metric system established?
6. With what do the fundamental units deal?
7. Discuss the metric system?
8. What is the unit of mass, and what is its standard?
9. What is the unit of time?
10. What is the approximate equivalent of 1 meter in the English system? One gram? One liter?
11. The temperature of a solution is 68 F. What is the equivalent temperature in centigrade units?
12. A patient is found to have a temperature of 40 C. What is his temperature in Fahrenheit units?
13. Find the temperature value which is the same in both the C and F systems. (There are two methods, graphic and algebraic.)

CHAPTER 3   THE PHYSICAL CONCEPT OF ENERGY

Force

ALL MATTER is endowed with a property called *inertia.* This may be defined as the tendency of a resting body (of matter) to remain at rest, and the tendency of a body moving at constant speed in a straight line to continue its state of motion. To set a resting object in motion, we must push or pull it; that is, we must apply a *force.* Conversely, to stop a moving object or change its direction, we must apply an opposing force. It should be pointed out that the *mass* of a body is a measure of its inertia.

Work and Energy

Whenever a force acts upon a body over a distance, *work* is done. For example, to lift an object, a force is required to overcome the force of gravity. This applied force multiplied by the distance through which the object is lifted equals the work done. If the object is lifted twice the distance, then twice as much work will be done. Thus,

$$\text{work} = \text{force} \times \text{distance}$$

In a physical sense, work results from the expenditure of energy. What, then, is energy? *Energy may be defined as the actual or potential ability to do work.* It is obvious that work and energy must be measured in the same units, since the work done must be equal to the available energy.

There are two main types of *mechanical* energy:

1. **Kinetic Energy.** Every moving body can do work because of its motion. The energy of such a moving body is called *kinetic*

31

energy (from the Greek *kinetikos* = motion). Thus, the energy of a moving car or a rolling ball is called its kinetic energy.

2. **Potential Energy.** A body may have energy because of its position or its shape. For example, a car is parked on a hill. If its brakes are released, the car will begin to roll down the hill. It is logical to suppose that at the instant before the brakes were released, the car possessed *stored energy*, which became actual energy of movement, or kinetic energy, when the car began to move downhill. A wound-up clock spring and a stretched rubber band are further examples of stored or potential energy.

Thus, there are two types of mechanical energy—kinetic energy (energy of movement) and potential energy (stored energy).

Law of Conservation of Energy

What is the relationship between potential energy and kinetic energy? This is readily demonstrated by a simple example. Consider a rock poised on top of a cliff. The rock has a certain amount of stored or potential energy. Where did this energy originate? A definite amount of work had to be done originally to lift the rock to its position on top of the cliff, either by man or by natural forces. Such work was done against the force of gravity and was stored in the rock as potential energy (see Figure 3.1). Now the rock is shoved over the edge of the cliff and, as it falls, its po-

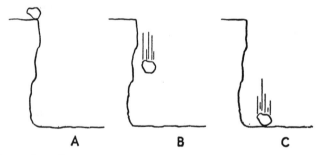

A   B   C

Figure 3.1. The relationship of potential energy to kinetic energy. In *A* the rock is at the top of the cliff, *external work* being required to lift it there against the force of gravity. This work is stored in the rock as *potential energy*. In *B* and *C* the rock is falling and the potential energy is thereby converted to *kinetic energy*. The external work = the potential energy = the kinetic energy.

tential energy is gradually converted to kinetic energy. As the potential energy decreases, the kinetic energy increases correspondingly, so that at the instant the rock reaches the ground its initial potential energy has been completely changed to kinetic energy.

<table>
<tr><td>initial energy
stored in rock
(P.E.)</td><td>=</td><td>final energy
of rock in motion
(K.E.)</td></tr>
</table>

Notice that the rock has different values of potential energy along its course; in other words, its potential energy gradually decreases as it falls toward the foot of the cliff. If the potential energy at any point above the ground is called an *energy level,* then it follows that the energy levels decrease at successively lower points. This important concept will be applied again later in the discussion of the energy levels of the electron shells in the atom. As exemplified in Figure 3.2, the *difference between any two energy levels equals the amount of potential energy that has been changed to kinetic energy. The rock will not of its own accord move from a lower to a higher energy level; external work is required.* This is another way of saying that *matter will move only from a higher to a lower energy level in the absence of outside work.*

It should be emphasized that the work required to lift the rock to the top of the cliff equals the potential energy of the rock at

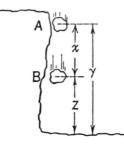

Figure 3.2. Potential energy at *A* — potential energy at *B* = kinetic energy at *B*; or, energy level at *A* — energy level at *B* = kinetic energy at *B*. If potential energy at $A = y$, and potential energy at $B = z$, then the difference in energy levels is $y - z = x$. Thus, x represents the amount of potential energy converted to kinetic energy when the rock drops from *A* to *B*.

the summit of the cliff; and this, in turn, equals the kinetic energy attained by the rock at the instant it reaches the ground. At the instant the rock strikes the ground its kinetic energy is converted to heat.

The foregoing discussion leads to the fundamental concept of the *law of conservation of energy*. This law states that *energy can be neither created nor destroyed, but various forms of energy can be changed to other forms. The total amount of energy in the universe is constant.*

What are the so-called *"forms of energy?"* They may be listed simply as mechanical, thermal (heat), light, electrical, chemical, atomic, molecular, and nuclear. These can be changed from one form to another. For example, an electric motor converts electrical energy to mechanical energy which is almost equal to the initial electrical energy, any difference being due to the waste of some of the energy due to the production of heat.

$$\frac{\text{electrical}}{\text{energy}} = \frac{\text{mechanical}}{\text{energy}} + \frac{\text{heat}}{\text{energy}}$$

Other examples of the transformation of energy from one form to another include: the conversion of chemical to electrical energy in a battery; the conversion of heat to mechanical energy in a steam engine; and the conversion of chemical to mechanical energy in a gasoline engine. Each of these transformations obeys the Law of Conservation of Energy; any energy disappearing in one form appears in some other form, and is never destroyed.

The most remarkable transformation of all, that of matter and energy, was formulated by Einstein in his famous equation

$$E = mc^2$$

in which E is energy, m is mass, and c is the speed of light in a vacuum. According to this equation, mass and energy are mutually convertible, one to the other. Thus, *matter cannot be destroyed, but can be changed to an equivalent amount of energy, and vice versa.* The conversion of even a minute amount of matter in this way releases a fantastically large amount of energy.

QUESTIONS

1. What is the physical concept of work?
2. Define energy; force.

3. Discuss kinetic energy and potential energy.
4. Name the various forms of energy.
5. Define the law of conservation of energy. Can you prove it?
6. What is the source of electrical energy in a battery?
7. Suppose that 1000 energy units of heat are applied to a steam turbine. What is the maximum number of energy units of electricity that can be obtained, neglecting the loss of energy due to friction?
8. Discuss the concept of energy levels.

CHAPTER 4 THE STRUCTURE OF MATTER

Speculation on the basic structure of matter has occupied the mind of man since ancient times. In fact, a number of Greek philosophers, including Epicurus, Democritus, and Aristotle, devoted much time to this problem. Democritus (about 400 B.C.) actually suspected that matter is composed of invisible and indivisible particles which he called *atoms* (from Greek *atomos* = uncut).

One of the best methods of studying something is to take it apart—the process of *analysis.* For instance, the most satisfactory way to ascertain the mechanism of a clock is to take it apart and study its individual parts and their relationships to one another. If this has been done successfully it should then be possible to reassemble the component parts into a perfectly operating clock. Such a recombination of parts is called *synthesis.*

Subdivisions of Matter

Let us now apply the process of analysis, based on logic and experiment, to the structure of matter. By definition, *matter is anything which occupies space and has inertia.*

Matter is usually found in nature in impure form, as a *mixture* of indefinite composition. An example is *rock salt,* a mixture of certain minerals (sodium chloride, calcium carbonate, magnesium chloride, calcium sulfate, etc) that can be separated in pure form. One of these constituents is ordinary salt—sodium chloride. If a piece of pure salt is now broken into smaller and smaller fragments, there will eventually remain the tiniest particle of salt that is still recognizable as such; that is, it will have the same properties of color, odor, taste, hardness, melting point, etc, as the original piece of pure salt. This particle, much too small to be visible, is called a *molecule* of salt. *A molecule is the smallest subdivison of a substance having the physical properties of that substance.*

The degree of attraction among the molecules of a given body of matter determines whether it will be a *solid,* a *liquid,* or a *gas.* Molecular attraction is strong in solids and relatively weak in liquids. The molecules of a gas actually repel each other.

How can a *molecule* of salt be further subdivided? Certainly not by ordinary physical means, such as crushing, because the salt molecule is made up of two *atoms*—one of sodium and one of chlorine—held together by strong electrochemical forces called *bonds* (see Figure 4.1). *The atom is the smallest particle of an element that has the characteristic properties of that element and can combine chemically with one or more atoms of another element (or elements).*

Two new terms have been introduced in the above discussion— *substance* and *element.* These will now be defined. *A substance is any material that has a definite, constant composition,* such as pure salt. (Wood and air are not substances because their composition varies; they are examples of mixtures.) Substances may be *simple* or *complex.* The simple ones, called *elements, cannot be decomposed to simpler substances by ordinary means.* Examples of elements are sodium, iron, lead, oxygen, hydrogen, and chlorine. There are in all ninety-two such naturally occurring elements.

The complex substances are *compounds,* formed by the *chem-*

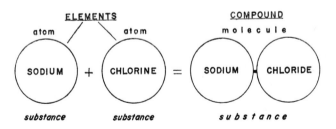

Figure 4.1. Atoms of certain elements are held together by strong electrochemical forces called bonds, to form compounds. Elements and compounds are both *substances* because they have a definite, constant structure.

ical union of two or more elements in definite proportions. Thus, as we have just seen, salt is a *compound* of equal numbers of atoms of the elements sodium and chlorine in chemical combination; every molecule of salt contains one atom of sodium and one atom of chlorine (see Figure 4.1). Note that *all elements and compounds are also substances* since they have a definite composition. Compounds may be made up of combinations of many atoms, ranging into the thousands.

Figure 4.2 illustrates the structure of matter down to the atomic level.

In summary then, rock salt is an indefinite *mixture* of *substances* each of which has a definite composition. Pure salt is both a *compound* and a *substance* because it has a definite composition, chemically combined atoms of sodium and chlorine in equal numbers. The smallest particle of salt (as well as other compounds) is a *molecule,* which can be separated chemically into simpler substances—*elements.* Thus, each molecule of salt consists of an atom of sodium and an atom of chlorine (see Figure 4.1). The *atom* is the smallest fragment of a particular element that is still recognizable as such. The atoms cannot be further subdivided by ordinary chemical or electrical methods, but can be broken down into smaller particles by special high energy,

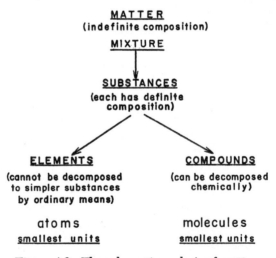

Figure 4.2. The schematic analysis of matter.

atom-smashing machines such as the nuclear reactor and the cyclotron.

Atomic Structure—the Electrical Nature of Matter

Let us now investigate the makeup of the atom itself. An atom is not a homogeneous particle, but consists of even smaller particles that are common to the atoms of all elements. Scientists have given us a highly satisfactory concept of atomic structure, based on extensive research, although no one has been, and no one probably ever will be, able to peer inside an atom. The modern *electrical* theory of atomic structure explains virtually all the observed natural phenomena in atomic physics. (A theory is a set of principles that attempts to correlate observed natural phenomena.) Note that the structure of the atom is represented by a diagram (not an actual picture) to help correlate certain experimental observations that have been treated mathematically.

The most widely accepted theory of atomic structure is that originally proposed by Niels Bohr in 1913, and thereafter extensively revised by others. We shall use the simplest concept, in which the atom is represented as a miniature solar system analogous to the sun with planets revolving about it (see Figure 4.3). In the center of the atom lies the positively charged core known as the *nucleus,* containing almost the entire mass of the atom. Revolving around the nucleus are the much lighter *orbital electrons,* each carrying a single *negative charge.* Each electron moves continuously in its own *orbit* or path with relation to the nucleus. However, the electrons whose orbits are at a particular distance from the nucleus (and hence at the same energy level) are grouped together and designated as belonging to a particular *shell* or *energy level.* These shells are identified by letters of the alphabet, the innermost being called the *K* shell, the next the *M* shell, and so on to *Q* (see Figure 4.3).

There is a specific maximum number of electrons that can occupy a particular shell, listed as follows:

K shell	2 electrons
L shell	8 electrons
M shell	18 electrons
N shell	32 electrons

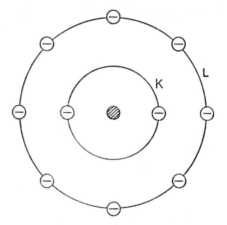

Figure 4.3. Greatly simplified atomic model according to the Bohr Theory. This shows the nucleus at the center surrounded by some of the shells of the electrons. (The rings do not represent orbits—see text.) The *K* shell can hold no more than 2 electrons, *L* no more than 8. The number of electrons that additional shells (up to *Q*) can hold has been determined. The electrons revolve very rapidly in their individual orbits and, at the same time, spin on their axes.

However, a shell may contain fewer than the maximum allowable. Furthermore, starting with the *M* shell, the outermost shell does not have to contain its maximum complement of electrons before electrons begin appearing in the next higher shell. In fact, an atom tends to be chemically stable when there are 8 electrons—OCTET—in the outer shell. For example, the element potassium has 2 *K* electrons, 8 *L* electrons, 8 *M* electrons, and 1 *N* electron, despite the fact that the *M* shell can hold a maximum of 18 electrons.

In a neutral atom the total number of orbital electrons exactly equals the total number of positive charges in the nucleus.

All of the atoms of a given element have the same total number of positive nuclear charges, but the atoms of different elements have different total nuclear charges. *An atom of a given element maintains its identity only if its nuclear charge is unaltered.* If the nuclear charge is altered, as by modern atom-smashing machines, the element is transmuted into an entirely different element.

Of what does the *nucleus* of an atom consist? We know that the nucleus contains two main kinds of particles—*protons* and *neutrons.* The proton carries a single positive charge (equal and opposite to the negative charge of an electron) and has an extremely small mass, about 1.6×10^{-24} gram. This is about 1,828 times *heavier* than the mass of an electron. Under ordinary conditions, protons are found only in the nucleus, but physicists can liberate them from the nucleus by means of special devices. Protons of all elements are identical, but *each element has its own characteristic number of nuclear protons.*

The other important constituent of the nucleus, the *neutron,* is a particle having about the same mass as a proton but carrying no charge; that is, the neutron is electrically *neutral.* Thus, protons and neutrons comprise virtually the entire mass of the nucleus but *only the protons contribute positive charges to the nucleus.* Other nuclear particles have been discovered but will not be considered here.

In summary, we can state that the atoms of various elements consist of the same *building blocks:* protons, neutrons, and electrons. Atoms of various elements differ from one another by virtue of different combinations of these same building blocks. The structure of an atom may be outlined as follows:

1. *Nucleus*—contains one or more:
   a. **Protons**—elementary positive particles, and
   b. **Neutrons**—elementary neutral particles having virtually the same mass as the proton
2. *Shells*—represents energy levels of one or more:
   a. **Electrons**—elementary negative particles in orbits

The number of electrons in the shells must equal the number of protons in the nucleus of a *neutral* atom.

Atomic Number

What determines the identity of an element, making it distinctly different from any other element? The answer lies in the number and arrangement of the orbital electrons, or what amounts to the same thing, the number of nuclear protons (see Figure 4.4). This number is characteristically the same for all

NUCLEUS

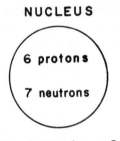

Atomic Number = 6

M a s s Number = 13

Figure 4.4. Nuclear composition (simplified).

atoms of a given element, and is different for the atoms of different elements. *The number of protons or positive charges in the nucleus of an atom is called its atomic number.* Thus, each element has a particular atomic number and, because of the corresponding specific number and arrangement of orbital electrons, its own distinctive chemical properties.

Mass Number

What determines the mass of an atom? We have already indicated that nearly the entire mass resides in the nucleus and, more specifically, in the nuclear protons and neutrons. These particles comprise the units of atomic mass. Therefore, the *total number of protons and neutrons in the nucleus of an atom is designated as its mass number* (see Figure 4.4). Protons and neutrons are often called *nucleons.*

Isotopes

If we were to examine a sample of a particular element, we would find that it consists of atoms that differ in mass number, even though they must all have the same atomic number. For example, the familiar element oxygen consists of a mixture of atoms of mass number 16, 17, and 18, although all these oxygen atoms have the same atomic number—8. Such atoms of the same element having different mass numbers are called *isotopes.* Any particular kind of atom, having a specific number of protons and neutrons, is called a *nuclide.*

Hydrogen, the simplest element, has three isotopes, as shown in Figure 4.5. All three hydrogen nuclides have atomic number 1, but they have different mass numbers—1, 2, and 3. The reason for the difference is immediately evident from the figure; *they differ only in the number of nuclear neutrons.*

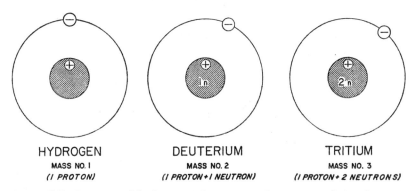

HYDROGEN   DEUTERIUM   TRITIUM
MASS NO. I   MASS NO. 2   MASS NO. 3
(I PROTON)   (I PROTON + I NEUTRON)   (I PROTON + 2 NEUTRONS)

Figure 4.5. Isotopes of hydrogen. These particular isotopes have the same atomic number, 1, but have different mass numbers. They are called *isotopes* of the same element, hydrogen. Their different mass numbers are due to the different numbers of neutrons in the nucleus. Tritium (pronounced trĭt'-ē-ŭm) is an artificial isotope and is radioactive.

The preceding discussion may be generalized as follows: *isotopes may be defined as atoms that have the same number of nuclear protons (equal to the atomic number of the element) but different numbers of nuclear neutrons.* This definition explains the difference in the mass numbers of isotopes.

Note that the atomic number and the mass number *together* specify the nuclear composition of any given atom, thereby designating a particular nuclide; the atomic number alone specifies the number of nuclear protons, and hence identifies the element.

The term *atomic weight* refers to the mass of any atom relative to the mass of an atom of oxygen nuclide 16. Because elements consist of mixtures of isotopes, the atomic weight is an average and is therefore almost never a whole number. However, since atomic weight and mass number are so nearly equal, these terms are often used interchangeably.

The Periodic System

The preceding discussion will be made clearer by actual examples. All of the elements can be arranged in an orderly series, called the *periodic table,* from the lightest to the heaviest (on the basis of atomic weight), or from the lowest atomic number to the highest (see Table 4.1). For instance, *hydrogen* is the simplest element, since it has only 1 proton and 1 orbital electron. The atomic number of hydrogen is therefore 1, and so is its mass number (see Figure 4.6.).

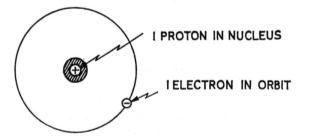

Figure 4.6. Atomic model of hydrogen.

The second element in the periodic series is *helium,* with two protons and two neutrons in the nucleus and two electrons in the K shell (see Figure 4.7). The atomic number of this element is 2 and the mass number is 4.

Lithium is the third element. Its atomic number is 3, and one of its isotopes has a mass number of 7 (see Figure 4.8). Note that the first shell has the maximum number of electrons, 2, and

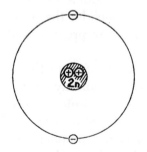

Figure 4.7. Atomic model of helium.

the third electron is in the next shell. Thus, the K shell of lithium has two electrons, and the L shell, one.

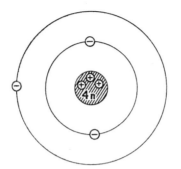

Figure 4.8. Atomic model of lithium.

Chemical Behavior

In tabulating the elements in order of increasing atomic number (periodic table), we find that there are *eight vertical groups* and *seven horizontal periods* (see Table 4.1). The *vertical groups* are subdivided, representing families of elements with surprisingly similar chemical properties. In other words, they participate in chemical reactions in a similar fashion. This relationship of the various elements was first discovered by Mendeleev, the eminent Russian chemist, in 1870. The *periods* (horizontal rows), on the other hand, consist of elements with the same number of electron shells, but with different chemical properties.

Why do the elements in any vertical group have similar chemical properties? These depend on the number of electrons in their *outermost shells*. For instance, lithium (Li), sodium (Na), and potassium (K) are seen in Table 4.1 to be members of the same family. Their atomic structure is shown in Figure 4.9. The neutrons are omitted here because they do not influence chemical behavior. The diagram shows that although these elements have different numbers of shells, they have one thing in common: *the outermost shell of each has one electron* which it can readily give up to assume a chemically stable octet configuration. This determines the similarity of their chemical behavior.

Consider now vertical group VII: fluorine (F), chlorine (Cl),

TABLE 4.1

ATOMIC PERIODIC TABLE

The numbers above each element represent atomic weight.
The numbers below each element represent atomic number.
Elements of particular interest to radiologic technologists are in dark type.
Transuranic elements (that is, beyond uranium) are man made.

Period	Group I	Group II	Group III	Group IV	Group V	Group VI	Group VII	Group VIII		
0	1.008 *Hydrogen* (*H*) 1									4.003 *Helium* (*He*) 2
1	6.940 Lithium (Li) 3	9.02 Beryllium (Be) 4	10.82 Boron (B) 5	12.01 *Carbon* (*C*) 6	14.008 Nitrogen (N) 7	16.000 *Oxygen* (*O*) 8	19.00 Fluorine (F) 9			20.183 Neon (Ne) 10
2	22.997 Sodium (Na) 11	24.32 Magnesium (Mg) 12	26.97 *Aluminum* (*Al*) 13	28.06 *Silicon* (*Si*) 14	30.98 Phosphorus (P) 15	32.06 Sulfur (S) 16	35.457 Chlorine (Cl) 17			39.994 Argon (A) 18
3	39.096 Potassium (K) 19	40.08 *Calcium* (*Ca*) 20	45.10 Scandium (Sc) 21	47.90 Titanium (Ti) 22	50.95 Vanadium (V) 23	52.01 *Chromium* (*Cr*) 24	54.93 Manganese (Mn) 25	55.85 *Iron* (*Fe*) 26	58.94 *Cobalt* (*Co*) 27	58.69 *Nickel* (*Ni*) 28
	63.57 *Copper* (*Cu*) 29	65.38 Zinc (Zn) 30	69.72 Gallium (Ga) 31	72.60 Germanium (Ge) 32	74.91 Arsenic (As) 33	79.00 Selenium (Se) 34	79.92 Bromine (Br) 35			83.7 Krypton (Kr) 36
4	85.48 Rubidium (Rb) 37	87.63 Strontium (Sr) 38	88.92 Yttrium (Y) 39	91.22 Zirconium (Zr) 40	92.91 Columbium (Cb) 41	96.0 Molybdenum (*Mo*) 42	99 *Technetium* (*Tc*) 43	101.7 Ruthenium (Ru) 44	102.9 Rhodium (Rh) 45	106.7 Palladium (Pd) 46
	107.88 *Silver* (*Ag*) 47	112.41 Cadmium (Cd) 48	118.70 Indium (In) 49	121.77 *Tin* (*Sn*) 50	127.6 Antimony (Sb) 51	126.93 Tellurium (Te) 52	126.92 *Iodine* (*I*) 53			131.3 Xenon (Xe) 54
	132.9 *Cesium* (*Cs*) 55	137.4 *Barium* (*Ba*) 56	Rare Earths 57–71	178.6 Hafnium (Hf) 72	180.9 Tantalum (Ta) 73	183.9 *Tungsten* (*W*) 74	186.3 *Rhenium* (*Re*) 75	190.2 Osmium (Os) 76	193.1 *Iridium* (*Ir*) 77	195.2 *Platinum* (*Pt*) 78

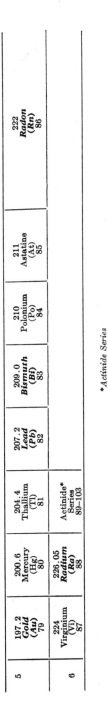

| 5 | 197.2 Gold (Au) 79 | 200.6 Mercury (Hg) 80 | 204.4 Thallium (Tl) 81 | 207.2 Lead (Pb) 82 | 209.0 Bismuth (Bi) 83 | 210 Polonium (Po) 84 | 211 Astatine (At) 85 | 222 Radon (Rn) 86 |
| 6 | 224 Virginium (Vi) 87 | 226.05 Radium (Ra) 88 | Actinide* Series 89–103 | | | | | |

*Actinide Series

227 Actinium (Ac) 89	232 Thorium (Th) 90	231 Protoactinium (Pa) 91	238.1 Uranium (U) 92
(237) Neptunium (Np) 93	(242) Plutonium (Pu) 94	(243) Americium (Am) 95	(247) Curium (Cm) 96
(247) Berkelium (Bk) 97	(251) Californium (Cf) 98	(254) Einsteinium (E) 99	(257) Fermium (Fm) 100
(256) Mendelevium (Md) 101	(254) Nobelium (No) 102	(257) Lawrencium (Lw) 103	

(Number in parenthesis is mass number of most stable nuclide.)

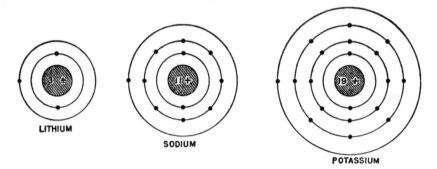

Figure 4.9. Elements with valence of +1.

and bromine (Br). As shown in Figure 4.10, these elements have different numbers of shells, but are similar in that they all have one less electron in the outermost shell than the number of electrons needed to saturate it to a chemically stable octet configuration. Therefore, these elements resemble each other in their chemical properties.

The number of electrons in the outermost shell determines *valence*. The family of elements represented by lithium, sodium, and potassium has one electron in the outermost shell, and the valence is +1, because these elements can *give* that electron to an element which needs an electron to saturate its outermost shell. On the other hand, the elements in the fluorine, chlorine, bromine family, as described above, have one less electron in the outermost shell than the number needed to saturate it. Their valence

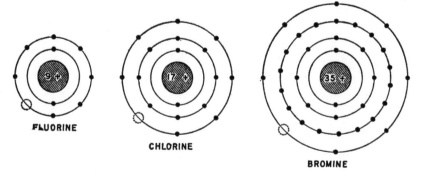

Figure 4.10. Elements with valence of −1.

is −1. These elements can *accept* an electron from the lithium family. The atom that gives up the electron now has one excess positive charge, whereas the atom that accepts the electron has one excess negative charge. Such charged atoms are called *ions,* and because of their opposite charges, firmly attract each other, forming a *compound.* The attraction between the two ions is a *chemical bond,* this type being called an *ionic bond* (see Figure 4.11).

Another type of bond is the *covalent bond* which involves the *sharing* of outer orbital electrons. This is exemplified by the water molecule which consists of two atoms of hydrogen bonded to one atom of oxygen by a sharing of the hydrogen valence electron as shown in Figure 4.12.

Suppose a particular element normally were to have its outer

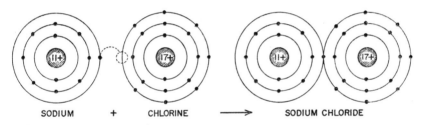

SODIUM   +   CHLORINE   ⟶   SODIUM CHLORIDE

Figure 4.11. Combination of sodium and chlorine chemically to form sodium chloride. The chemical bond is *ionic,* common to the formation of almost all inorganic compounds.

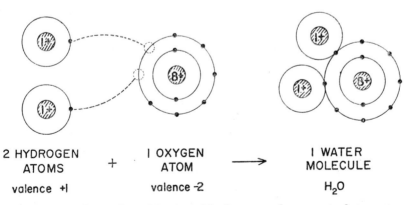

| 2 HYDROGEN ATOMS | + | I OXYGEN ATOM | ⟶ | I WATER MOLECULE |
| valence +I | | valence -2 | | H_2O |

Figure 4.12. Chemical combination of hydrogen and oxygen to form water, by sharing of two electrons—that is, a *covalent bond.*

shell saturated with electrons, or in octet configuration. It could not enter into a chemical reaction, since it would be unable to give up, accept, or share an electron. Referring again to Table 4.1, notice that there is such a family of elements known as the *inert elements.* They include helium, neon, argon, etc., as shown in Figure 4.13. (It has been possible, recently, to bring about chemical union of inert elements in very small quantities.)

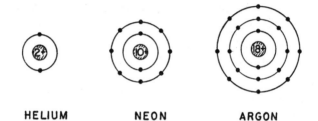

HELIUM   NEON   ARGON

Figure 4.13. Three members of the family of inert elements, with valence of 0. The outer shell in each case is completely filled with electrons. Therefore, these elements do not enter into chemical reactions.

Valence explains why elements always combine in *definite proportions.* Thus, sodium chloride is always formed by the union of one atom of sodium and one atom of chlorine. Water is always formed by the combination of one atom of oxygen with two atoms of hydrogen.

We may conclude from the foregoing discussion that *the chemical properties of an element depend on its valence,* which in turn depends on the number of electrons in the outermost shell. On the other hand, protons and neutrons do not participate in chemical reactions, although they go along for the ride, as it were.

Ionization

If an electron is removed from one of the shells of a neutral atom, or if an electron is added to a neutral atom, the atom does not lose its identity because it retains the same charge on its nucleus so that its atomic number is unaltered. However, the atom is now electrically charged. Thus, if an electron is *removed,* the atom becomes *positively charged* because one of its nuclear charges is now unbalanced and the atom has an excess of positive electricity. If an electron is *added* to a neutral atom, the atom now has an excess of negative electricity and is *negatively charged*. Such charged atoms will drift when placed in an electric field, as will be discussed later, and are called *ions* (from the Greek root "go"). The process of converting atoms to ions is termed *ionization*.

Note that *ionization is produced exclusively through the addition or removal of orbital electrons;* the protons do not participate in this process, their role being entirely passive.

1. **X-ray Bombardment of Matter.** X-ray energy displaces electrons from the atoms of matter lying in its path. Such atoms become positively charged, that is, positive ions. The removed electrons may ionize other atoms, combine with neutral atoms to produce negative ions, or recombine with positive ions to form neutral atoms.

2. **Electron Stream Bombardment of Matter.** The fast-moving electrons in an electron beam, or the electrons released from atoms by x rays as in (1) may collide with orbital electrons of other atoms and, in turn, displace them. An atom, upon losing an electron in this manner, becomes positively charged, but it is soon neutralized by recapturing an electron. Ionization by x rays or by fast electrons is of extremely short duration; the ions formed in this way become neutralized almost instantaneously.

3. **Spontaneous Breakdown of Radioactive Nuclides.** The radiation emitted by the radium series, for example, can produce ionization.

4. **Light-ray Bombardment of Certain Elements.** When light strikes the surface of certain metals such as cesium or potassium, electrons are liberated from the atoms and emitted from the surface of the metal. This principle is utilized in the photoelectric cell, which is the main component of the x-ray phototimer.

5. **Chemical Ionization.** If two neutral atoms such as sodium and chlorine are brought together, the sodium atom with valence +1 gives up its outermost or valence electron to the chlorine atom to fill its outermost shell. The sodium atom becomes a positive sodium ion (deficiency of one electron), while the chlorine becomes a negative chlorine ion (excess of one electron). When

the compound sodium chloride is dissolved in water its ions separate so that the water now contains sodium ions (Na^+) and chlorine ions (Cl^-). If the poles of a battery are connected to electrodes and these are immersed in the solution, the Na^+ ions move toward the negative electrode (cathode) and are called *cations,* while the Cl^- ions move toward the positive electrode (anode) and are called *anions,* as shown in Figure 4.14. On arriving at the *cathode,* each Na^+ ion picks up an electron and be-

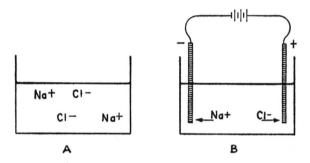

A   B

Figure 4.14. Chemical ionization and electrolysis. In *A* is shown a solution of ordinary salt (sodium chloride = NaCl) whose ions have separated to form equal numbers of Na^+ an Cl^- ions. In *B* a pair of electrodes has been immersed in the solution; the Na^+ ions move toward the negative electrode (cathode) and the Cl^- ions drift toward the positive electrode (anode), a process called *electrolysis.*

comes a neutral sodium atom. Each Cl^- ion gives up its extra electron to the *anode* and becomes a neutral chlorine atom. Separation of ions by an electric current is called *electrolysis.*

6. **Thermionic Emission.** When a metal is heated to incandescence (glowing hot), electrons are released from its surface. These electrons are the loosely bound, outermost orbital electrons of the atoms of the metal; the heating of the metal imparts sufficient kinetic energy to the electrons to increase the likelihood of their escaping from the surface, analogous to the evaporation of water molecules from a body of water. The ejection of electrons in this manner is called *thermionic emission,* or *thermionic effect;* it is essential to the operation of a modern x-ray tube.

Ionization is an extremely important process. Many chemical reactions take place between ions in solution. Ionization of air by x rays underlies the modern measurement of the exposure rate of an x-ray beam. Ionization of body tissues is believed to be the fundamental mode of action of x and gamma rays in therapy.

QUESTIONS AND PROBLEMS

1. Define element; compound; substance.
2. State briefly the Bohr concept of atomic structure.
3. Draw a model of the hydrogen atom and label the parts.
4. What constitutes the nucleus of an atom?
5. What is meant by orbit? Shell?
6. How does the mass of a proton compare with an electron?
7. A neutral atom has twelve electrons revolving in the orbits around the nucleus. What is its atomic number?
8. An atom has a nucleus containing six protons and two neutrons. What is its atomic number? Mass number?
9. Define isotope, and give an example. Define nuclide.
10. Why do elements combine chemically in fixed or definite proportions?
11. What is an ion? By what methods can ionization be produced?
12. Explain the similarity of chemical behavior of certain elements.
13. Discuss two types of chemical bonds. Give an example of a compound formed by each type.
14. Where are the following located in the atom: proton, neutron, electron? Why are they called the building blocks of matter?

CHAPTER 5 ELECTROSTATICS

Definition

THE TERM *electrostatics* is defined as *the branch of physics that deals with stationary or resting electric charges.* Another name for resting charges is *static electricity.*

We have previously pointed out that all atoms contain electrons revolving about a nucleus. If one or more of these electrons are removed, the atom is left with an excess of positive charge. Should the removed electron become attached to a neutral atom, then the latter will become negatively charged. Thus, *there are only two kinds of electricity, positive and negative.*

The same principle holds for a body of matter (made up of atoms) from which electrons may be removed, or to which electrons may be added. This process is called *electrification.* Consequently, an electrified or charged body has either an excess or a deficiency of electrons, being negatively charged in the first instance, and positively charged in the second.

Methods of Electrification

How can we electrify a body? There are three available methods, one or more of which you may have experienced in the past.

1. **Electrification by Friction.** You have probably observed, after walking over a woolen carpet, that touching a metal door knob will cause a spark to jump between your hand and the metal. Again, you may have noted that in combing your hair, a crackling sound is heard as the comb seems to draw the hair to it. These are examples of *electrification by friction,* that is, the removal of electrons from one object by rubbing it with another of a different kind. This phenomenon was known to the ancient

Greeks who observed that amber, after being rubbed with fur, attracts small bits of straw or dried leaves. Since the Greek word for amber is *elektron,* the origin of the root word for *electricity* is obvious.

Based on the two kinds of charges developed by friction, we customarily call one *negative* and the other *positive.* When a glass rod is rubbed with silk, some electrons are removed from the rod so that it becomes positively charged; this is *positive electrification.* The silk, of course, now has an excess of electrons and is negatively charged.

If amber is rubbed with fur, the amber picks up electrons and becomes negatively charged; this is *negative electrification.* It should be emphasized that friction is the simplest and most fundamental method of electrification.

2. **Electrification by Contact.** When a body is charged by friction and is then allowed to touch an uncharged object, the latter also acquires a charge. If the first object is negatively charged, it will give up some of its electrons to the second object, imparting to it a negative charge. If, on the other hand, the first object has a positive charge, it will remove electrons from the second object, leaving it positively charged. *We can conclude that a charged object confers the same kind of charge on any uncharged body with which it comes in contact.*

3. **Electrification by Induction.** Matter can be roughly classified on the basis of its ability to allow electrons to move freely through it. The materials in the first group are called *non-conductors* or *insulators* because electrons do not flow freely in them; examples are plastics, hard rubber, and glass. The materials in the second class are called *conductors* because they allow a free flow of electrons; they are exemplified by *metals,* such as silver, copper, and aluminum. Electrification by induction utilizes metallic conductors.

Surrounding every charged body there is a region in which a force is exerted on another charged body. This region or zone is called an *electric field.* If an uncharged metallic object is brought into the electric field of a charged object, there will be a shift of the electrons in the uncharged metal; this is the method of *electrification by induction.* Note that *only the electrons move.* In

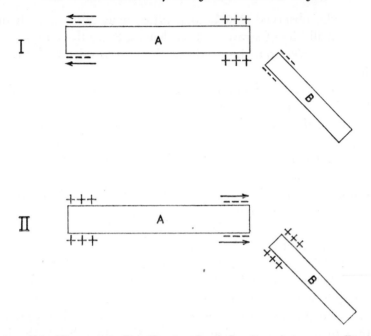

Figure 5.1. Temporary electrification by induction. *A* is a metallic rod. *B* is a charged body. In I electrons on *A* have been repelled to the left by the negative charge on *B*. In II the electrons on *A* have been attracted to the right by the positive charge on *B* (see Figure 5.3).

Figure 5.1, if the negatively charged rod, *B*, is brought *near* the originally uncharged metal body *A*, but not touching it, the negative field near *B* will repel electrons on *A* distally, leaving the end of *A* nearest *B* positively charged. The reverse is also true; if *B* were a positively charged rod, the end of *A* nearest *B* would become negatively charged. As *B*, in either case, is withdrawn from the electric field of *A*, the electrons on *A* are redistributed to their original position and *A* is no longer charged.

It is possible to charge a piece of metal by induction so that it will *retain its charge* even after the charging body has been removed. In Figure 5.2, *A* is shown first as it is affected by the electric field of *B*. Connecting the negative end of *A* temporarily to a water pipe (ground) causes the excess electrons to pass down to ground. If the ground connection is broken and *B* is *then* removed, *A* remains positively charged. It must be emphasized

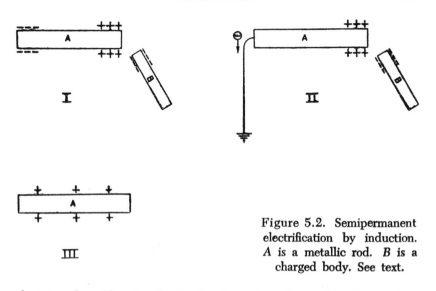

Figure 5.2. Semipermanent electrification by induction. *A* is a metallic rod. *B* is a charged body. See text.

that *in electrification by induction, the charged body confers the opposite kind of charge on the metallic body which is placed in its field.*

What is meant by *ground?* Since the earth is essentially an infinite reservoir of electrons, any charged body can be neutralized (same number of positive and negative charges) if it is grounded; that is, connected to the ground by a conductor. Thus, if a positively charged body is grounded, electrons will move up from the ground to neutralize it. On the other hand, if a negatively charged body is grounded, its excess electrons will move to ground and the body will again become neutral. In both instances, the charged body was neutralized by being brought to the same electrical potential energy level as ground. *Ground potential equals zero.* The symbol for ground is ⏚.

Laws of Electrostatics

There are five fundamental laws of electrostatics.

1. **Like Charges Repel Each Other; Unlike Charges Attract Each Other.** As noted previously, every charge has around it an electric field of force. This field consists of lines of force which start at a positive charge, pass through space, and end on a negative charge (see Figure 5.3). There is a resulting force of attrac-

tion between such oppositely charged bodies due to the electric field of force. If, on the other hand, we have two similarly charged bodies, their electric lines of force repel each other as in Figure 5.4 and so there is a repelling or separating force between them. Figure 5.5 shows this law demonstrated experimentally.

2. **The Electric Force Between Two Charges Is Directly**

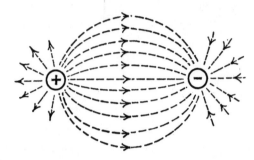

Figure 5.3. Electric fields. The electric lines of force are assumed to start on a positive charge, pass through space, and converge on a negative charge. A negative charge, such as an electron, placed in the field would drift toward the positive charge.

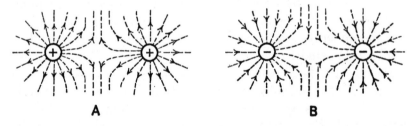

A   **B**

Figure 5.4. When like charges are brought near each other, their electric fields exert a force of repulsion or separation between the like charges.

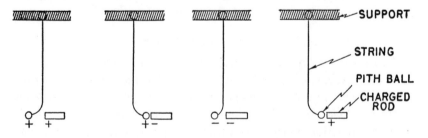

Figure 5.5. Like charges repel. Unlike charges attract.

Proportional to the Product of Their Magnitudes, and Inversely Proportional to the Square of the Distance Between Them. The force may be one of attraction or repulsion, depending on whether the charges are different or alike. For example, if we have two unlike charges and the strength of one is doubled, the force of attraction between them is doubled. If both are doubled, the force of attraction between them is $2 \times 2 = 4$ times as great. If the distance between the orginal charges is doubled, the force between them is $\frac{1}{2} \times \frac{1}{2} = \frac{1}{4}$ as great. If the distance is halved, the force is $2 \times 2 = 4$ times as great.

3. **Electric Charges Reside Only on External Surfaces of Conductors.** Thus, if a hollow metal ball is charged, all of the charges are on the outside surface, while the inside remains uncharged, as is evident from Figure 5.6. This is explained by the mutual

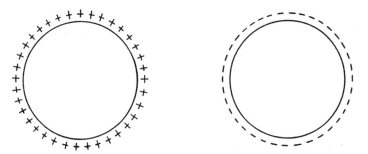

Figure 5.6. Electric charges tend to distribute themselves on the outer surface of a hollow metallic sphere.

repulsion of like charges which tend to move as far away from each other as possible. The greatest distance obviously exists between them if they move to the outer surface of the object.

4. **The Concentration of Charges on a Curved Surface of a Conductor Is Greatest Where the Curvature Is Greatest.** This is shown in Figure 5.7. The greatest possible concentration of charges is on a point, which obviously has an extremely high degree of curvature; in fact, charges become so crowded on a point that they easily leak away.

5. **Only Negative Charges (Electrons) Can Move in Solid Conductors.** The positive charges are actually charged atoms which do not drift in solids.

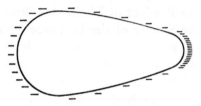

Figure 5.7. Electric charges tend to concentrate at the most pointed part of a metallic object.

Electroscope

A simple type of electroscope is shown in Figure 5.8. It is a device used to detect electric charges.

An electroscope may be charged either by contact or by induction.

1. **Charging an Electroscope by Contact.** Figure 5.9 shows how this is done. If the negatively charged rod is allowed to touch the knob, as in *A*, electrons move from the rod to the knob, stem, and leaves of the electroscope. Since the leaves now have like charges, they repel each other and spread, and the electroscope is negatively charged. On the other hand, if a positively charged rod is brought into contact with the knob of the electroscope, some electrons will move from the electroscope to the rod leaving the electroscope with a positive charge; the leaves will diverge just as they did when they were negatively charged. The leaves repel each other whenever they are similarly charged.

2. **Charging an Electroscope by Induction.** In Figure 5.10, a

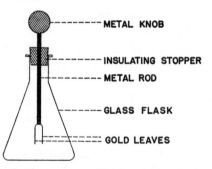

Figure 5.8. Gold leaf electroscope. When an electric charge is imparted to the electroscope, the leaves repel each other and spread apart.

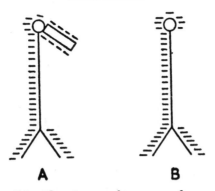

Figure 5.9. Charging an electroscope by contact.

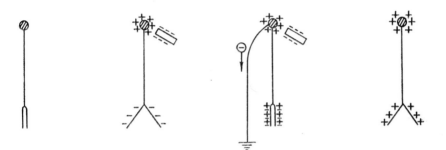

Figure 5.10. Charging an electroscope by induction. The successive steps are indicated in the diagrams from left to right.

negatively charged rod is brought *near* (but not touching) the knob. Its negative field repels electrons of the knob down to the leaves causing them to spread. If the knob is grounded, electrons pass to ground and the leaves collapse because they have lost their charge. If the ground connection is next broken and the rod is *then removed,* there is a deficiency of electrons in the electroscope (since some have passed to ground), and the electroscope is positively charged; the leaves consequently diverge. In general, the larger the charge, the greater will be the spread of the leaves.

The electroscope can be used not only to detect electric charges, but also to determine whether they are positive or negative. When a positively charged body is brought near the knob of a negatively charged electroscope, some of the electrons will be attracted from the leaves up to the knob, and the leaves will come together. But when a negatively charged body is brought near the knob of a negatively charged electroscope, some of the electrons from the knob will be repelled down to the already negative leaves and they will diverge even further. We can therefore deduce that a charged body brought near a similarly charged electroscope causes the leaves to diverge farther. But a charged body brought near an electroscope that has an opposite charge causes the leaves to come together.

A charged electroscope can be used to detect the presence of x rays because of the ionization they produce in air. The resulting ions neutralize the charge on the electroscope, causing the leaves to collapse.

Static Discharge

Suppose we have two oppositely charged bodies separated by an air space or insulator. The natural tendency is for the electrons on the negatively charged body to move over to the positively charged body, but this is prevented by the air space or insulator. However, if the electrons continue to be piled up on the negatively charged body, the electric field becomes correspondingly stronger. Eventually, if this process continues, the insulator cannot withstand the electrical stress or pressure and the electrons jump from the more negative to the less negative body, producing a *spark*.

There is a difference of electrical potential energy between two oppositely charged bodies, or between two bodies having similar charges of different sizes. The point where electrons are in excess is at a higher negative potential than the point at which they are deficient. Therefore, the electrons move down hill, so to speak, from the point of higher to the point of lower negative potential. This is analogous to the difference in mechanical potential energy in the example cited on pages 32-34.

The above discussion can best be brought home by considering the familiar *lightning discharge.* A thunder cloud may have a positive charge on its upper side and a negative charge on its lower side, or *vice versa,* built up by its motion through the air. The earth beneath develops an opposite charge by electrostatic induction. When the difference of electrical potential between a negatively charged cloud and the earth exceeds a certain critical value, electrons rush from the cloud to the earth in the form of a bolt of lightning. Since this usually involves an excessive movement of electrons to the earth, they may rush back toward the cloud, then back to the earth, repeating the process many times within a few millionths of a second.

QUESTIONS

1. What is meant by electrification and by what three methods can it be accomplished?
2. What happens if two similarly charged bodies are brought close together?
3. How many kinds of electricity are there? Which one is capable of drifting in a solid conductor?
4. If a negatively charged object is touched to a neutral object, what kind of charge does the latter acquire?
5. State five laws of electrostatics.
6. Of what value is the electroscope in radiology? How does it work?
7. What is meant by an electric field?
8. Describe electrification by contact; by induction.
9. Explain static discharge and give an example.
10. What is a conductor? an insulator? Give examples.

CHAPTER 6   *ELECTRODYNAMICS –* *THE ELECTRIC CURRENT*

Definition

IN THE PRECEDING chapter, we described the characteristics of stationary electric charges. Under certain conditions, electric charges can be made to drift through a suitable material, known as a *conductor. The science of electricity in motion is known as electrodynamics, or current electricity.*

The Nature of an Electric Current

Fundamentally, *an electric current consists of a flow of charged particles.* This may occur under the following conditions:

1. **In a Vacuum**—electrons may jump a gap between two oppositely charged electrodes in a vacuum tube. This will be discussed in detail later in connection with x-ray tubes.

2. **In a Gas**—two oppositely charged electrodes placed in a gas will cause positively charged ions in the gas to drift toward the negative electrode (cathode), and negatively charged ions to drift toward the positive electrode (anode). This occurs in an ordinary neon tube.

3. **In an Ionic Solution**—as already described on page 51, when a salt is dissolved in water, its ions separate. The positive ions move toward the cathode, while the negative ions move toward the anode, when electrodes are immersed in the solution and connected to a source of direct current. This flow of ions in opposite directions characterizes the flow of an electric current in an ionic solution (see Figure 4.14).

4. **In a Metallic Conductor**—this will be our main concern in the present chapter, and in later ones as well. *An electric current*

in a solid conductor consists of a flow of electrons only. In the atoms of metallic conductors, the valence (outermost) electrons are "free" and under certain conditions, these free electrons can be made to drift through the conductor. An electron does not move instantaneously from one end of a wire to the other, the individual electron drift being relatively slow—less than 1 mm per sec. However, the electrical impulse or *current* moves with

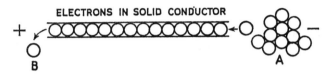

ELECTRONS IN SOLID CONDUCTOR

Figure 6.1. Diagrammatic representation of electron flow in a solid conductor. As an electron enters the conductor at *A* where there is an excess of electrons, another electron leaves at *B* where there is a deficiency of electrons.

tremendous speed, approaching that of light in a vacuum (3×10^{10} cm per sec). To make electrons flow in this manner, the conductor must have at one end an excess of electrons, and the other end a relative deficiency. The net result is that electrons will then flow from the point of excess (higher negative potential) to the point of deficiency (lower negative potential), just as a body will fall from a point of higher potential energy to a point of lower potential energy (see pages 32-34 and Figure 6.1).

An *electric circuit is defined as the path over which the current flows.* Electrons will not flow in a nonconductor such as glass because they are firmly bound in the atomic shells, and they flow only under certain conditions in semiconductors. Figure 6.2 shows diagrammatically the components of a simple electric circuit.

Sources of Current Electricity

There are two main types of devices that can pile up electrons at one point and simultaneously remove them at another point, thereby setting up a difference of electrical potential and causing an electric current to flow in a conductor. Since current electricity is a form of energy, it can be produced only by conversion of

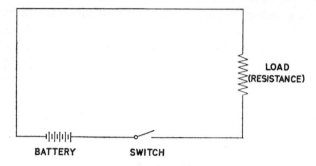

Figure 6.2. Simple resistance circuit. The load represents the total resistance of electrical devices in the circuit.

some other form of energy as exemplified in the following devices:

1. **Chemical** includes cells or batteries which convert chemical energy into electrical energy.

2. **Mechanical** includes the dynamo or generator which converts mechanical energy into electrical energy.

Other sources include *solar energy* (sunlight) and *atomic energy*. However, these are still incompletely developed and are not widely used as sources of electricity.

The Factors in the Simple Electric Circuit

There are three main factors in a simple electric circuit which is carrying a *steady direct current;* that is, a current of constant strength flowing always in the same direction. These factors can best be understood by comparing them with those involved in the flow of water through a pipe.

1. **Potential Difference.** We may define potential difference as the *difference in electrical potential energy between two points in an electric circuit,* due to an excess of electrons at one point relative to the other. A piling up of electrons at one point and simultaneous removal at another point are brought about through the use of a battery or generator. The potential difference actually represents the amount of *work* expended in causing a specified quantity of electricity to flow in a circuit. This resembles the situation shown in Figure 6.3, in which there are two containers of water at different levels connected by a pipe closed

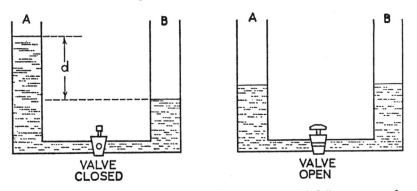

Figure 6.3. Water pressure analogy of electrical potential difference or voltage. With the valve closed, there is a difference of pressure between the water in arms *A* and *B*, proportional to the difference in water levels, *d*. (This is analogous to the potential difference between the poles of a battery on open circuit with no current flowing.) When the valve is opened, the water flows from arm *A* into arm *B* until there is no difference of pressure. (In a closed electric circuit, current continues to flow as long as potential difference is applied.)

by a valve. There is a difference of pressure between the water in both containers, proportional to the difference in height, *d*, of the water levels, when the valve is closed. As soon as the valve is opened, the difference in pressure causes the water to flow from container *A* to container *B* until their levels become equalized and the pressure difference disappears.

Similarly, an excess of electrons at one end of a conductor causes a difference of electrical pressure or a *potential difference* to be set up, *even if the conductor is interrupted by an open switch* as in Figure 6.2. As soon as the switch is closed, electrons rush from the point of excess (higher negative potential) to the point of deficiency (lower negative potential). Thus, just as with the water analogy, it is *the difference of potential that causes the electrons to flow.* If the difference of potential can be maintained by a battery or generator, electron current will be maintained as long as the switch is closed and the circuit is complete.

As the current flows through the circuit, there is a gradual fall of electrical potential along the circuit. Similarly, with water flowing through a pipe, the water pressure decreases gradually along the pipe.

It would be well to review the section in Chapter 3 on the difference of potential energy between two points at different levels above ground, and the movement of a rock from a higher to a lower energy level. The analogy with the difference in water pressure and the difference of electrical potential is self-evident.

The term *electromotive force* (emf) applies to the maximum difference of potential between the terminals of a battery or generator. As noted above, this is not really a force but rather a quantity of work or energy needed to move a unit quantity of electricity through the circuit.

The unit of potential difference (or emf) is the *volt, defined as that potential difference which will cause a current of one ampere to flow in a circuit whose resistance is one ohm.*

2. **Current.** The second factor in an electric circuit is *current, defined as the amount of electricity flowing per second.* (In a water pipe, there is a definite amount of water flowing per second past a given point.) Since an electric current consists of a flow of electrons, the more electrons flowing per second the "stronger" is the current. Similarly, the more water flowing per second in a pipe, the stronger is the current of water.

The unit of current is the *ampere, which may be defined as one coulomb quantity of electricity flowing per second* (ie, 6.3 $\times$ 10^{10} free electrons per sec). A current of one ampere will deposit a standard weight of silver per second from a solution of silver nitrate by electrolysis (to be exact, 0.001118 g per sec).

3. **Resistance.** The third factor in an electric circuit is *resistance,* a property of the materials making up the circuit itself. *Electrical resistance is that property of the circuit which opposes or hinders the flow of an electric current.* As an analogy, there is resistance to the flow of water in a pipe because of the friction of the molecules of water against the molecules of the pipe material. Note that the resistance of a nonconductor is so great that current cannot be made to flow in it.

The unit of electrical resistance is the ohm (symbol Ω), defined by the International Electrical Congress of 1893 as follows: *one ohm is the resistance of a standard volume of mercury under standard conditions* (to be exact, it is the resistance to a steady current by a column of mercury weighing 14.45 g, having a length of 106.3 cm and a uniform cross-sectional area, at O C).

On what does the resistance of a conductor depend? It depends on four things: material, length, cross-sectional area, and temperature.

a. *Material.* Certain materials are *conductors* of electricity, best exemplified by metals such as copper and silver. Copper is not quite so good a conductor as silver, but since it is much cheaper it is very widely used in electric wires. Other materials, such as glass, wood, and plastic, are nonconductors, since they have virtually no free electrons and hence offer tremendous resistance to the flow of electricity. Such nonconducting materials are called *insulators* or *dielectrics.* Intermediate in resistance are the *semiconductors* (see pages 217-221).

b. *Length.* A long conductor, such as a long wire, has more resistance than a short one. In fact, the *resistance is directly proportional to the length of the conductor.*

c. *Cross-sectional Area.* A conductor with a large cross-sectional area has a lower resistance than one with a small cross-sectional area. A more precise way of expressing this idea is: *the resistance of a conductor is inversely proportional to the cross-sectional area.* Similarly, a water pipe having a large cross-sectional area offers less resistance than a small pipe.

d. *Temperature.* With metallic conductors, the *resistance becomes greater as the temperature rises.*

Batteries or Cells

Let us turn now to some of the *chemical devices* that are used as sources of electric current. There are two main types, the *dry cell* and the *wet cell.*

1. **Dry Cell.** A typical dry cell consists of a carbon rod immersed in a paste composed of ammonium chloride, manganese dioxide, cellulose, and a small amount of water. These ingredients are placed in a zinc can, the top of which is closed by a layer of asphalt varnish (see Figure 6.4). The dry cell is actually *moist.* Note that the carbon rod nowhere comes in contact with the zinc can. A chemical reaction sets up a potential difference of 1.5 volts between the battery terminals or electrodes; that is, between the zinc and the carbon. The zinc becomes the negative

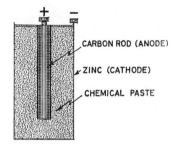

Figure 6.4. Construction of dry cell.

electrode (cathode), while the carbon becomes the positive electrode (anode). Dry cells are used in flashlights.

2. **Wet Cell.** An example of the wet cell is the lead storage cell used in automobiles. It consists of a hard rubber or plastic case containing sulfuric acid in which are immersed two electrodes; one, lead, and the other, lead oxide. As the result of a chemical reaction with the sulfuric acid, the lead becomes negative while the lead oxide becomes positive. Each cell in a storage battery produces a potential difference of about 2 volts. Automobile batteries usually have 3 or 6 wet cells.

Elementary Electric Circuits

Certain fundamental principles having been presented, we can now examine the simplest type of electric circuit; that is, one supplied by a *battery* as the source of potential difference. The current is a *steady direct current* that flows in *one direction only.* Since the only hindrance to the flow of current in such a circuit is resistance, it is often termed a *resistance circuit.*

Essential Parts. The resistance circuit has three essential parts: (1) battery, (2) conductor, and (3) load or resistance. The circuit may be represented by a diagram which shows, incidentally, some of the more common electrical symbols (see Figure 6.2).

Current flows only when the switch is *closed.* If the circuit is broken at any point, it is then called an *open circuit.* When the circuit is completed, it is called a *closed circuit.* However, keep in mind that there is a potential difference even on open circuit.

Polarity of Circuits. The current in a resistance circuit has a definite direction, and is therefore said to have *polarity.* Most physics books have again come to regard the direction of flow of the current to be from positive to negative. However, for the sake of simplicity, we shall assume the *direction of the current to be the same as that of the flow of electrons; that is, from the cathode (—), through the external circuit, to the anode (+)* (see Figure 6.5).

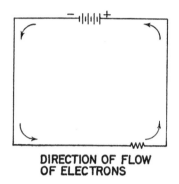

**DIRECTION OF FLOW
OF ELECTRONS**

Figure 6.5. The direction of *electron flow* determines the direction of an electric current. The *conventional direction* is opposite to this.

Connection of Meters. The polarity of circuits makes it necessary for *direct current* measuring devices or meters to be connected correctly in the circuit. Such devices have *polarity;* thus the two terminals on a meter are labeled (+) and (—) and, to avoid damage, must be so connected that the (+) terminal is on the (+)side of the circuit and the (—) terminal on the (—) side.

A *voltmeter* measures in volts, the potential difference between any two points in a circuit. Therefore, the voltmeter is always connected in *parallel;* that is, it is placed in a small circuit which is a branch of the main circuit, as in Figure 6.6. Obviously, if a voltmeter is connected successively between various pairs of points in the circuit where the resistance may differ, the voltage drops will be different. Voltage readings on the voltmeter will be larger across the higher resistance and smaller across the lower resistance. Similarly, the drop in pressure between any pair of

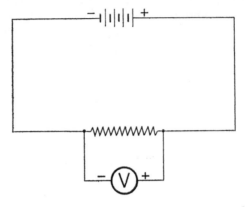

Figure 6.6. Connection of a voltmeter—in parallel.

points in a water pipe will be different if the bore of the pipe is not uniform, there being a greater fall in pressure where the pipe is narrower.

An *ammeter* (ampere + meter) measures, in amperes, the quantity of electricity flowing per second. It is connected directly into the circuit, that is, in *series* (see Figure 6.7). When so connected, *an ammeter will read the same no matter where it is placed in a series circuit* (see page 74).

Ammeters and voltmeters look alike externally and consist of a box enclosing the mechanism, with a glass window in which a pointer moves over a calibrated scale, as shown in Figure 6.8. The instrument is always labeled "voltmeter" or "ammeter." The construction of these devices will be described later.

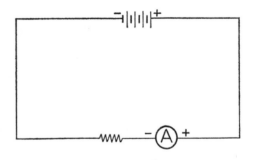

Figure 6.7. Correct connection of an ammeter—in series.

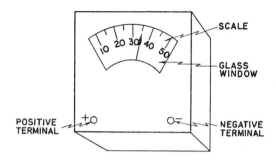

Figure 6.8. Voltmeter or ammeter.

Ohm's Law

Ohm, a German physicist, discovered that when a steady direct current is flowing in a resistance circuit, there is a definite relationship between the voltage, amperage, and resistance; ie, a circuit in which resistance is the only obstacle to the flow of current). This is expressed in *Ohm's Law*, which states that *the value of the current in a resistance circuit supplied by steady direct current is equal to the potential difference divided by the resistance:*

$$I = \frac{V}{R} \tag{1}$$

where I = current in amperes
V = potential difference in volts
R = resistance in ohms

If any two of these values are known, the third may be found by substituting in the equation and solving for the unknown.

Two rules must be observed in applying this law:

1. When Ohm's Law is applied to a *portion of a circuit,* the current in that portion of the circuit equals the voltage across that portion of the circuit divided by the resistance of that portion of the circuit.

2. When Ohm's Law is applied to the *whole circuit,* the current in the circuit equals the total voltage across the circuit divided by the total resistance of the circuit.

Series and Parallel Circuits

Electric appliances (toasters, flatirons, fans, etc) and electric sources (batteries, generators) may be connected in a circuit in two principal ways. One is called *a series circuit, which may be defined as an electric circuit whose component parts are arranged in a row, so that the current passes consecutively through each part* (see Figure 6.9). The other arrangement is called a *parallel circuit, wherein the component parts are connected as branches of the main circuit so that the current is divided among them* (see Figure 6.10).

1. **Series Circuit.** When electric appliances and other *current-consuming devices* are connected in series as shown in Figure 6.9,

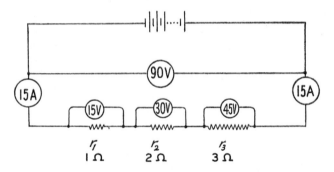

Figure 6.9. Resistors in series. Ohm's Law applies to the entire circuit, and to any portion of it. The total voltage drop, 90 V, is equal to the sum of the voltage drops across all the resistors, 15 V + 30 V + 45 V. The current, 15 amp, is the same everywhere in the circuit. The voltage drop across each resistor is proportional to the resistance, because $V = IR$ and I is constant in this circuit.

the total resistance is equal to the sum of the separate resistances. Thus, if R = total resistance of the whole circuit, and r_1, r_2, and r_3 represent the resistances of various devices connected in series, then,

$$R = r_1 + r_2 + r_3 \tag{2}$$

Suppose in this case, $r_1 = 1$ ohm, $r_2 = 2$ ohms, and $r_3 = 3$ ohms, and the ammeter reads 15 amp. (Remember that the current is the same everywhere in a series circuit.) What is the voltage across the *entire* circuit? First substitute the given values of resistance in equation (2),

$$R = 1 + 2 + 3 = 6 \text{ ohms}$$

Then apply Ohm's Law

$$I = V/R$$
$$\text{or}$$
$$V = IR \tag{3}$$

and substitute the values of I and R (obtained above)

$$V = 15 \text{ amp} \times 6 \text{ ohms} = 90 \text{ volts}$$

There is another approach to this problem—the application of Ohm's Law to *each part* of the circuit. Since $V = IR$ in general, the voltage across each resistor in Figure 6.9 is equal to the current times the resistance of each resistor:

$$v_1 = Ir_1 = 15 \times 1 = 15 \text{ volts}$$
$$v_2 = Ir_2 = 15 \times 2 = 30$$
$$v_3 = Ir_3 = 15 \times 3 = 45$$

since I is the same throughout the circuit. The total voltage is then

$$V = v_1 + v_2 + v_3$$
$$V = 15 + 30 + 45 = 90 \text{ volts}$$

which is the same result obtained above by applying Ohm's Law to the entire circuit.

What is the real significance of the equation $V = IR$? It actually states that the *voltage across a resistor equals the current times the resistance.* But, besides this, it indicates that there is a fall or drop in potential across a resistor, and as the resistance increases so does the potential drop. Sometimes this is called a voltage drop or, what is the same thing, an IR drop. Keep in mind that the terms potential difference, voltage, voltage drop, and IR drop are used interchangeably.

In summary, then, voltage is a measure of the potential difference across a resistor. The voltage drop across the resistor increases as the resistance increases. It is obvious from the above examples that the voltage drop across an entire series circuit must equal the sum of the voltage drops across each resistor.

If *current-producing devices* such as electric cells are connected in series, **the total voltage equals the sum of the individual voltages.** In such a circuit, the positive pole of one cell is connected to the negative pole of the next, as in Figure 6.10. If

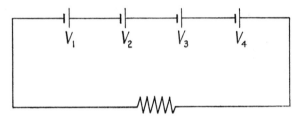

Figure 6.10. A circuit containing batteries in series.

V is the total voltage, and v_1, v_2, v_3, and v_4 are the voltages of the individual cells, then

$$V = v_1 + v_2 + v_3 + v_4 \qquad (4)$$

The total current flowing in such a circuit can be determined from the total voltage and the total resistance simply by applying Ohm's Law.

2. **Parallel Circuit.** Figure 6.11 shows a parallel circuit; careful comparison of A and B reveals that the circuits are essentially identical, although they have been drawn somewhat differently.

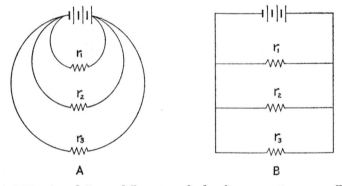

A   B

Figure 6.11. A and B are different methods of representing a parallel circuit. Note that they are fundamentally identical.

This is due to the fact that the conductors themselves have so little resistance that it may be disregarded. In a parallel circuit the *voltage drop is the same across all branches.* Thus the voltage drop across r_1 is the same as that across r_2 and that across r_3, and the voltage drop in each case is therefore equal to V, the voltage supplied by the battery.

What is the resistance of a circuit with several parallel branches? It is obvious that in such a circuit, the current divides among the several branches or paths, so that the total resistance must be less than that in an undivided path. As more resistors are added in *parallel,* there is an increase in their total cross-sectional area so that the total resistance decreases.

The total resistance of a *parallel circuit* is expressed by the following equation:

$$1/R = 1/r_1 + 1/r_2 + 1/r_3 + \dots 1/r_n \qquad (5)$$

where R is the total resistance and r_1 r_2 etc. are the resistances of the separate branches.

Thus, the total resistance of a parallel circuit can be found by the rule based on equation (5): *the reciprocal of the whole resistance equals the sum of the reciprocals of the separate resistances in a parallel circuit.*

A typical example is presented in Figure 6.12 to clarify this rule. If the resistances in the branches of a parallel circuit are

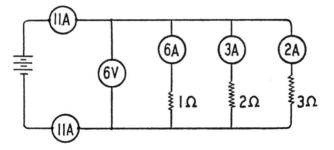

Figure 6.12. Resistors in parallel. The voltage drop, 6 V, is the same across the whole circuit and across each branch. The current is different in the branches, the smallest current flowing in the largest resistance. The total current, 11 amp, equals the sum of the branch currents: 6 amp + 3 amp + 2 amp. The total resistance is found by $1/R = 1/r_1 + 1/r_2 + 1/r_3$.

1, 2, and 3 ohms respectively, what is the total resistance of the circuit? Substituting the values of r in equation (5), we find that the total resistance is less than any of the resistances in the individual branches.

$$1/R = 1/1 + 1/2 + 1/3$$

$$1/R = \frac{6 + 3 + 2}{6}$$

$$1/R = 11/6$$

Inverting both sides of the equation

$$R = 6/11 = 0.55 \text{ ohm}$$

Suppose, in the above example, that the potential difference across the circuit is 6 volts. How do we find the current in each branch and in the main lines? Since this is a parallel circuit, the voltage drop across each branch must be the same—6 volts. Then, applying Ohm's Law to each branch,

$$i = v/r$$

$$i_1 = 6/1 = 6 \text{ amp}$$

$$i_2 = 6/2 = 3 \text{ amp}$$

$$i_3 = 6/3 = 2 \text{ amp}$$

The *current in the main lines of a parallel circuit equals the sum of the branch currents,*

$$I = i_1 + i_2 + i_3 \qquad (6)$$

Substituting the values of the branch currents just obtained,

$$I = 6 + 3 + 2 = 11 \text{ amp}$$

To verify the answer, Ohm's Law is applied to the entire circuit. Since the total current is 11 amp and the total resistance is 0.55 ohm,

$$V = IR$$
$$V = 11 \times 0.55 = 6 \text{ volts}$$

As expected, the current divides among the branches so that the *smallest current flows in the branch having the greatest re-*

sistance. Since V is equal in all branches, if r in a given branch is greater, i must be smaller, and *vice versa,* in accordance with $V = IR$.

With *current-producing devices* connected in parallel, the total voltage is the same as that of a single battery, if all the batteries are alike. Figure 6.13 shows this type of connection. The total current (ie, in the main line) is equal to the sum of the currents provided by all the batteries.

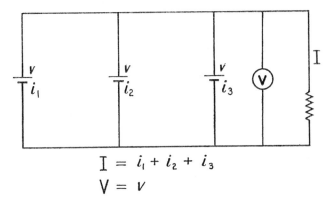

$$I = i_1 + i_2 + i_3$$
$$V = v$$

Figure 6.13. A circuit containing batteries connected in parallel. Total current, $I = i_1 + i_2 + i_3$. Voltage across entire circuit, $V = v$.

The parallel circuit is used in electrical wiring of homes, factories, and other buildings. With a series, circuit electrical failure of one appliance, such as a light bulb, causes all other lights in the circuit to go out. Besides, as more appliances are added in a series circuit, each appliance receives less voltage, since the total voltage drops must be divided among more appliances; thus, as more electric lights are added, all become dimmer.

The main advantage of the parallel circuit is that failure of one appliance does not prevent operation of the others, since they are on branches of the main circuit. As more branches are added, the total resistance decreases and the total amperage increases while the voltage remains unchanged. Thus, the more appliances added, the greater is the amperage in the *main circuit.* When too many appliances are added, the amperage in the main line may become excessive and the circuit *overloaded.* Under these conditions the wiring system could become hot enough to ruin one or more appliances, or start a fire. This danger can be avoided by connecting protective devices such as circuit breakers or fuses in the circuit. A *fuse* contains a wire which melts more easily than the wires of the circuit being protected (see Figure 14.3). When the amperage exceeds the safe maximum, the fuse melts, thereby opening the circuit and stopping the flow of current before the circuit can become overheated. Such a fuse, in which the protective wire has melted, is called a *blown fuse*. When two non-insulated wires carrying a current touch each other, a *short circuit* results; there is a marked fall in resistance and a corresponding rise in amperage which is usually sufficient to blow a fuse. The *circuit breaker,* another type of protective device, will be discussed later.

Electric Capacitor (Condenser)

We have already noted that electrical energy can be obtained from a chemical reaction in a dry or wet cell; in a sense, this may be regarded as *stored electricity.* There is another method of storing electrical energy that does not depend on chemical reaction. You should recall from the chapter on Electrostatics that when a conducting body is charged, the entire charge is distributed on the surface of the body. Such an electrically charged body *stores* electrical energy, provided it is insulated so that the charge does not "leak" away; it is an *electric capacitor* or *condenser.*

One of the simplest types is the *parallel-plate capacitor,* consisting of a pair of flat metallic plates arranged parallel to each other and *separated* by a small space containing air or special insulating material. Capacitor operation is shown in Figure 6.14. When the capacitor is connected to a source of direct current, such as a battery, electrons move from the negative terminal of the battery to the plate to which it is connected. At the same time, an equal number of electrons passes from the opposite plate to the positive terminal of the battery. This is indicated in Figure 6.14A. *No electrons pass from one plate to the other across the space between them,* unless there is a breakdown of the insulator due to application of excessively high voltage. When the capaci-

CAPACITOR

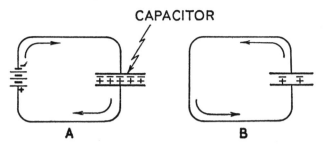

A   B

Figure 6.14. Parallel plate capacitor. In A the capacitor is connected to a battery and is being charged. In B the capacitor has been disconnected and its plates connected by a wire, whereupon it loses its charge. The arrows indicate the direction of electron flow in the wire.

tor is fully charged, there is a difference of potential (voltage) across its plates equal and opposite in direction to that of the current which charged it; the current now ceases to flow in the wires.

If the capacitor is then disconnected from the battery, it retains its electrical charge until the plates are connected by a conductor as in Figure 6.14B, whereupon the capacitor discharges and once more becomes neutral. The direction of the discharging current is *opposite* to that of the original charging current, and stops flowing as soon as the capacitor is completely discharged. (With an alternating current the capacitor plates repeatedly change polarity and the current continues to flow.)

How much electricity can a capacitor store? The quantity of electricity in coulombs stored per volt applied to the capacitor is called its *capacitance,* the unit of capacitance being the *farad,* a very large unit; a more practical unit is the microfarad (millionth of a farad). The larger the area of the plates, and the smaller the space between them, the greater is the capacitance or electrical storing ability. The insulator or, as it is more often called, the *dielectric* between the plates also determines the capacitance for a given plate size and spacing; if the *dielectric constant* (insulating ability) of air is taken as 1, then wax paper has a dielectric constant of 2, and glass a constant of about 7. These materials will increase the capacitance of a given capacitor two times, and seven times, respectively, as compared with an air dielectric. In practice, a capacitor usually is made up of multiple parallel plates to increase the capacitance.

The Work and Power of an Electric Current

Electrical energy, just as any other form of energy, is capable of performing work. According to the Law of Conservation of Energy, a given amount of electrical energy is convertible to a definite amount of work. The power of a steady direct current, or the *amount of work the current can do per second,* is simply the product of the current and voltage:

$$P = IV \qquad (7)$$

Power   = amperage × voltage
(in *watts*)

This relationship is called the *power rule.* The unit of power is the *watt,* 746 watts being equal to 1 horsepower.

It is well known that an electric current produces heat in the circuit. The greater the *resistance* of the circuit or any portion of the circuit, the greater is the rate of heat production. To find the amount of electrical power lost in heat production, we must convert equation (7) to a form which contains the term *resistance.* This may be done by the use of Ohm's Law.

$$I = V/R$$
$$V = IR$$

Substituting *IR* for *V* in equation (7),

$$P = I \times IR$$

Power Consumed = I²R watts $\qquad$ (8)
(*as heat*)

where *I* is the current in amperes and *R* the resistance in ohms.

Two conclusions may be drawn from equation (8). First, the power loss due to the heating effect of the current is proportional to the resistance; thus, that portion of the circuit having the greatest resistance will have the greatest amount of heat produced in it.

Second, the power loss in the form of heat increases very rapidly with an increase in the current. Thus, *the power loss is proportional to the square of the current.* In other words, if the current is doubled, the power loss in heat formation is 2^2 or 4 times as great. If the current is tripled, the power loss is 3^2 or 9 times as great. The importance of this principle will be discussed later.

QUESTIONS AND PROBLEMS

1. How does an electric current flow along a wire? In a salt solution? Under what other conditions do electrons flow?
2. What are the two main sources of electric current?
3. Define potential difference. What is the unit of measurement? What is the difference between potential difference, electromotive force, and voltage?
4. What is meant by electrical resistance, and on what does it depend?
5. Define electric current strength and state its unit.
6. In what direction does an electric current flow? What is meant by the "conventional" direction?
7. State Ohm's Law.
8. An electric circuit has, in series, appliances with these resistances: 4 ohms, 10 ohms, 6 ohms, 40 ohms. What is the value of the current when the applied potential difference is 120 volts?
9. How many volts are required to produce a current of 30 amperes in a circuit having a resistance of 5 ohms?
10. What type of electric circuit is used in ordinary house wiring? Why?
11. A circuit has the following resistances connected in *parallel:* 3 ohms, 6 ohms, 18 ohms. What is the total resistance of the circuit? If 12 volts are applied across the circuit, what is the amperage in each branch and in the main lines?
12. What is the power formula? What is the unit of power?
13. A circuit has a current of 20 amperes flowing under an impressed electromotive force of 110 volts. What is the power of the circuit?
14. What is an electric capacitor? Describe its mode of operation, with the aid of a diagram. Contrast its behavior in a direct current with that in an alternating current.

CHAPTER 7 MAGNETISM

Definition

Magnetism *may be defined as the ability of certain materials to attract iron, cobalt, or nickel.* In general, any material that attracts these materials is called a *magnet.* The various types of magnets will be described in the next section.

Classification of Magnets

There are three main types of magnets: *natural magnets, artificial permanent magnets,* and *electromagnets.*

1. Natural magnets include, first, the *earth* itself, a gigantic magnet which, in common with all others, can deflect the needle of a compass. The other type of natural magnet is the ore *lodestone,* consisting of iron oxide (magnetite) that has become magnetized by lying in the earth's magnetic field for an extraordinarily long period of time. In some parts of the world there are actually magnetic mountains consisting of such ores. The ancient Greeks recognized the magnetism of lodestone more than 2500 years ago.

2. Artificial permanent magnets usually consist of a piece of *hard steel* in the shape of a horseshoe or bar which has been *artificially* magnetized so that it is capable of attracting iron (see Figure 7.1). The *magnetic compass* is simply a small permanent magnetic needle swinging freely on a pivot at its center, used to detect the presence of magnetic materials and to locate the earth's magnetic poles. A widely used alloy of aluminum, nickel, and cobalt—*alnico*—is an extremely powerful permanent magnet that has many times the lifting ability of a steel magnet. It is interesting to note, however, that aluminum itself is not attracted by a magnet.

BAR   MAGNET   **HORSESHOE MAGNET**

Figure 7.1. Permanent magnets.

3. **Electromagnets** are usually temporary magnets produced by means of an electric current, as will be described later.

Laws of Magnetism

The *three fundamental laws of magnetism* are as follows:

1. *Every magnet has two poles,* one at each end. This concept is based on the observation that when a magnet is inserted into a container of iron filings and then removed, iron particles will cling to its ends or *poles,* as illustrated in Figure 7.2. When

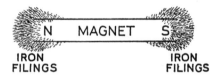

**IRON   IRON
FILINGS   FILINGS**

Figure 7.2. Polarity of a magnet. This is demonstrated by immersing first one end, then the other, of a magnet into a box of iron filings which cling mainly at the ends or *poles.*

allowed to swing freely (as in a magnetic compass), one pole of the magnet points north and the other end south. The poles are therefore called, respectively **north** and **south** poles.

2. *Like magnetic poles repel* each other; *unlike poles attract* each other. Thus, if two bar magnets are brought together with their south poles facing each other as in Figure 7.3A, there is a force which tends to push them apart. If, on the other hand, the magnets are brought together with opposite poles facing each other, there is a force which pulls them together, as in Figure 7.3B.

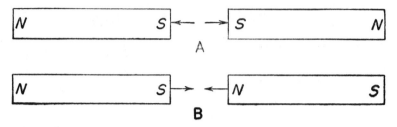

Figure 7.3. (A) like poles repel. (B) unlike poles attract.

3. *The force of attraction (or repulsion) between two magnetic poles varies directly as the strength of the poles, and inversely as the square of the distance between them.* Thus, doubling the strength of one pole doubles the force between the poles. Tripling the strength triples the force. On the other hand, doubling the distance between two poles reduces the force between them to $(\frac{1}{2})^2$ or $\frac{1}{4}$. Tripling the distance reduces the force to $(\frac{1}{3})^2$ or $\frac{1}{9}$.

Nature of Magnetism

Why are some materials magnetic (iron-attracting) whereas others not? The most widely accepted explanation is an outgrowth of Weber's theory of magnetism, first proposed in the early nineteenth century. Weber's theory was based on observations such as the following:

1. Breaking a magnet results in each fragment becoming a *whole magnet* with its own north and south poles, as in Figure 7.4.

2. Heating or hammering a piece of steel while it lies near a magnet causes the steel to become a magnet.

3. Stroking an iron bar repeatedly with one end of a magnet causes it to become a magnet.

4. Gently jarring a test tube of iron filings near a magnet causes it to become a magnet. If the test tube is now shaken vigorously, it loses its magnetism.

These observations suggested that elements such as iron, nickel, and cobalt consist of extremely small, discrete magnets. But only with the advent of modern atomic science has the true nature

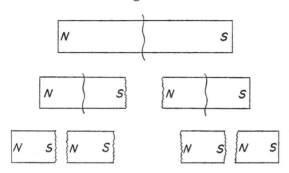

Figure 7.4. When a magnet is broken, each part becomes a whole magnet with a north and a south pole.

of such elements been fully explained. According to atomic theory, the orbital electrons in magnetic elements spin predominantly in one direction, resulting in the formation of *magnetic poles in the atom as a whole.* These atomic magnets are believed to form groups called *magnetic domains.*

A *magnetic element* can exist either in the magnetized or non-magnetized state. In a *nonmagnetized* piece of iron, the domains are in disorderly arrangement, as shown in Figure 7.5A. There-fore, the iron does not have magnetic poles. On the other hand, in *magnetized* iron the domains are aligned in orderly fashion, all their north poles pointing in one direction, and their south poles in the opposite direction, as shown in Figure 7.5B.

In *nonmagnetic elements* such as copper, just as many electrons spin in one direction as in the other, so their opposing magnetic

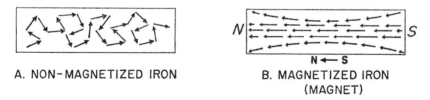

A. NON-MAGNETIZED IRON B. MAGNETIZED IRON (MAGNET)

Figure 7.5. Magnetic domains in a magnetic substance. In *A*, when a ferro-magnetic substance such as iron is not magnetized, the domains are ar-ranged haphazardly and there is no net polarity. In *B* the iron has been magnetized and the domains are arranged in orderly fashion to produce opposite poles.

effects cancel out. Therefore, in such elements the atoms are not magnetic, no domains are formed, and they *cannot* be magnetized.

Magnetic Fields

What is there about a magnet that causes it to attract iron? Surrounding a magnet there is a zone of influence called a *magnetic field,* representing a summation of the fields of all the domains. The magnetic field can be demonstrated by placing a piece of cardboard over a magnet and sprinkling iron filings on the cardboard. If the latter is gently tapped, the filings arrange themselves into a pattern resembling the one in Figure 7.6. The lines along which the filings arrange themselves are called *lines of force* or *magnetic flux.* In other words, the magnetic field is made up of lines of force, and the more concentrated these lines, the stronger is the field. In fact, the *strength of a magnetic field is proportional to the number of lines per square centimeter.*

The lines of force can be mapped out by means of a small magnetic compass. Starting at the north pole of a bar magnet, one simply moves the compass around it toward its south pole and observes the behavior of the needle which acts as though it were following a line through space from the north to the south pole of the magnet, as shown in Figure 7-6.

Characteristics of Lines of Force

Certain facts are known or assumed about lines of force.

1. They are assumed to leave the north pole, travel through space in a curved path, and enter the magnet at its south pole (see Figure 7.6).

2. Lines of force seem to repel each other when they travel in the same direction, and attract each other when they travel in opposite directions (see Figure 7.7).

3. The field is distorted by magnetic materials, but is not affected by nonmagnetic materials (see Figure 7.8).

Magnetic Induction (Magnetization)

If a piece of ordinary, nonmagnetized iron is brought near one pole of a magnet, the end of the iron nearest the pole assumes the

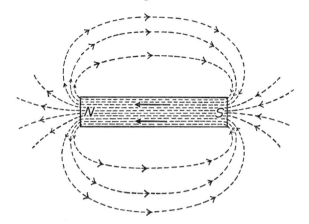

Figure 7.6. The magnetic lines of force about and in a magnet. The lines have been mapped by a small magnetic compass held successively in the indicated positions outside the magnet. They can also be shown by means of iron filings. (The direction of the lines within the magnet are not mapped, but are assumed.)

opposite polarity, a phenomenon called *magnetic induction.* This is explained as follows: the magnetic field of the pole exerts a force on the magnetic domains of the iron, causing them to align themselves as in Figure 7.9A. The iron thereby becomes magnetized and behaves as a magnet while it lies in the field. When the iron is removed from the magnetic field, its domains again become jumbled, as in Figure 7.9B, because of molecular motion, thereupon losing its magnetism.

In summary, then, whenever a magnetic material (iron, cobalt, or nickel) is attracted by a magnet, it first becomes magnetized (magnetic induction), its end nearer the magnetic pole acquiring unlike polarity. These unlike poles then experience a force of attraction, resulting in attraction by the magnet.

Magnetic Permeability and Retentivity

Some materials are more susceptible than others to magnetic induction; that is, their magnetic domains can be readily lined up when placed in a magnetic field. The ease with which a given material can be magnetized in this way is designated as its *magnetic permeability.*

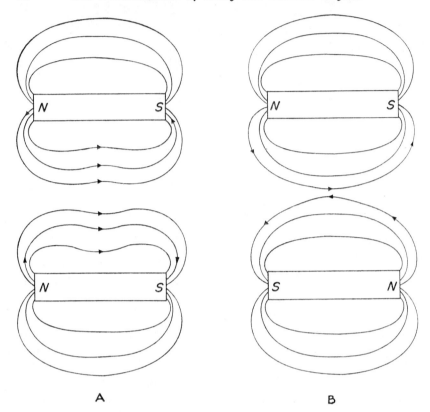

A   B

Figure 7.7. Magnetic lines of force act upon one another, depending on their relative directions. In A the lines of force are in the same direction and the resultant force tends to separate the magnets. In B the lines of force are in opposite directions and the magnets attract each other.

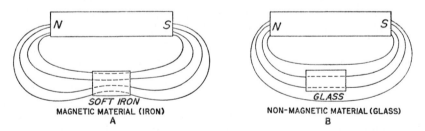

Figure 7.8. The effect of magnetic and nonmagnetic materials on a magnetic field. In A, the soft iron concentrates the lines of force because of its great magnetic permeability. In B, the glass has no effect on the magnetic lines of force.

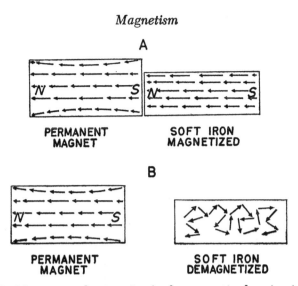

Figure 7.9. Magnetic induction. In *A*, the magnetic domains in the soft iron are orderly and the iron acts as a magnet. When removed from the field of the magnet, as in *B*, the iron loses its magnetism because the domains have become disarranged.

Some materials such as steel do not readily become demagnetized after they have been removed from the magnetizing field; that is, once the magnetic domains have been aligned they tend to remain so. The ability of a magnet to resist demagnetization is called *retentivity.*

Examples of the above include:

1. Soft iron—high permeability, low retentivity.
2. Hard steel—low permeability, high retentivity.

Thus, a metal which is easily magnetized is also easily demagnetized (ordinary iron). On the other hand, a metal which is difficult to magnetize is also difficult to demagnetize (hard steel).

Magnetic Classification of Matter

All matter can be classified on the basis of its behavior when brought into a magnetic field.

1. **Ferromagnetic materials** (ordinarily called magnetic materials) are strongly attracted by a magnet because of their extremely high magnetic permeability; that is, their great suscepti-

bility to magnetic induction. The ferromagnetic elements are *iron, cobalt, and nickel.* Ferromagnetic substances are also capable of being *permanently* magnetized under certain conditions. For example, steel is a form of iron that has been hardened by special treatment, as a result of which its magnetic permeability decreases while its retentivity increases; it is therefore susceptible to permanent magnetization. Another example is the alloy *alnico,* composed of aluminum, nickel, and cobalt, which, when magnetized, is capable of lifting many times its own weight. It is employed in medicine in the removal of magnetic foreign bodies such as nails and bobbie pins from the stomach, under fluoroscopic control.

2. **Paramagnetic materials** are feebly attracted by a magnet. Platinum, an example of such a material, has extremely low magnetic permeability.

3. **Nonmagnetic materials** are not attracted by a magnet because they are not susceptible to magnetic induction. In such materials, there are as many electrons spinning in one direction as in the other so that the atoms are not magnetized and therefore do not form magnetic domains. Examples of nonmagnetic materials are wood, glass, and plastic.

4. **Diamagnetic materials** are actually repelled by a magnet, although this action is feeble. Only a few elements exhibit this property, among them beryllium and bismuth.

Detection of Magnetism

We can detect magnetism most easily by using a *magnetic compass.* This consists of a horizontal magnetized steel needle which swings freely about a vertical axis at its center. When brought near a magnet the needle, since it is a magnet, is deflected according to the laws of magnetism.Thus, the south pole of the compass needle turns toward a magnetic north pole, while the north pole of the compass needle turns toward a magnetic south pole. Notice that the north pole of a compass (or any other magnet) is really a geographic "north-seeking" pole, since it points north when allowed to swing freely. Hence the earth's geographic "north" pole is actually a south magnetic pole, whereas the earth's geographic "south" pole is a north magnetic pole.

QUESTIONS

1. Define magnetism in your own terms.
2. Name and discuss briefly the three main types of magnets.
3. What is believed to be the intimate structure of a magnet? Show by diagram. State three facts which support this theory.
4. State the laws of magnetism.
5. Explain magnetic induction. How does a magnet attract a piece of iron?
6. Compare ordinary iron and hard steel with regard to their magnetization and demagnetization.
7. What is meant by magnetic field? Magnetic flux?
8. What happens to a magnetic field when a piece of ordinary iron is placed in it?
9. In what direction do the magnetic lines of force travel in space? Within the magnet?
10. How is magnetism detected?
11. State the magnetic classification of matter.
12. Define the following terms pertaining to magnetism:
    a. Domain
    b. Permeability
    c. Retentivity
    d. Flux
    e. Field
    f. Poles
    g. Compass

CHAPTER *8* ELECTROMAGNETISM

Definition

EVERY ELECTRIC CURRENT is accompanied by magnetic effects. *Electromagnetism is defined as the branch of physics that deals with the relationship between electricity and magnetism.*

Electromagnetic Phenomena

In 1820, Oersted, a Danish scientist, discovered that a magnetic compass needle turns when placed near a wire carrying a direct electric current. Since this indicated the presence of a magnetic field, Oersted concluded that *a magnetic field always surrounds a conductor in which a current is flowing.* Actually, a magnetic field surrounds any stream of charged particles, whether in a metallic conductor or in space.

What is the direction of this magnetic field? It is specified by the *thumb rule: if the wire is grasped in the left hand so that the thumb points in the direction of the electron current [(−) to (+)], the fingers encircling the wire will indicate the direction of the magnetic lines around the current* (see Figure 8.1).

These lines of force can be demonstrated by Davy's Experiment. A wire is pushed through a hole in a piece of cardboard and connected to a source of direct current. If iron filings are sprinkled on the cardboard, and the cardboard is then gently tapped, the iron particles arrange themselves in concentric circles around the wire. This is illustrated in Figure 8.2, in which the concentric circles represent the magnetic lines of force, or *magnetic flux,* surrounding the electric current in the wire.

Note carefully this very important fact: *the magnetic field exists around the wire only while the current is flowing.* When

94

Figure 8.1. Left hand thumb rule. The thumb points in the direction of the electron current. The fingers designate the direction of the magnetic field around the conductor.

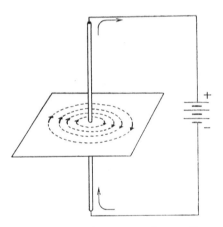

Figure 8.2. The magnetic lines of force around a conductor carrying a direct current. The dotted lines represent the magnetic flux when the electron current is in the indicated direction.

the circuit is opened, the magnetic field around the wire disappears.

A magnetic field is present also around an alternating current, but because of the rapid reversal of field direction, it cannot be demonstrated by a compass or by Davy's Experiment.

The Electromagnet

If a wire is fashioned into a coil or *helix*, as in Figure 8.3, and an electric current is passed through it, the coil responds in a

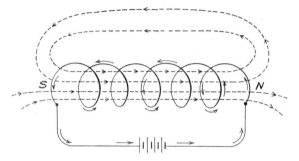

Figure 8.3. A helix carrying an electric current is called a *solenoid*. This shows the magnetic field associated with a solenoid.

most interesting manner. One end of the coil behaves as though it were the north pole and the opposite end the south pole of a magnet. By applying the thumb rule to each turn of the wire in the coil, we can predict not only the direction of the magnetic field inside the coil, but also which end of the coil will be the north pole. Such a coiled helix carrying an electric current is known as a *solenoid.* In modern x-ray equipment, solenoids are used to lock various movable parts, such as tube stand or Bucky tray, in almost any desired position.

As we have already pointed out (page 89), certain metals, such as iron, possess a high degree of magnetic permeability. Placing an iron rod inside a solenoid increases the strength of the associated magnetic field because the iron becomes magnetized by the magnetic field of the solenoid. *Such a solenoid with an iron core is called an electromagnet.* Figure 8.4 shows the main features of a simple electromagnet.

The electromagnet has many practical applications, the best known being the electric bell and the telegraph. In x-ray equipment, electromagnets are used in remote control switches and relays of various kinds. These will be discussed later under the appropriate headings.

Electromagnetic Induction

You have just learned that electrons moving in a conductor are surrounded by a magnetic field. Under suitable conditions the reverse is also true; *moving magnetic fields can produce elec-*

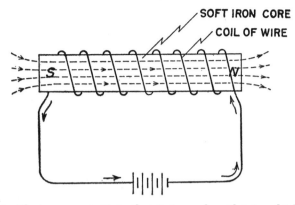

Figure 8.4. Electromagnet. Note that it is a solenoid into which has been inserted an iron core to increase the strength of the magnetic field. If the current were reversed in the illustration, the magnetic poles would be reversed. If the current were interrupted, the electromagnet would lose its magnetism because of the low retentivity of the soft iron core.

tric currents. In 1831, Michael Faraday discovered, on rapidly passing a closed wire between the poles of a horseshoe magnet, that an electric current was generated in the wire. This is the principle of *electromagnetic induction,* defined more broadly as follows: *whenever a conductor cuts across magnetic flux, or is cut by magnetic flux, an electromotice force—emf—is induced in the conductor.* The practical unit of emf is the volt, which is also the unit of potential difference. Thus, the terms emf, voltage, and potential difference are used interchangeably, the common unit being the volt. The magnitude or size of the induced emf depends *on the number of lines of force cut per second* and is determined by four factors:

1. The *speed* with which a wire cuts the magnetic lines of force or magnetic flux; *the more magnetic flux cut per second, the higher the induced emf.*

2. The *strength* of the magnetic field, or the degree of crowding of the magnetic lines of force: *the stronger the magnetic field the higher the induced emf.* A moment's thought will reveal that this is essentially the same as statement (1); the closer the spacing of the magnetic lines, the more flux cut per second.

3. The *angle* between the conductor and the direction of the

lines of force. As this angle approaches 90° progressively more lines of force are cut per second, and therefore a correspondingly larger emf is induced as shown in Figure 8.5.

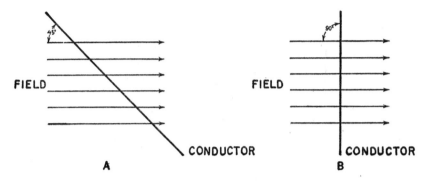

Figure 8.5. In *A*, where the conductor cuts the magnetic field at a 45° angle (down through this page) the induced emf is less than in *B* where the wire cuts the field at 90°.

4. The number of *"turns"* in the conductor if it is wound into a coil. The induced emf is directly proportional to the number of turns in the coil.

Bear in mind that *an emf is induced only while the wire is moving across the field.* If the motion of the wire is interrupted, the induced emf drops to zero. But the motion of the wire need only be relative; thus, *if the wire is stationary and the magnetic field passes across it, an emf will be induced in it* just as well, and the above four principles apply. Finally, *if the wire is stationary and the magnetic field varies in strength an emf will be induced in the wire.*

Even when a wire moving relative to a magnetic field does not form a closed circuit, an emf is induced, but no current flows. On the other hand, when the circuit is closed, the induced emf causes an induced current to flow.

In summary, we may state that there are three ways in which an emf or voltage can be induced in a wire by electromagnetic induction:

1. The wire may move across a stationary magnetic field, or
2. The magnetic field may move across a stationary wire, or

3. The magnetic field may vary in strength while a stationary wire lies in it. This principle is extremely important in the electric transformer, to be described later.

We may conclude that *whenever a conductor and a magnetic flux move with relation to each other, an emf is induced in the conductor, the magnitude of the induced emf being directly proportional to the number of lines of force per second crossing the conductor.*

Direction of Induced Electric Current

The induced electric current resulting from the induced emf has a definite direction depending on the relationship between the motion of the wire and the magnetic field. This is stated in the *left hand rule* or *dynamo rule: first, hold the thumb and first two fingers of the left hand so that each is at right angles to the others, as in Figure 8.6. Next, let the index finger point*

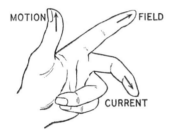

Figure 8.6. Left hand dynamo rule. This is based on the direction of electron current flow.

in the direction of the magnetic lines of force, and the thumb point in the direction the wire is moving. The middle finger will then point in the direction of the induced electron current.

The relationships among the direction of the magnetic field, the motion of the wire, and the induced current are summarized in Figure 8.7. These principles apply to the electric *generator* which will be discussed in detail in the next chapter.

Self-induction

Suppose we have a coil in a direct current circuit whose switch is open. At the instant the switch is closed a magnetic field is

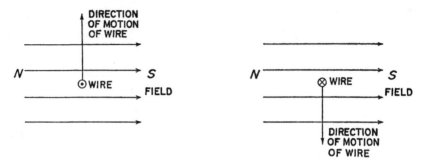

Figure 8.7. The direction of the electron current induced in a wire. The wire is shown in cross section. ⊗ indicates that the current is moving into this page. ◉ indicates that it is moving out of this page, toward the reader.

set up about the coil, building up rapidly to a maximum, where it remains *while the primary current continues to flow.* During the *brief interval* in which the magnetic field "grows up" around the coil, immediately after closure of the switch, the magnetic flux is really expanding, thereby *cutting the coil itself.* By the rules of electromagnetic induction, it induces an *emf* in the coil in a direction *opposite* to the *applied emf,* a phenomenon known as *self-induction.* Thus, at the instant the switch is closed, the self-induced current tends to *oppose* the flow of the primary current in the coil.

While the switch is closed and the primary current is flowing steadily in one direction, the surrounding magnetic field is also constant and therefore does not move relative to the coil. Consequently, *the self-induced current is now zero.*

If the switch is then opened, the magnetic field collapses, sweeping across the coil in a reverse direction from that occurring when the switch was closed. As a result, the self-induced emf reverses and tends to maintain the current in the coil momentarily after the switch has been opened. Thus, when the switch is closed, there is a momentarily induced emf opposing the applied emf; and when the switch is opened, the self-induced emf is momentarily reversed, being in the same direction as the applied emf. Figure 8.8 describes self-induction.

An iron core inserted into the coil produces an increase in the self-induced emf which opposes or *bucks* the applied emf. The

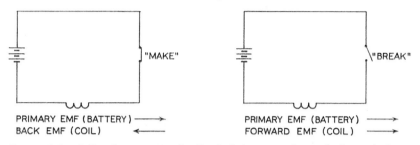

PRIMARY EMF (BATTERY) ⟶
BACK EMF (COIL) ⟵

PRIMARY EMF (BATTERY) ⟶
FORWARD EMF (COIL) ⟶

Figure 8.8. Self-induction. On the "make" (instant of switch closure) there is a *momentary* self-induced emf bucking or opposing the primary battery emf. On the "break" (instant of switch opening), the self-induced emf is reversed momentarily while the battery emf is unchanged.

size of this bucking effect depends on the depth of insertion of the core.

In actual practice self-induction devices such as the choke coil operate only with *alternating current* (ac) as described on page 132. The rapid reversal in direction and magnitude of ac produces an effect similar to the manual opening and closing of a switch with dc.

QUESTIONS

1. Define electromagnetism in your own words. How can you determine its presence in a very simple manner?
2. What is the thumb rule? How is it applied?
3. Describe an electromagnet. In what two ways can you determine its polarity?
4. What is electromagnetic induction?
5. What are the four factors that determine the magnitude of the induced emf? Name two terms used interchangeably with emf.
6. In the following diagram, what is the direction of the electromagnetically induced current?

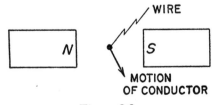

WIRE

N   *S*

MOTION
OF CONDUCTOR

Figure 8.9.

7. Two coils of wire are lying parallel to each other. If one is connected to the terminals of a battery, what will happen in the second coil? What happens when the switch is opened? Closed?

8. What is the difference between a helix, a solenoid, and an electromagnet?

9. What type of current is usually employed to produce self-induction?

CHAPTER 9 ELECTRIC GENERATORS AND MOTORS

ELECTRIC GENERATOR

Definition

An electric generator, dynamo, or turbine is a device that converts mechanical energy to electrical energy by electromagnetic induction.

Essential Features

As already described in the last chapter, electromagnetic induction is the setting up of an emf in a conductor as it cuts, or is cut by, magnetic flux. In actual practice, an electric generator has two components to bring about the continual cutting of magnetic flux by a conductor:

1. *A powerful electromagnet* to set up the necessary magnetic field. This is called the *field magnet.*

2. *An armature,* consisting of a coil of wire that is rotated mechanically in the magnetic field. The mechanical energy needed to turn the armature is most often obtained from a waterfall as in Figure 9.1. Sometimes, other forms of energy are used to rotate the armature. For example, steam may be forced against the vanes of the turbine causing it to turn; or a windmill may produce the same effect.

Simple Electric Generator

An *elementary form* of electric generator may be represented by a single loop of wire (armature) rotated mechanically between the poles of a magnet (see Figure 9.2). What are the size and direction of the current produced by such a generator?

103

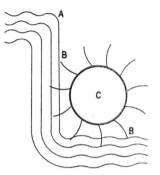

Figure 9.1. The waterfall at *A*, dropping on the blades *B* of turbine, rotates it counterclockwise about the axis *C*. If an armature is connected to axis *C*, it will rotate with the turbine.

Examination of the relationship between the armature and the magnetic field in which it is rotated should give the answer. In Figure 9.2 the armature is represented by the wire loop *AB* whose ends are *connected separately* to the metal *slip rings E* and *F* which rotate with the armature in the direction indicated by *C*. Metal brushes firmly touched each slip ring, at the same time letting the rings rotate. In this way the circuit is completed through the external circuit containing *R* (resistance, representing load or current-consuming devices). Magnetic flux passes from the *N* pole to the *S* pole.

When the armature is rotating clockwise, as in Figure 9.2A, with *A* passing up through the magnetic field, an emf is induced in it in the direction indicated by the straight arrow (according to the left hand rule). At the same time, *B* is passing down through the field, so that its emf is in the direction of the straight arrow near *B*. The net effect is that the current leaves *A* through slip ring *E*, passes through the external circuit *R*, and enters slip ring *F*.

As the armature coil rotates, *A* and *B* exchange positions, as shown in Figure 9.2B. Now *A* is moving down through the field while *B* is moving up, and the current *leaves slip ring F* and *enters slip ring E*. Note that the current is now *reversed in both the coil and the external circuit.* This reversal takes place repeatedly as the armature continues to rotate. Such a current

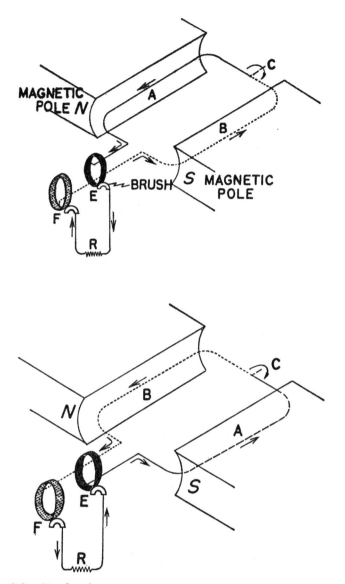

Figure 9.2. Simple alternating current generator. Armature *AB* rotates clockwise in the magnetic flux (which passes from *N* to *S*). In the upper diagram, side *A* moves up through the field while side *B* moves down. Arrows show the direction of the induced electron current which leaves the generator through slip ring *E*, passes through external circuit *R*, and completes the circuit through slip ring *F*. As the armature continues to rotate, sides *A* and *B* reverse positions as in the lower diagram, and the current is now reversed in the armature and in the external circuit *R*.

which periodically reverses its direction is called an **alternating current,** usually abbreviated **ac.**

Let us now examine the magnitude of the induced voltage *from instant to instant* as the coil turns in the field. In Figure 9.3, the generator is simply shown end-on, *A* and *B* representing the cross section of the wire armature, rotating in the direction of the arrows around axis *C.* At the instant the armature is in position (1) it is moving parallel to the magnetic field and is therefore not cutting the magnetic flux. No emf is induced at this instant, so the graph shows the voltage in the armature to be zero. As the armature moves toward position (2) it cuts progressively more magnetic flux per second until, at position (2), it is moving perpendicular (90°) to the magnetic lines. Now the maximum emf is induced in the armature, as represented by the peak of the curve. At (3) the armature is again moving parallel to the field and the induced emf is zero. At (4), with the armature again cutting the flux at an angle of 90° but in the opposite direction, the induced emf is reversed, and the peak of the curve is below the horizontal axis of the graph. Finally, the cycle is completed when the armature returns to its original zero position.

The alternating current curve embodies certain useful information. Such a curve is presented in Figure 9.3 to show the

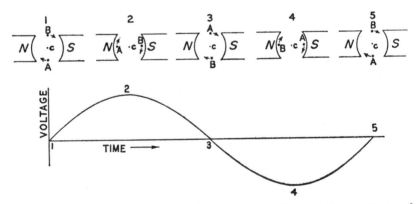

Figure 9.3. The relationships of the various positions of the rotating coil (armature) of an alternating current generator to the alternating voltage curve. *A* and *B* represent the wire of the coil in cross section. The voltage depends on the angle at which the wire cuts the magnetic field at that particular instant. *C* is the axis of rotation of the coil armature.

relationship of the alternating current to the successive positions of the generator armature as it rotates between the magnetic poles. (This is a *sine curve* because it depends on the sine of the angle, from instant to instant, between the plane of the armature and the plane perpendicular to the magnetic field.)

In Figure 9.4 the ac curve is shown again in its usual form. The

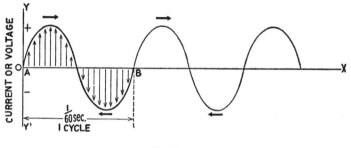

TIME

Figure 9.4. Sine curve representing a 60-cycle alternating current. The thick horizontal arrows show the direction of the *current* during different parts of the a.c. cycle. Note that the current reverses its direction every ⅟₁₂₀ sec. The thin vertical arrows indicate the variation in voltage or amperage from instant to instant.

YY' axis (vertical) may represent either the induced emf or the current while the X axis (horizontal) represents time. The distance between two successive *corresponding points* on the curve, for example AB, represents *one cycle* of the alternating current, and indicates that the armature has made one complete revolution. In common use in the United States is 60-cycle ac, consisting of 60 complete cycles such as AB in each second, and corresponding to 60 revolutions of the armature per second. *The number of cycles per second is called the frequency of an alternating current.* Note that *each cycle consists of two alternations;* that is, the voltage starts at zero, rises gradually to a maximum in one direction (shown above the X axis), gradually returns to zero, increases to a maximum in the opposite direction (shown below the X axis), and finally returns to zero. Thus, there are 120 alternations per second with a 60-cycle current, which means that the *current flows back and forth in the conductor 120 times in each second.* The single sine curve represents a *single phase ac.*

Since the voltage and amperage of an alternating current vary from instant to instant, they cannot be expressed in the same way as in a direct current which flows steadily in one direction (see Figure 9.5). Instead, we must use the concept of *effective current,* defined as follows: *the effective value of an alternating current is equal to that direct current which has the same heating effect in a resistor as the alternating current in question.* For example, if a steady direct current of 1 ampere produces the same heating effect in a 1 ohm resistor as does a given alternating current, then the alternating current may be said to be 1 ac ampere in strength. If we have a pure sine wave alternating current as in Figure 9.4, then the relationship between the peak and the *effective* or *root mean square value* of the current is as follows:

$$\text{maximum current} = 1.41 \times \text{effective current}$$
$$\text{or}$$
$$\text{effective current} = 0.707 \times \text{maximum current}$$

The *effective voltage* has the same relationship to the *maximum voltage:*

$$\text{maximum voltage} = 1.41 \times \text{effective voltage}$$
$$\text{or}$$
$$\text{effective voltage} = 0.707 \times \text{maximum voltage}$$

Properties of Alternating Current Circuits

In Chapter 6 we found that a circuit carrying a *steady direct current* has a property called resistance, which tends to hinder the flow of current. If a coil of wire is introduced into the circuit, the flow of current is further hindered at the time the switch is closed, by the back emf resulting from the self-inductance of the coil (see section on self-induction). However, the direct current immediately rises to a plateau where its value is constant (see Figure 9.5) and the associated magnetic field is also constant. Now there is no back emf. There still remains the true electrical resistance of the coil wire. If a *capacitor* is introduced into a dc circuit, the current flows only until the capacitor is fully charged, and then ceases.

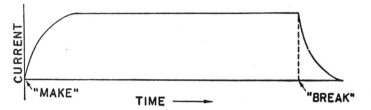

Figure 9.5. Curve representing a steady direct current from a battery. The circuit is closed at "make." The current rises rapidly to a maximum and maintains that value at a constant level until the circuit is opened at "break," when it rapidly drops to zero.

With an *alternating current,* on the other hand, the rapid change in magnitude and direction of the current causes a corresponding fluctuation in the strength and direction of the magnetic field set up around a coil in the circuit. The coil is said to have *inductance,* defined as the opposition to a *change* in the amount of current flowing. The persistently changing magnetic field around the coil supplied by an alternating current sets up a *back emf* in a direction opposite the applied alternating voltage, first in one direction and then in the other. This "bucking" tendency of inductance is called *inductive reactance,* and is measured in ohms because it opposes the flow of current.

If a *capacitor* is added to an ac circuit, it periodically reverses polarity as the current reverses. This offers a certain amount of hindrance to the flow of current, known as *capacitive reactance.*

The apparent total resistance of an ac circuit is called its *impedance,* and this depends on the inductive reactance, capacitive reactance, and true electrical resistance. The calculation of the impedance from these three factors is *not* that of simple addition, and is too complex for our discussion here.

It is interesting to note that Ohm's Law applies to alternating currents if the above discussed properties are taken into consideration,

$$\text{effective current} = \frac{\text{effective voltage}}{\text{impedance}}$$

$$I = \frac{V}{Z}$$

where I is the effective current measured in amperes, V is the effective voltage measured in volts, and Z is the impedance measured in ohms.

Direct Current Generator

The principle of the ac generator also applies to the dc generator, but in the latter a *commutator* instead of slip rings introduces the generated current into the external circuit. Whenever A and B segments of the armature exchange positions as shown in Figure 9.6, the corresponding half of the commutator always

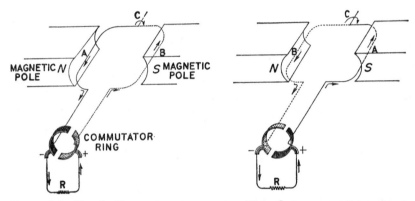

Figure 9.6. Simple direct current generator. Note that a commutator ring is used, with the result that the current is always supplied to the external circuit in the same direction. (Compare with Figure 9.2, which illustrates the ac generator.)

introduces the current to the external circuit in the same direction even though there is an alternating current in the armature itself. The commutator actually consists of a metal ring split in two, with the halves separated by an insulator, each half being permanently fixed to its corresponding wire.

The wave form of the current obtained from such a generator is represented in Figure 9.7. Note that the part of the sine wave which, with alternating current, would be below the horizontal line lies above the line. This is a *pulsating direct current.* In actual practice, the dc generator consists of numerous coils and a commutator which is divided into a corresponding number of segments. Since the coils are so arranged that some are always

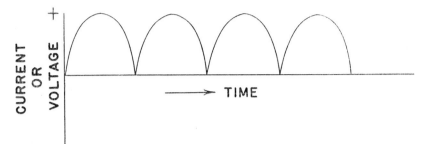

Figure 9.7. Wave form of a direct current curve obtained from a simple dc generator such as that shown in Figure 9.6. Note that the curve differs from a sine curve in that the portions of the curve that lie below the horizontal axis are turned up to lie above the axis. This is a *pulsating direct current*, in contrast to the steady direct current supplied by a battery.

cutting magnetic flux during rotation of the armature, emf is always being induced. At no time does the voltage fall to zero, as in the simple one-coil dc generator illustrated above. Consequently, the voltage is steadier than that generated by a single-coil armature. Figure 9.8 shows the dc curve obtained with a multiple coil generator.

Advantages of Alternating Current

For at least two reasons, alternating current is employed much more frequently than direct current.

1. An alternating current is needed to operate transformers which are essential to many phases of industry, and particularly to radiology. A transformer will not operate adequately on voltage with the wave form shown in Figure 9.8.

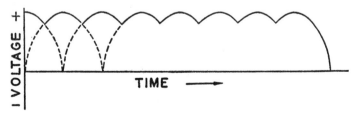

Figure 9.8. Direct current wave form produced by a multiple-coil generator. The solid line curve represents the voltage variation from instant to instant. This fluctuation is much less than that produced by any one of the coils alone (shown by the broken line).

2. The electric current generated at the power plant is often transmitted over a great distance. It has been shown earlier that with **direct current** the power loss (in the form of heat) is related to the current and resistance as follows:

$$P = I^2R$$

where $P =$ loss of power in the form of heat, measured in watts
$I =$ current in amperes
$R =$ resistance in ohms

According to this equation, the power loss is proportional to the *square of the amperage.* It is obvious, therefore, that if a current is transmitted at a very *low* amperage there will be a much smaller loss of electric power than if transmitted at high amperage. An example will illustrate this point.

Suppose we wish to transmit 50,000 watts of electric power over a distance of 10 miles (a total of 20 miles of wire), the wire having a resistance of 0.15 ohm per mile. The total R is $0.15 \times 20 = 3$ ohms. If we use *dc* with a voltage of 500 and an amperage of 100, $(500 \times 100 = 50,000$ watts), the loss of power is obtained by the power equation:

$$Power\ loss = I^2R$$
$$= 100 \times 100 \times 3$$
$$= 30,000\ watts$$

Thus, ⅗ of the power is wasted as a result of the heating effect of the current in the transmission wires!

If, instead, we used *ac* at 5000 volts and 10 amperes (again 50,000 watts), the power loss would be much less. The above equation for power consumption cannot be applied accurately to ac without a correction known as the *power factor,* but it does give a rough approximation. Neglecting the power factor,

$$Power\ loss = I^2R$$
$$= 10 \times 10 \times 3$$
$$= 300\ watts$$

as contrasted with a power loss of 30,000 watts with dc transmitted at lower voltage and higher amperage.

Upon reaching its destination the high voltage (often called high tension) must be reduced to useful values, ordinarily 115 volts for lighting and 115 or 230 volts for most x-ray equipment. This is accomplished by means of a *transformer,* an electromagnetic device which will be described in detail in the next chapter.

ELECTRIC MOTOR

Definition

An electric motor is a device that converts electrical energy to mechanical energy. In other words, electricity can be made to do useful work by means of a motor.

Principle

Whenever a wire carrying an electric current is placed in a magnetic field, there is a force on the wire tending to push it out of the field. It should be emphasized that two conditions must obtain: *the wire must carry a current, and it must be located in a magnetic field.* The conductor experiences a force that is directly proportional to the length or number of turns of the conductor, the strength of the magnetic field, and the magnitude of the current flowing in the conductor.

A magnetic field acts on a current-carrying wire, as shown in Figure 9.9, where the wire is represented in cross section. The following conditions exist:

1. There is an electric current in the wire directed away from the reader.

2. Magnetic flux surrounds the wire as shown by the small arrows (associated with the current in the wire).

3. Magnetic flux passes to the right between the poles of the magnets (called the *field magnets*).

Now look at the upper side of the wire; at that point, the magnetic flux surrounding the wire is directed opposite to that of the flux between the field magnets. Such oppositely directed lines of force "attract" each other. At the same time, note that on the lower side of the wire the surrounding magnetic flux at that point is in the same direction as the flux between the field magnets. Lines of force in the same direction "repel" each other. The net effect is for the wire to be thrust upward by the interaction of the two magnetic fields, as shown by the vertical arrow.

This principle is embodied in the *right hand or motor rule* which mirrors the left hand rule. *The right hand or motor rule states that if the thumb and first two fingers of the right hand are held at right angles to each other; and if the index finger indicates the direction of the magnetic field, and the middle finger the direction of the electron current in the wire, then the thumb will point in the direction that the wire will move.* Applying this rule to Figure 9.9, we find that the same result is obtained as before: the wire is thrust upward.

FORCE

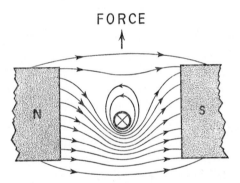

Figure 9.9. The dark circle represents a wire in cross section carrying a current "into the page." As explained in the text, the wire is forced upward by the weaker magnetic field above (lines of force in opposite directions) and by the stronger magnetic field below (lines of force in the same direction). Adapted from *The New Physics and Chemistry Dictionary and Handbook* by Robert W. Marks. Copyright© 1967 by Bantam Books, Inc.

The Simple Electric Motor

The diagrams shown in the section on the electric generator apply also to the electric motor, except that current is *introduced* at the brushes instead of being withdrawn. An ac motor has *slip rings,* whereas a dc motor has a *commutator.*

In its elementary form, an electric motor may be represented, as in Figure 9.10, by a cross-sectional diagram of a single coil armature carrying an electric current and lying in a magnetic field. Applying the *right hand rule,* we find that as wire *A* is

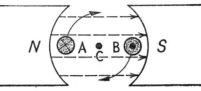

Figure 9.10. Cross section of an elementary motor. *A* and *B* are the coil wire seen end-on. *C* is the axis of rotation. The curved arrows show the direction in which the coil armature rotates when the proper current is fed to the armature.

thrust upward *B* is thrust downward. The net result is that the armature turns in a clockwise direction about axis C.

At this point we should mention the *back emf* of a motor. When the coil of a motor rotates in the magnetic field, an emf is induced in the coil according to the principles of electromagnetic induction. This self-induced emf is in a direction *opposite* to the direction of the emf already applied to the coil and is therefore called *back emf.* Thus, an electric motor acts at the same time as a generator, sending some current back to the main lines. This is one of the reasons why a motor consumes less current than a heating device.

Types of Electric Motors

There are two main types of electric motors:

1. **Direct current motor.** This is the reverse of a direct current generator.

2. **Alternating current motors.** There are two kinds of alternating current motors, *synchronous* and **induction.**

The *synchronous motor, supplied* by ac, is built like a single-phase ac generator. However, current is fed to the armature, causing it to rotate in the magnetic field. Certain conditions are necessary for the operation of a synchronous motor. Its armature must rotate at the same speed (revolutions per minute) as the armature of the ac generator supplying the current, or at some fixed multiple of that speed. In other words, the speed of this type of motor *must be synchronous with the speed of the generator.* Special devices are required to bring the speed of the motor armature up to the required number of revolutions per minute. The synchronous motor is used in electric clocks and synchronous timers. The main *disadvantage* of the synchronous motor lies in the limited speeds with which it will operate, that is, it must be in step with the generator. At the same time, a fixed speed may be of *advantage* in certain types of equipment, such as x-ray exposure timers and electric clocks.

The *induction motor* has a *stator* which is made up of an even number of stationary electromagnets distributed around the sides of the motor. In the center is a rotating part called a *rotor,* consisting of bars of copper arranged around a cylindrical iron core, and resembling a squirrel cage. There are no slip rings, commutators, or brushes because this type of motor works on a different principle from other types of motors and generators. The stator is supplied by a *multiphase* current, which means that the ac supplying one opposite pair of coils is out of phase (out of step) with that supplying the next pair of coils. Thus, successive pairs of coils set up magnetic fields which successively reach their peak strength, acting as though they were rotating, and dragging along the rotor. The reason for this drag on the rotor lies in the fact that currents are induced in its copper bars by the "moving" magnetic field. Each copper bar, now carrying an induced current while lying in a magnetic field, experiences a push or force according to the principles discussed earlier. This force causes rotation of the rotor.

Figure 9.11 illustrates the principle of a *two-phase induction motor.* Its power supply is a two-phase ac generator whose armature consists of two coils arranged perpendicular to each other. The current produced by such a generator may be represented by two curves, one of which is ¼ cycle ahead of the other as shown in the upper part of Figure 9.11. This two-phase current is applied to the *stator* to induce a rotating magnetic field. (Actually, a three-phase stator is more often used.) The *rotor* is shown in cross section at the center.

The induction motor is used to turn the anode of a rotating anode tube, to be described later.

Current-measuring Devices

An interesting and important application of the *motor principle* is in the construction of instruments for the measurement of current and voltage. The basic device employing this principle is

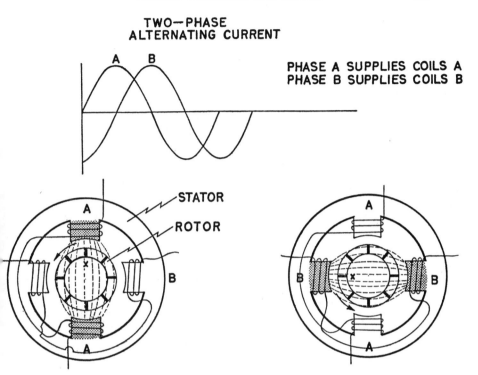

Figure 9.11. Principle of the induction motor. In phase *A* (left diagram) the magnetic flux is directed downward as shown. In the next ¼ cycle, phase *B* activates its stators and the magnetic flux is now directed from right to left (the flux in the preceding ¼ cycle having disappeared). In the next phase, the magnetic flux is directed upward, and finally from left to right. This process continues as long as current is supplied. In effect, then, we have a magnetic field which rotates counterclockwise, dragging the rotor with it and making it rotate.

the *moving coil galvanometer*, which consists essentially of a coil of fine copper wire suspended between the poles of a horseshoe field magnet. One end of the coil is fixed and the other is attached to a hairspring helix. When a *direct current* is passed through the coil as it lies in the magnetic field of the permanent magnet, the coil moves according to the motor principle. In this case, the construction of the instrument, as shown in Figure 9.12, is such that the coil must rotate through an arc that is usually less than one-half circle (180°). Rotation takes place against the spring, so that when the current is interrupted, the coil returns to the zero position. Since the degree of rotation of the coil is proportional to the current, the attached pointer moving over a *calibrated scale* measures the current as shown in Figure 9.12.

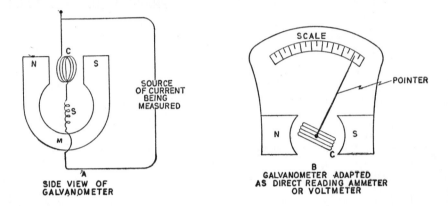

Figure 9.12. A. Moving coil galvanometer. The coil, *C*, is suspended between the poles of a horseshoe magnet *M*. One end of the coil is attached to the spring, *S*. When a current is passed through the coil it rotates, twisting the spring. When the current ceases, the spring returns the coil to the initial position. B. A direct current meter provided with a pointer and a scale. After initial calibration as an ammeter or voltmeter, it may be used to measure current or voltage.

If a galvanometer, provided with a low resistance in parallel, is connected in series in the circuit being measured, it indicates *amperage* and constitutes an *ammeter.* If a galvanometer, protected by a high resistance in series, is connected in parallel with the circuit being measured, it determines *voltage* and constitutes a *voltmeter.* Thus, modifications of the same basic instrument allow the measurement of voltage or amperage in a dc circuit, provided the instrument has been previously calibrated against known values of voltage or amperage.

Further modification is necessary to measure the voltage and amperage of an *alternating current,* because the coil would "freeze" due to the rapid alternation of the current in the coil. (Recall that a 60 cycle alternating coil sustains 120 reversals of direction per second.) Instead of using a horseshoe magnet, one must provide a magnetic field which reverses its direction in step with the current entering the coil. To do this, the ac being measured is, at the same time, used to supply the coils of a pair of field electromagnets. Since the current alternates simultaneously in the rotating coil and in the field coils, the rotation of the coil and hence the deflection of the pointer are unaffected by the alternations of the current, but are still affected by the size of the current. This device, called an *electrodynamometer*, can be calibrated to read ac voltage or amperage. The same type of instrument can be calibrated to measure electric power in a circuit, reading the values directly in watts.

QUESTIONS

1. What is an electric generator? The generator rule?
2. What are some of the commonly used sources of energy in a generator?
3. Describe, with the aid of a diagram, a simple electric generator. Show how it produces an alternating current and relate it to the ac sine curve. Label the curve in detail.
4. How does an ac generator differ from a dc generator?
5. Compare the current curves of single phase ac, single phase dc, and an electric battery.
6. Why is ac universally preferred to dc? Prove by means of an example.
7. Define an electric motor. What is the motor principle?
8. What is the right hand or motor rule?
9. Under what conditions will a wire, placed in a magnetic field, tend to move up through the field?
10. What are the two main types of motors?
11. Describe the principle and construction of an induction motor.
12. What is a synchronous motor? How does it differ from an induction motor?
13. What is the importance of the synchronous motor and the induction motor in radiographic equipment?
14. State the underlying principle of the galvanometer. Show with the aid of a simple diagram the construction of a dc voltmeter.

CHAPTER *10* PRODUCTION AND CONTROL OF HIGH VOLTAGE

TRANSFORMER

For most ordinary purposes, electricity is brought to the consumer at 115 or 230 volts. However, this is inadequate for the direct operation of an x-ray tube, since many thousands of volts are required to impart sufficient speed to the electrons in the x-ray tube to generate x rays of satisfactory quality. Such high voltages can readily be obtained by means of a *transformer,* which is one of the major components of an x-ray unit. The high voltage transformer is often called the *x-ray generator.*

Principle

A transformer is an electromagnetic device which changes an *alternating current* from low voltage to high voltage, or from high voltage to low voltage, without loss of an appreciable amount of energy (usually less than 10 per cent). The transformer transfers electrical energy from one circuit to another *without the use of moving parts or any electrical contact between the two circuits,* employing the principle of *electromagnetic mutual induction.*

The simplest type of transformer is the *air core* transformer, consisting of two highly insulated coils of wire lying side by side. One of these, the *primary coil,* is supplied with an alternating current (ac). The other or *secondary coil* develops ac by mutual induction. This is shown in Figure 10.1. The primary coil is the *input side* of the transformer, whereas the secondary coil is the *output side.*

120

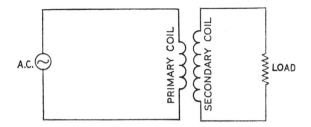

Figure 10.1. Simplest type of transformer. An electromotive force is set up in the secondary coil by mutual induction.

How does a transformer function? An ac flowing in the *primary* coil sets up around it a magnetic field that varies rapidly in direction and strength, just as does the ac itself. As this varying magnetic flux cuts the *secondary* coil, it induces in it an ac having the same wave form as the primary ac in accordance with the principle of *mutual induction.*

Note here, again, the important principle of electromagnetic induction: the emf induced in a coil is directly proportional to the number of turns in the coil that cuts, or is cut by, a given magnetic flux. This also applies to the transformer. Suppose the primary and secondary coils each have one turn. The ac in the primary coil sets up a magnetic field that cuts the same number of turns (1) in the secondary coil; therefore, the emf is the same in both coils (neglecting any loss of energy during the process). But suppose the primary coil has one turn and the secondary has two turns. Now the *same* magnetic field set up by the ac in the single-turn primary coil cuts the double-turn secondary coil, *doubling the emf in the latter.* Similarly, if the primary coil has two turns and the secondary one turn, the secondary emf will be half the primary emf.

The preceding discussion is summarized in the *transformer law: the emf induced in the secondary coil is to the emf in the primary coil, as the number of turns in the secondary coil is to the number of turns in the primary coil, expressed simply in practical units in the following equation:*

$$\frac{V_s}{V_p} = \frac{N_s}{N_p}$$

where V_s = voltage in secondary coil
V_p = voltage in primary coil
N_s = no. of turns in the secondary coil
N_p = no. of turns in the primary coil

Stated in words, this means that if the number of turns in the secondary coil is twice the number in the primary coil, then the voltage in the secondary will be twice the voltage in the primary. If the number of turns in the secondary coil is three times that in the primary coil, then the voltage induced in the secondary will be three times as great as in the primary, etc. Such a transformer, having more turns in the secondary coil than in the primary, puts out a *higher* voltage than is supplied to it and is therefore a *step-up transformer.*

If the secondary coil has fewer turns than the primary coil, the output voltage will be *less* than the input voltage, and the transformer is a *step-down transformer.*

What happens to the value of the *current* in a transformer? According to the Law of Conservation of Energy, there can be no more energy coming out of the transformer than is put in, and similarly, the power output (energy per unit time) can be no greater than the power input. Since the power in an electric circuit equals voltage multiplied by amperage, then (neglecting the power factor),

$$I_s V_s = I_p V_p$$

where I_s = current in amperes in the secondary coil
I_p = current in the primary coil
V_s = voltage in the secondary coil
V_p = voltage in the primary coil

Rearranging this equation, we get the proportion:

$$\frac{I_s}{I_p} = \frac{V_p}{V_s}$$

In other words, if the voltage is increased, as in a step-up transformer, the amperage is decreased; and if the voltage is decreased, as in a step-down transformer, the amperage is increased. Thus, a *step-up transformer increases the voltage, but decreases the amperage in an inverse ratio.* The power output of an x-ray transformer is rated in kVA (kilovolt-amperes).

In roentgen *diagnostic* equipment the step-up transformer takes 115 or 230 volts and multiplies this voltage to 30,000 to 150,000 volts (30 to 150 kilovolts), thereby providing the high voltage necessary to drive the electrons through an x-ray tube. At the same time it decreases the current to thousandths of an ampere (milliamperes).

Construction of Transformers

Four main types of transformers will now be described.

1. **Air Core Transformer.** This consists simply of two insulated coils lying side by side as described on page 120.

2. **Open Core Transformer.** An iron core inserted into a coil of wire carrying an electric current causes a marked intensification of the magnetic field in and around the coil because of the magnetization of the core. Therefore, a transformer becomes more efficient if each *insulated* coil has an iron core, as shown in Figure 10.2. This is known as an *open core transformer.* Although more efficient than an air core transformer, the open core type is still subject to an appreciable waste of power due to loss of magnetic flux at the ends of the cores, a condition termed *leakage flux.*

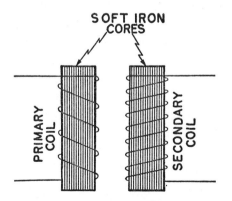

Figure 10.2. Open core transformer. There is some waste of power because the magnetic flux is partly lost in the air at the ends of the iron cores.

3. **Closed Core Transformer.** Here the heavily insulated coils are wound around a *square or circular iron "doughnut"* as in Figure 10.3. With this type of transformer, the closed core pro-

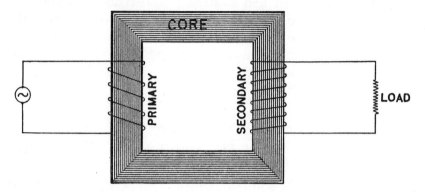

Figure 10.3. Closed core transformer. The doughnut type core concentrates the magnetic flux and, at the same time, provides a continuous path for the magnetic flux.

vides a continuous path for the magnetic flux, so that only a small fraction of the magnetic energy is lost by leakage. The core is *laminated,* that is, it is made of layers of metal plates. In a solid core, eddy currents would be set up by electromagnetic induction during transformer operation. Such eddy currents take energy from the current in the transformer coils, causing a waste of power known as *transformer loss.* Lamination of the core hinders the formation of these eddy currents and thereby increases the efficiency of the transformer; in other words, there is less power wasted. Silicon steel is often used in laminated cores because its high electrical resistance further reduces eddy current power loss. Since the doughnut type of closed core transformer is much more efficient than the open core type, it is the *one most commonly used in x-ray-generating equipment.* The entire transformer is submerged in a metal box containing a special type of oil for maximum insulation. Furthermore, the oil helps to keep the transformer cool, since a certain amount of heat is unavoidably produced during its operation.

4. **Shell-type Transformer.** This, the most advanced type, is used as a commercial or power transformer. Here, too, a laminated

core is used, consisting of a pile of sheets of silicon steel, each having two rectangular holes, as illustrated in Figure 10.4. The

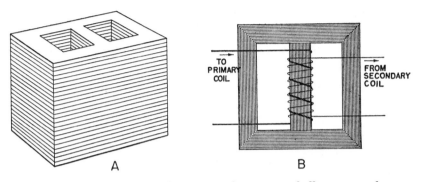

Figure 10.4. *A*, core of shell-type transformer. *B*, shell-type transformer, top view. Note the windings of the primary and secondary coils around the *same* central section of the core. Very heavy insulation is required.

primary and secondary coils are both wound around the central section of the core for maximum efficiency, as shown in Figure 10.4. This winding is possible because in any type of transformer *the coils must be highly insulated from each other,* and when the coils are placed close together, as in the shell-type transformer, efficiency is increased. Insulation is achieved by coating the wires of both coils with special insulating material, and also by immersing the transformer in a container filled with a special type of oil for maximum insulation and cooling.

Transformer Efficiency and Power Losses

The efficiency of a transformer is the ratio of the power output to the power input:

$$\% \text{ efficiency} = \frac{\text{power output}}{\text{power input}} \times 100$$

In the ideal situation, the power input and power output would be equal, and the efficiency 100 per cent. However, in practice the efficiency is more likely to be 90 to 95 per cent; that is, there is 5 to 10 per cent less power output than power input, the lost energy appearing as heat. In contrast to other types of electrical and mechanical equipment, this is a small waste of energy, and the transformer is therefore regarded as a highly efficient device.

We shall now summarize the various kinds of *losses of electrical power* in a transformer and the methods of minimizing them in order to improve efficiency.

1. **Copper Losses.** These include mainly the loss of electrical power due to the *resistance* of the coils. Such loss of power can be reduced by using copper wire of adequate diameter. In a step-up transformer the primary coil carries a current of high amperage, so its wire must be thicker than the wire of the secondary coil which carries a small current (milliamperes). Recall that a thick wire, having less resistance than a thin one, can carry more current with less waste of energy in the form of heat (*Power Loss* $= I^2R$).

2. **Eddy Current Losses.** The fluctuating magnetic field, set up in the transformer *core* by the ac in its coils, induces electrical *eddy currents* (swirling currents) in the core itself by electromagnetic induction. The eddy currents, in turn, produce heat in the core because electrical power loss due to heat $= I^2R$. Hence, electric power is wasted as heat, and transformer efficiency is impaired. Eddy currents can be minimized by the use of *laminated (layered) silicon steel plates*, highly insulated from each other by a special coating. Lamination and high-resistance silicon steel increase the electrical resistance of the core, thereby decreasing the size of the eddy currents according to Ohm's Law.

3. **Hysteresis Losses.** Since the transformer operates on and puts out ac, the tiny magnetic domains in the *core* are repeatedly rearranging themselves as the core is magnetized first in one direction and then the other by the ac in the coils. This rearrangement of the domains produces heat in the core, thereby wasting electrical power. Such a loss of power, called *hysteresis loss*, is reduced by the laminated silicon steel core.

CONTROL OF HIGH VOLTAGE

Properly designed x-ray equipment must include adequate control of voltage to permit the choice of a variety of kilovoltages to be applied to the x-ray tube. This allows the technologist to obtain x rays of suitable penetrating power for a particular technic. Without flexible voltage control, modern radiographic procedures would be seriously hampered. The main device for high voltage control is the *autotransformer*. Another type of voltage regulator is the variable resistor or *rheostat,* which is widely used to regulate filament voltage and current and, ultimately, tube current (mA). The rheostat also helps control high voltage in orthovoltage therapy equipment. A third device, the *choke coil,* was formerly used to control filament current and voltage.

Autotransformer

According to the description of the step-up transformer, you should realize that there is a *fixed ratio* of voltage output to voltage input. Furthermore, the voltage input is fixed at 115 or 230 volts, depending on the power supply and type of equipment. Without any further modification, we would have available only a single kilovoltage, a situation that would seriously limit the range of radiography, since various kilovoltages must be applied to the x-ray tube to obtain x-ray beams with a variety of penetrating abilities.

How can we modify the basic equipment so as to obtain the required range of kilovoltages? This is accomplished simply by *varying the voltage input to the transformer primary.* Thus, if the ratio of the step-up transformer is 500 to 1 and the input to the primary coil is 115 volts, then the transformer will put out $500 \times 115 = 57,500$ volts (57.5 kV). If the input is 180 volts, the output will be $500 \times 180 = 90,000$ volts (90 kV). To obtain the required variety of input voltages we connect a device known as an *autotransformer* between the source of ac and the primary side of the transformer. As you will see, the autotransformer is really a variable transformer.

Construction. The autotransformer is made up of a coil of insulated wire wound around a *large iron core.* At regular intervals along the core, insulation is interrupted and the bare points connected or tapped off to metal buttons, as shown in Figure 10.5. A movable contactor, *C*, varies the number of turns included in the secondary circuit of the autotransformer, thereby varying its output voltage.

Principle. The autotransformer is an electromagnetic device which operates on the principle of *self-induction. A single coil*

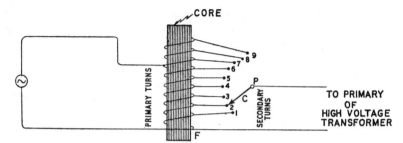

Figure 10.5. Diagram of an autotransformer. Only a single coil is used. The *autotransformer* secondary voltage stands in the same ratio to the primary voltage, as the ratio of the number of tapped turns (in this case, 2) is to the number of primary turns (7); that is, $\frac{2}{7}$, the voltage being stepped down. If *C* is turned above 7, the ratio exceeds 1 and the voltage is stepped up.

serves as both the primary and secondary coil, the number of turns being adjustable. (This is in sharp contrast to the transformer, wherein the number of turns cannot be changed.) By turning the contactor to include more or fewer turns on the load side, we can vary at will the ratio of the number of secondary to the number of primary turns, thereby varying the ratio of the voltage output to the voltage input. This relationship is very similar to that in the transformer and is embodied in the *autotransformer law:*

$$\frac{\text{autotransformer secondary voltage}}{\text{autotransformer primary voltage}} = \frac{\text{number of tapped turns}}{\text{number of primary turns}}$$

Thus, in Figure 10.5, if the primary voltage is 230,

$$\frac{\text{autotransformer secondary voltage}}{230} = \frac{2}{7}$$

autotransformer secondary voltage $= 65.7$ volts

The secondary voltage of the autotransformer is applied to the primary side of the main step-up transformer in x-ray equipment.

For those desiring a more detailed description of the operation of an autotransformer, a wiring diagram is shown in Figure 10.6. Because of the large inductance, there is a large voltage drop across the primary side. We may regard this voltage drop as being related to the number of turns; that is, *volts per turn.* Suppose that the applied potential is 230 volts and there are 7 primary turns. We would then have $230/7 = 33$ volts per turn.

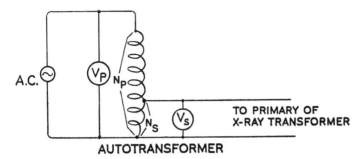

AUTOTRANSFORMER

Figure 10.6. Further simplification of wiring diagram of autotransformer. Due to self inductance, *the voltage drop is proportional to the number of turns.* If there are 7 primary turns N_p, and the applied potential is 230 volts, then there would be 230/7 = 33 volts per turn. On the secondary side there are 2 tapped turns N_s; therefore the voltage across the secondary side $V_s =$ 2 turns × 33 volts per turn = 66 volts. So the voltage output to input = 66/230 = 2/7, the same result as that obtained with the autotransformer law.

Now, looking at the secondary side of the autotransformer, we see that there are 2 tapped turns; therefore, the voltage drop across this side is 2 turns × 33 volts per turn = 66 volts. Accordingly, the ratio of the voltage output to input of the autotransformer is 66/230 = 2/7, the same result as that obtained by means of the autotransformer law.

Note that an autotransformer can be used only where there is a relatively small difference between its input and output voltage. Furthermore, *the position of the contactor cannot be changed while the exposure switch is closed because sparking may occur between the metal buttons, thereby damaging them.*

Rheostat

Another device used for voltage control is the *rheostat,* a variable resistor which allows us to vary the resistance of a circuit. It can operate on either alternating or direct current.

Principle. In the dicussion of Ohm's Law, we found that there is a fall of potential, or voltage drop, across a resistance in a circuit, as shown in the following equation:

$$V = I \times R$$

Voltage Current Resistance
Drop in Amps in Ohms

The rheostat consists of a series of resistance coils and a sliding tap which can be moved so as to increase or decrease the amount of resistance in the circuit. When placed in series in the *primary circuit,* it can be used to vary the voltage that is available to the primary of the transformer. This can be readily seen from a study of Figure 10.7.

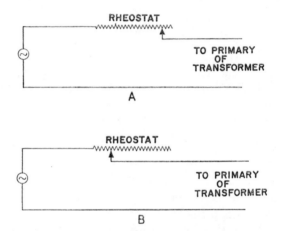

Figure 10.7. Principle of the rheostat or variable resistor. In *A* most of the resistor is in the circuit, producing a large voltage drop; hence, only a small voltage is available to the primary of the *transformer.* In *B* only a small segment of the resistor is in the circuit, producing a small voltage drop; now a large voltage is available to the primary of the transformer. Furthermore, according to Ohm's Law, the current in *A* is smaller than in *B.*

It must be emphasized that the rheostat controls voltage by converting more or less electrical energy into heat. This is a wasteful process, reducing the efficiency of the equipment.

Another disadvantage is that the voltage drop in a circuit controlled by variable resistance depends also on the amperage, according to the equation, $V = IR$. When the exposure switch is closed and current flows through the x-ray tube, there is a momentary surge of current in the primary circuit of the transformer. Consequently, the *IR* or voltage drop across the rheostat increases, leaving less voltage across the primary of the transformer. This is reflected in a smaller kV across the secondary of the transformer and the x-ray tube. Hence, voltage regulation is not so stable with a rheostat as with an autotransformer which allows more current to flow in the primary circuit as needed, without excessive decrease in kV across the tube. For this reason, the rheostat has been abandoned as the chief method of voltage control in modern x-ray equipment, and has been superseded by the autotransformer type of control.

In most orthovoltage x-ray machines (200-250 kV), the tube must be warmed before application of the full voltage in order to avoid damage. A rheostat is used for this purpose, since it can be varied while the tube is in operation; whereas an autotransformer cannot because sparking may occur, charring the contacts. Thus, when the main switch is first closed, all the resistance is *in* the circuit, *in series with the primary coil* of the x-ray transformer; the resulting large voltage drop across the rheostat leaves a small voltage across the primary coil, inducing a low kV in the secondary coil. As the rheostat resistance is now slowly reduced, manually or automatically, the voltage drop across it decreases, thereby gradually increasing the voltage drop across the primary and inducing a higher kV in the secondary until the maximum kV is applied to the x-ray tube. An autotransformer is included with the rheostat for kV control in orthovoltage therapy units.

CONTROL OF FILAMENT CURRENT
AND TUBE CURRENT

In addition to the control of high voltage for the purpose of varying the penetrating power of an x-ray beam, it is equally important to be able to regulate tube *current* (milliamperage). The construction and operation of an x-ray tube will be explained in detail later. At present it is enough to say that an x-ray tube has two circuits: (1) a filament circuit carrying the current needed to heat the filament and (2) the tube circuit, itself, carrying the tube current which passes between the electrodes of the x-ray tube and produces x rays. Of utmost significance is the fact that *a small change in filament current produces a large change in tube current*. Therefore, by regulating the filament current we can control the x-ray output in terms of quantity of x rays per minute.

Three devices are available for controlling filament current: choke coil, rheostat, and saturable reactor.

Choke Coil

Principle. The choke coil is an electromagnetic device operating on the principle of *self-induction*. It will therefore operate only on *alternating current*.

In Chapter 8 we learned that an electric current in a coil is associated with a magnetic field so that one end of the coil is a north pole and the opposite end a south pole. If the magnetic field varies in direction and strength, as when it is generated by an *alternating current*, the magnetic flux cuts the coil itself and induces a voltage which *opposes* the voltage already applied across the coil. This is the *back emf of self-induction*.

The introduction of an *iron core* into the coil intensifies the magnetic field and increases the inductive reactance of the coil (analogous to, but not identical with, resistance), producing the following effects:

1. A larger voltage drop across the choke coil.
2. A smaller remaining voltage for the rest of the circuit.
3. A *decrease in current* due to the larger inductance of the circuit.

These effects become more pronounced as the core is inserted deeper into the coil. The principle of the choke is explained further in Figure 10.8.

The choke coil was previously used to regulate filament current in the x-ray tube, in order to control tube current (mA). However, it has been largely replaced by the rheostat and, more recently, by the saturable reactor.

Rheostat

We shall not repeat the entire discussion of the rheostat (variable resistor) which has already been presented in this chapter. However, as already mentioned, the choke coil has been largely replaced by the rheostat for the control of filament current. A variable resistor controls not only the voltage but also the current. According to Ohm's Law, $I = V/R$, and if the applied volt-

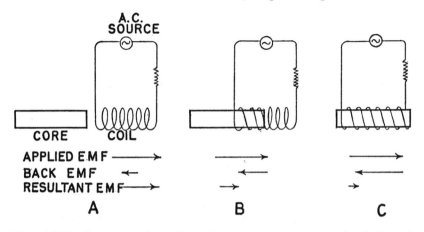

Figure 10.8. Operation of a choke coil. Note that ac is required. The length of the arrows indicates the emf in the specified direction. In *A*, with the core completely out, the back emf is minimal and the voltage drop across the coil is small; hence, the resultant emf is only slightly less than the applied emf. In *B*, with the core partly inserted, there a larger back emf and the voltage drop across the coil is larger than in *A*. In *C*, with the core completely inserted, the back emf of self-induction is maximal, producing a large voltage drop across the coil with a small resultant emf available for the rest of the circuit; *at the same time, the current is decreased*. The position of the core is adjustable, so that the resultant emf and current can be varied as needed.

age across the *entire* circuit is kept constant, an increase in *R* must necessarily cause a proportional decrease in *I*, the current.

Saturable Reactor

In more advanced types of modern radiographic equipment, a new device, the **saturable reactor,** is coming into use for control of filament current. It depends on the electromagnetic principle, not discussed heretofore, that if an iron core within or near a coil is saturated with magnetic flux by an independent source of direct current, the inductance of the coil increases as the degree of saturation of the core increases. This variable degree of core saturation is accomplished by varying the size of the applied dc. As usual, an increase in the inductance of a circuit causes a decrease in the current, *I*. Figure 10.9 shows a simplified version of a saturable reactor.

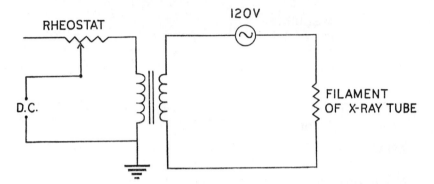

Figure 10.9. Current control by a saturable reactor. The dc voltage in the left coil saturates the iron core with a steady magnetic flux whose strength can be changed by changing the voltage by means of a rheostat. The higher the saturation of the core, the greater will be the inductance of the right coil and the lower will be the current in the filament circuit.

QUESTIONS AND PROBLEMS

1. Why is a high voltage needed in x-ray equipment?
2. What device is used to change low voltage to high voltage?
3. What is the transformer law? A transformer is so constructed that there are 100 turns in the primary coil and 100,000 turns in the secondary. If the input is 110 volts, what is the output voltage?
4. Show by diagram a doughnut transformer, first as a step-up transformer, second as a step-down transformer.
5. How are eddy currents reduced in a transformer? Why should they be kept to a minimum?
6. Why is a core used in a transformer?
7. What two types of insulation are used in a transformer and why?
8. What are the purpose and principle of an autotransformer? How does it operate? What is the autotransformer law?
9. Why is the autotransformer preferred in the control of high voltage in an x-ray machine?
10. The main electrical supply to an x-ray machine is 230 volts. If the x-ray transformer has a step-up ratio of 1,000 and we wish to obtain 100 kv, what will be the step-down ratio of the autotransformer? (Step-up and step-down ratios

refer to the quotient of the number of turns in the secondary coil by the number of turns in the primary coil.)

11. What is a rheostat? In what type of x-ray equipment is it most widely used today? How does it operate?

12. Describe the principle of the choke coil. Where is it often used in x-ray equipment?

13. What type of current (ac or dc) is needed for the operation of choke coil; autotransformer; rheostat?

14. Why is it necessary to regulate the filament current of an x-ray tube?

15. What purpose is served by the saturable reactor? How does it operate?

CHAPTER 11 RECTIFICATION

Definition

As we have pointed out before, an alternating current (ac) periodically reverses its direction and varies in magnitude. While an x-ray tube can operate on *any* high voltage ac, it *operates most efficiently when supplied by high voltage direct current* (dc)—a current that is *unidirectional,* flowing always in the same direction. *Rectification may be defined as the process of changing alternating current to direct current.*

There are two possible ways of rectifying the high voltage ac leaving the secondary side of the step-up transformer: (1) by *suppressing* that half of the ac cycle represented by the portion of the ac curve that lies below the line (that is, the negative half cycle) or (2) by changing the negative half cycle to a *positive* one. The resulting curves are shown in Figure 11.1.

In order to impart a better understanding of rectification, we shall summarize here the construction and theory of x-ray tubes, although Chapter 13 contains a detailed discussion of this subject.

An *x-ray tube* consists of a glass bulb from which the air has been evacuated as completely as possible. Sealed into the ends of the bulb—or *tube,* as it is better known—are two terminals or electrodes: the *cathode* and the *anode.* The cathode consists of a thin tungsten filament which, when heated to incandescence (white hot) by a *low voltage current,* "boils off" electrons. These form a cloud or *space charge* near the filament. If, now, a *high voltage* from the secondary of the transformer is applied between the cathode and anode so as to produce a large negative charge on the cathode and a large positive charge on the anode, the resulting strong electric field drives the space charge electrons toward the anode at tremendous speeds. When the electrons

136

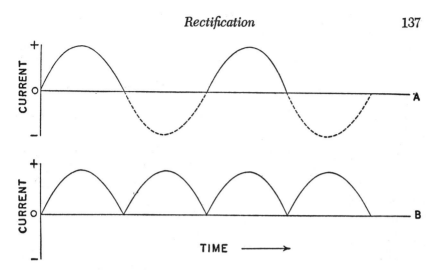

Figure 11.1. Two fundamental types of rectified circuits. In *A*, with half wave rectification, the negative half lying below the line is *suppressed*. In *B* the negative half of the alternating current is *inverted* so that it lies above the line and becomes positive. With either method, one obtains a pulsating direct current.

strike the anode their *kinetic energy is converted to heat and x rays.* Under ordinary conditions of operation, *the high voltage current can pass in only one direction, from cathode to anode.* If the applied high voltage ac has been rectified before reaching the tube so that its direction is always from cathode to anode, the tube can withstand larger energy loading ("exposures") than if the current is not rectified.

Methods of Rectifying an Alternating Current

There are two main systems of rectification in use today: (1) self-rectification and (2) vacuum tube or solid state diode rectification. In this chapter we shall describe the use of vacuum tube diodes, called *valve tubes,* in rectification, since they are still the most widely used. However, since solid state rectifiers are gaining acceptance, they will be discussed on pages 217-228. (A third form, mechanical rectification, is now obsolete.) *All rectifying systems are located between the secondary coil of the transformer and the x-ray tube.*

1. **Self-rectification.** In this, the simplest type of rectification,

the high voltage is applied *directly* to the terminals of the x-ray tube. Under ordinary conditions, an x-ray tube allows passage of electrons only from the cathode to the anode during the positive half cycle of the ac curve, when the anode is positively charged. This half of the voltage cycle is the *useful voltage* or *forward bias.*

During the negative half cycle the "anode" is negative and the "cathode" is positive, but despite the presence of a high voltage across the tube (although in the wrong direction) no current will *normally* flow because there is no space charge near the anode. This reverse voltage, called *inverse voltage* or *reverse bias,* is actually higher than the useful voltage.

The chief disadvantage of self-rectification is that it lowers the energy rating (that is, maximum exposure factors) of the x-ray tube. In a self-rectified circuit the anode must never be heated to the point of electron emission because then, during the negative half cycle, the inverse voltage would drive the electrons in the wrong direction—toward the filament—causing the filament to melt, and ruining the tube. Therefore, with self-rectification technical factors of kV, mA, and time must be limited to lower values than with full-wave rectification wherein both halves of the ac cycle are utilized and there is no inverse voltage.

An important point to remember is that the mA, as registered by the milliammeter, is an *average* value. In a self-rectified circuit, because only one-half the cycle is used, the peak mA and consequent anode heat must be greater, for the same milliammeter reading, than in a full-wave rectified circuit in which both halves of the cycle are utilized.

Figure 11.2 describes the current through the tube in a self-rectified circuit. This type of rectification is known as *self-half-wave rectification.*

Self-rectification is used in small, mobile x-ray apparatus, but it is being gradually replaced by more sophisticated methods of rectification, to be described next.

2. **Valve-tube Rectification.** A valve tube is a *thermionic diode tube* which resembles, in general, an x-ray tube, allowing passage of current *in one direction only; that is, from cathode to anode.* However, it is so designed that it normally does not produce x

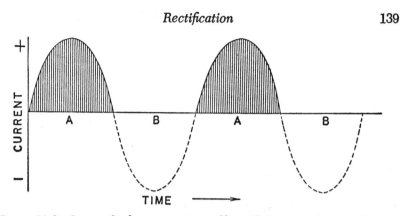

Figure 11.2. Curve of tube current in a self-rectified circuit. During the first half cycle, *A*, the current passes through the tube; the applied voltage is the *effectual* or *useful voltage*. During the second half, *B*, no current flows from anode to cathode and now the applied voltage is *ineffectual* or *inverse voltage*. Note that the inverse voltage is greater than the useful voltage.

rays. Details of construction will be considered later. Valve tubes can be introduced into the high voltage x-ray circuit in such a way as to produce either *half-wave* or *full-wave rectification*. Solid state rectifiers, to be described later, may be used for the same purpose.

One or two valve tubes can be used to achieve half-wave rectification. A single valve tube can be connected as shown in Figure 11.3. During the *positive half* of the ac cycle, the polarity of the secondary coil of the transformer is that shown in the figure, the direction of the current being from cathode to anode. The current flows through both the valve tube and the x-ray tube as indicated by the arrows. During the *negative half cycle,* the polarity of the transformer secondary is reversed and a strong negative charge is placed on the anode of the valve tube (instead of the x-ray tube, as would occur in a self-rectified circuit). Thus, the *valve tube suppresses the inverse voltage,* diminishing the possibility of reverse flow of electrons, should the anode become hot enough during operation to be a source of electron emission. Such a reverse flow of electrons could easily destroy the filament. Therefore, the valve tube allows greater loading of the x-ray tube than is possible with self-rectification, although half-wave rectification results in either case (see Figure 11.2).

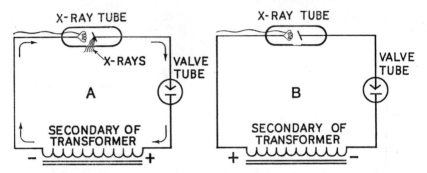

Figure 11.3. Half-wave rectification with a single valve tube. In *A* electrons flow readily in both the x-ray tube and valve tube; this is called *forward bias.* In *B,* during the next half cycle, the polarity of the transformer is reversed and current cannot flow normally from anode to cathode in the valve tube, thereby protecting the x-ray tube from the inverse voltage, also called *reverse bias.*

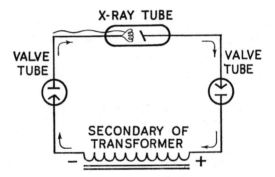

Figure 11.4. Half-wave rectification with two valve tubes. When the polarity of the transformer reverses, no current flows in the circuit.

If *two valve tubes are used,* as in Figure 11.4, the high voltage during the negative half cycle is divided between the two valve tubes, increasing the efficiency of the system and improving the loading capacity of the x-ray tube. The wave form of this type of rectified current resembles that of self-rectification (see Figure 11.2), but notice again that the current is rectified by the valve tubes rather than by the x-ray tube.

Four valve tubes can be so connected that the negative half of the ac cycle is reversed to provide *full-wave rectification.* Figure 11.5 indicates the circuitry of a four-valve-tube rectifier. In

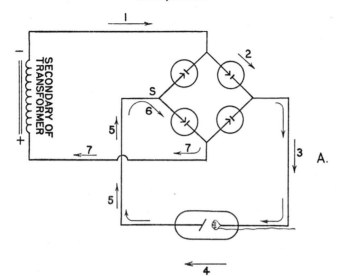

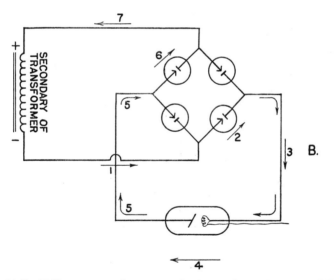

Figure 11.5. Full-wave rectification with four valve tubes. By following the numbered portions of the circuits in A and B, you can see that regardless of the polarity of the transformer during the different halves of the ac cycle, the current always reaches the x-ray tube in the same direction; that is, the alternating current is fully utilized.

A, the current passes from the negative pole of the transformer through the successively numbered portions of the circuit to the cathode of the x-ray tube, and finally returns to the positive end of the transformer. (Remember that each valve tube permits the current to flow only in one direction, from cathode to anode.) In *B,* the polarity of the transformer reverses due to alternation of the current. The current again passes through the successively numbered portions of the circuit to the cathode of the x-ray tube and finally reaches the positive end of the transformer. Note that at some time during its course through the circuit, the current arrives at the junction of two valve tubes at their cathodes, as at point *S* in *A;* the current passes through one of the two valve tubes because passage through the other valve tube would lead it back to a point of higher potential and this is physically impossible (analogy: a ball will not, of itself, roll uphill to a point of higher potential energy).

You should have no difficulty learning the proper connection of the valve tubes once you have mastered the following simple rules:

(1) First connect four valve tubes without indicating anode or cathode.

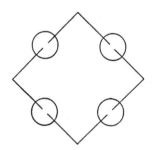

Figure 11.6.

(2) Then insert two *cathodes* so that they are connected together.

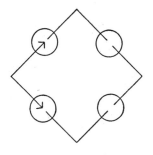

Figure 11.7.

(3) Now connect *these cathodes* to the *anode* of the x-ray tube.

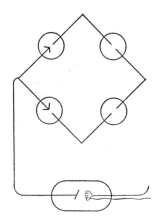

Figure 11.8.

(4) Insert two *anodes* in the opposite valve tubes, and join them to the *cathode* of the x-ray tube.

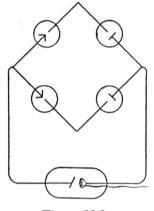

Figure 11.9.

(5) Now designate the remaining electrodes in the valve tubes.

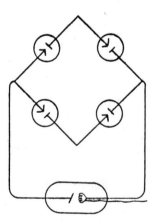

Figure 11.10.

(6) Finally, connect the remaining valve tube ends to the terminals of the transformer.

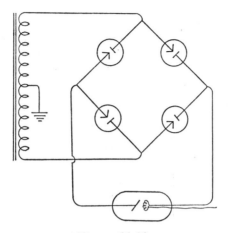

Figure 11.11.

The above rules are necessary because there is only one correct way to connect the valve tubes to each other and to the x-ray tube and transformer. If this relationship is altered in any way, the system will fail to rectify. Notice that in any particular half cycle, one parallel pair of valve tubes is conducting (see Figure 11.5).

There is one disadvantage in four-valve-tube rectification. Utilization of the entire ac wave results in the production of a high percentage of low energy x rays at lower kilovoltages (see Figure 11.12). However, these can be minimized either by using ap-

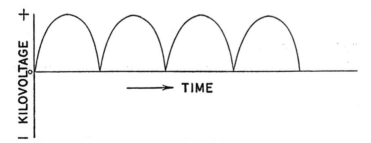

Figure 11.12. Wave form of a current rectified by four valve tubes. Note that all the values of kV are represented from zero to the peak value.

propriate filters or by boosting the low points in the ac wave to higher voltages by means of capacitors, as in *constant-potential* equipment.

The use of full-wave rectification increases considerably the *tube rating* or *heat loading capacity,* which means that larger exposures can be used without damaging the tube. Table 11.1

TABLE 11.1

COMPARATIVE TUBE RATINGS OF STATIONARY ANODE TUBE WITH 2.2 MM FOCUS OPERATING ON 60 CYCLES. MAXIMUM PERMISSIBLE EXPOSURE TIME WITHOUT DANGER OF DAMAGING THE TUBE IS INDICATED FOR THE SELECTED VALUES OF kV AND mA

Type of Rectification	kV	mA	Maximum Safe Exposure Time
Full-wave	80	40	20 sec
Half-wave	80	40	10 sec
Self-rectified	80	40	2 sec

shows the comparative maximum values of kV, mA, and exposure time with different types of rectification. It is obvious that with the same factors of kV and mA, in this particular example, the maximum safe exposure is increased five times with half-wave rectification, and ten times with full-wave rectification, as compared with self-rectification. We must emphasize that this table applies to a selected tube; the rating for any particular tube must be obtained from the manufacturer. However, with any type of equipment, full-wave rectification increases the tube rating, thereby making possible modern high speed radiography.

QUESTIONS

1. Define rectification.
2. Why is it desirable to rectify the current for x-ray generation?
3. Name two main types of rectification.
4. How are x rays produced?
5. Show with the aid of a diagram: (a) single-valve-tube rectification and (b) four-valve-tube rectification. Draw the shapes of the current waves produced by each.

6. What is meant by a constant potential unit? Under what conditions, and why, is it desirable?

7. Draw and label the current curves in a self-rectified circuit; a single-valve-tube rectified circuit; a four-valve-tube rectified circuit. Which are half-wave rectified? Full-wave rectified?

8. Define inverse voltage; reverse bias; useful voltage. What is their importance?

9. What property of valve tubes makes them suitable for use in rectification?

10. In what type of equipment is self-rectification most often used?

CHAPTER *12*   X RAYS (ROENTGEN RAYS)

How X Rays Were Discovered

For a number of years before the discovery of x rays physicists had been observing high voltage discharges in vacuum tubes. In 1895 Wilhelm Konrad Roentgen, a German physicist, while studying these phenomena in a Crookes tube operated at high voltage in a darkened room, suddenly noticed the *fluorescence* —glowing—of a piece of barium platinocyanide lying several feet from the end of the tube. He promptly realized that the fluorescence of the barium platinocyanide was caused by a hitherto unknown type of invisible radiation. Moreover, he soon found that this radiation could pass completely through solid matter such as paper, cardboard, and wood, since they did not prevent fluorescence when placed between the tube and the barium platinocyanide. However, these rays could be stopped by denser materials such as lead.

Thus, by the sheerest accident, Roentgen had discovered a new type of radiation which, upon leaving the Crookes tube, was capable of passing through certain solids that are opaque to ordinary light.

Roentgen quickly discovered the potentiality of the new radiation in medicine when he placed his hand between the tube and a piece of cardboard coated with barium platinocyanide. Imagine his excitement on seeing the bones of his hand depicted on this, the first fluoroscopic screen!

Roentgen gave the name *x rays* to this invisible, penetrating radiation because the letter *x* represents the unknown in mathematics, although within the next few months he had discovered most of the properties of x rays.

This monumental discovery has had a tremendous impact on pure science, medicine, and industry. Roentgen's name is often directly linked with the rays in that they are also called *roentgen rays*. *Roentgenology is that branch of medicine dealing with the use of roentgen rays in diagnosis and treatment.* Radiology is a broader term which includes also radium and radioactive nuclides.

Roentgenography or *radiography* deals with the art and science of recording x-ray images on photographic film.

Roentgenoscopy or *fluoroscopy* is the observation of x-ray images on a screen coated with fluorescent material.

What Are X Rays?

An x-ray beam consists of a group of rays which are fundamentally of the same nature as white light, ultraviolet, infrared, and other similar types of radiant energy. They are all *electromagnetic waves*—wavelike disturbances associated with *vibrating electric charges.* Most kinds of waves are transmitted by some medium; for example, you have probably all seen waves on the surface of the water, wherein water is the transmitting material. When a stringed instrument is plucked, waves are set up in the string, and here the string is the transmitting material. But what transmits electromagnetic waves? Strangely enough, no one knows exactly what carries them. At one time scientists assumed that empty space really consisted of a nonmaterial medium called the ether, which was supposed to conduct these waves, but this concept has been abandoned.

X rays are a class of electromagnetic waves traveling with the same *constant* speed as light, 186,000 miles per second (3×10^{10} cm per sec) in a vacuum. The waves are believed to have the form shown in Figure 12.1. All electromagnetic waves (including radio, heat, light, ultraviolet, and x rays) have the same form and travel with the same speed, but differ in *wavelength*—the distance between two successive crests in the wave, such as A to B in Figure 12.1. The number of crests or cycles per second is the *frequency* of the wave, the unit of frequency being the *hertz* $= 1$ cycle per sec. You can see, by comparing the upper and lower waves in Figure 12.1, that if the wavelength is decreased,

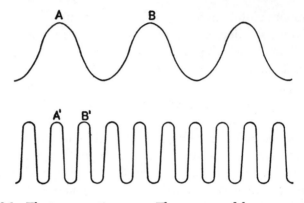

Figure 12.1. Electromagnetic waves. The upper and lower waves differ in wavelength, that is, the distance between two crests such as *A* to *B* and *A' to B'*. The lower wave thus has a shorter wavelength and therefore more crests or cycles in a given period of time, than does the upper wave, because their speeds are identical. Thus, the wave with the shorter wavelength vibrates faster, that is, it has greater frequency. Another way of saying this is that in a given interval of time, the wave with shorter wavelength will have greater frequency—number of cycles per second.

the frequency must increase correspondingly. This can also be shown in another way. The speed with which the wave travels is equal to the frequency multiplied by the wavelength:

$$c = \nu\lambda \qquad (1)$$

where c = speed of x rays in vacuum or air (3×10^{10} cm per sec)

ν (Greek *nu*) = frequency in hertz (cycles per sec)

λ (Greek *lambda*) = wavelength in cm

Since c, the speed of all electromagnetic waves, is constant in a given material, an increase in frequency (ν) must always be accompanied by a corresponding decrease in wavelength (λ), and conversely, a decrease in frequency by an increase in wavelength. In other words, the *frequency is inversely proportional to the wavelength.*

The wavelengths of x rays are extremely short; for instance, in *ordinary radiography,* the useful range of x-ray wavelengths is about 0.1 to 0.5 Å (recall that 1Å $= 10^{-8}$ or 1 one-hundred-millionth cm!). Because of the extremely short wavelengths of

x rays their frequencies are enormous—3 × 10¹⁹ to 6 × 10¹⁸ hertz!
The various types of electromagnetic waves, ranging from radio
waves with wavelengths measured in several thousand meters,
down to gamma rays with wavelengths in thousandths of an Å,
are shown with their wavelengths in Figure 12.2.

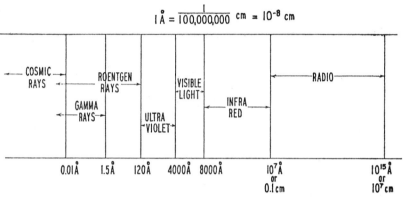

Figure 12.2. The electromagnetic spectrum.

As we shall attempt to explain in more detail later, each electromagnetic wave is associated with an extremely tiny amount of
energy that depends on the frequency of the wave. Such an energy "fragment" is called a *photon* or *quantum*.

Source of X Rays

The x radiation used in roentgenography and therapy is produced by man-made machines. *These rays arise whenever a
stream of fast-moving electrons suddenly undergoes a reduction
in speed.* Gamma rays, which are physically identical with x rays,
are emitted by the *nuclei* of certain radioactive elements.

The X-ray Tube

A man-made device in which x rays are produced is called an
x-ray tube. The details of construction will be discussed later,
but the principles of construction must be outlined now in order
to appreciate how x rays are produced.

We should mention, out of historic interest, that the early x-ray
tubes were cold cathode gas tubes. Such a tube consisted of a

glass bulb in which a partial vacuum had been produced, leaving a small amount of gas. In it were sealed two electrodes, one negative (cathode) and the other positive (anode). The cathode terminal was *not* heated. Application of a high voltage across the terminals caused ionization of the gas in the tube with release of a stream of electrons which produced x rays upon striking the positive terminal (anode). The gas tube is now obsolete not only because of its inefficiency but also because tube mA could not be changed independently of kV, making it difficult to control x-ray quantity and quality.

In 1913 W. D. Coolidge, at the General Electric Company Laboratories, invented a new type of x-ray tube based on a radically different principle; this was the *hot cathode diode tube,* which revolutionized radiographic technic because it made possible the independent control of mA and kV. It is still the most satisfactory device for the production of x rays. The principles of construction and operation are surprisingly simple and must be clearly understood by all those concerned with the use of x rays. The Coolidge tube consists of the following (see Figure 12.3):

1. A *glass envelope* or *tube* from which the air has been evacuated as completely as possible. Air must be removed not only from the interior of the tube, but from the glass and metal parts as well, by prolonged baking before the tube is sealed. This process is called *degassing.* A high vacuum is necessary for two reasons:

a. To prevent collision of the high speed electrons with gas molecules, which would cause significant slowing of the electrons (see below).

b. To prevent oxidation and burning-out of the filament.

2. A *hot filament* supplied with a separate low voltage heating current. The filament becomes the *cathode,* or negative electrode, when the high voltage is correctly applied.

3. A *target* which becomes the *anode,* or positive electrode, when high voltage is correctly applied.

4. A *high voltage* applied across the electrodes, charging the filament negatively and the target positively.

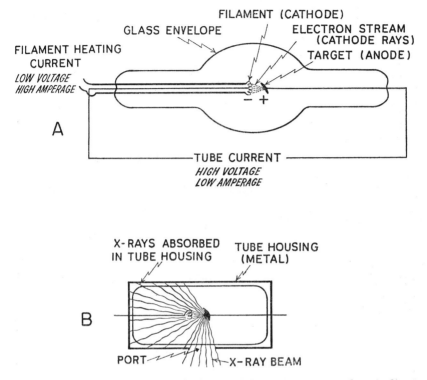

Figure 12.3. Model of *hot cathode* (Coolidge type) x-ray tube. *A* shows the two circuits and the general construction. The inset at *B* shows that x rays are emitted from the target in all directions, but only those approaching the port (window) pass out as the *useful beam*, the remaining x rays being absorbed in the protective housing.

Notice that there are *two circuits in the modern x-ray tube:* one, a low voltage heating circuit through the cathode itself; and the other, a high-voltage circuit between the cathode and anode to drive the electrons (see Figure 12.3).

Conditions Necessary for the Production of X Rays

Whenever *rapidly moving electrons are suddenly stopped, x rays are produced.* The x-ray tube is a device for obtaining free electrons, then speeding them up, and finally stopping them. In addition to these basic processes, the electrons must also be concentrated on a small area of the anode known as the *focus.* These four conditions required for the production of x rays in a hot cathode tube will now be described in some detail because of their extreme importance.

1. **Separation of Electrons.** In common with all other atoms, the tungsten atoms in the filament have orbital electrons circulating around a central nucleus. How can these electrons be liberated from their atoms? The *filament current* supplying the filament causes it to become glowing hot or *incandescent*, with resulting separation of some of its outer orbital electrons. We may regard such electrons as "boiling off" the filament to form a small cloud or *space charge.* The electrons liberated in this manner are called *thermions*, and the process of their liberation through the heating of a conductor by an electric current is called *thermionic emission.*

2. **Production of High Speed Electrons.** If, now, high potential difference is developed between the filament cathode and the target anode by applying sufficient kV between them, space charge electrons that have been separated by the preliminary heating of the filament rush toward the anode at high speed. The reason for this is as follows: since the filament cathode is given a very high *negative charge* and the target anode a similarly high *positive charge* by the applied kV, the resulting strong electrical field causes electrons to rush *at an extremely high speed through the tube from cathode to anode,* this stream of electrons constituting the *cathode rays* or *tube current. The speed of these electrons approaches one-half the speed of light,* and even more in some modern equipment.

3. **Concentration of Electrons.** The electron stream in the tube is confined to a very narrow beam and is concentrated on a small spot on the anode face—the *focus*—by a negatively charged molybdenum collar surrounding the filament; the narrower the electron beam the smaller the focus and the sharper the x-ray images.

4. **Sudden Stopping of the Electron Stream.** Upon striking the target, the electron stream in the x-ray tube is stopped abruptly and the *kinetic energy* of the electrons undergoes conversion to some other forms of energy (Law of Conservation of Energy). With equipment ordinarily used in clinical radiology, about 99.8

per cent of this energy is changed to heat, while *only 0.2 per cent is converted to x rays.* In fact, only about one part in a thousand of the kinetic energy of the electrons eventually results in x rays that are useful in roentgenography! As shown in Figure 12.3B, x rays are emitted in *all* directions, but only those leaving the window of the x-ray tube comprise the *useful beam.*

You must avoid confusing the electrons flowing in the tube, with the x rays emerging from the tube. Imagine a boy throwing stones at the side of a barn. The stones are analogous to the electron beam in the x-ray tube, whereas the emerging sound waves are analogous to the x rays that come from the target when it is struck by electrons.

Further Consideration of Production of X Rays

When the electron stream encounters the tube target, x rays are produced by the following two main processes:

1. **Brems Radiation.** Upon approaching the strongly positive nuclear field of a target atom, the oppositely charged high-speed electron is deviated from its initial path because of the attraction between these opposite charges. As a result, the electron is slowed down or *decelerated.* In slowing down, the electron loses some of its kinetic energy, the *lost kinetic energy being radiated as an x ray of equivalent energy.* The term *bremsstrahlung* or *braking radiation* has been applied to this process. Brems radiation is *heterogeneous;* that is, nonuniform in energy and wavelength because the amount of braking or deceleration varies among electrons according to their speed and how closely they approach the nucleus. With each different deceleration, a corresponding amount of kinetic energy is converted to x rays of equivalent energy; just as the deceleration varies, so does the energy of the x rays. A fraction of the electrons approaches the nucleus head-on and is completely stopped by the nuclear electrostatic field. In this special case of the brems effect, all of the kinetic energy of the electron is converted to an x ray of equivalent energy. The deceleration of electrons depends also on a third factor—the atomic number of the target. Thus, targets of higher atomic number are more efficient producers of brems

radiation (for example, rhenium with atomic number 75, vs tungsten with atomic number 74).

2. **Characteristic Radiation.** An electron with a sufficient minimum kinetic energy may interact with an *inner orbital electron* (for example, in K or L shell) of a target atom, ejecting it from its orbit (see Figure 12.4). To free this electron, work has to be

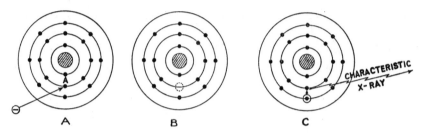

Figure 12.4. Production of characteristic radiation. In *A*, the incoming *electron* collides with an inner shell electron. In *B*, the atom is in an "excited state," electron A now being displaced from its shell. In *C*, an electron jumps from an outer shell to replace electron A; this is accompanied by characteristic radiation, as the atom returns to its normal state.

done against the attractive force of the nucleus. The atom is now unstable, being *ionized* (electron missing from atom) and in an *excited state* (electron vacancy in a shell). Immediately, the space or "hole" vacated by the electron is filled by an electron dropping into it from one of the outer shells. Since, in the first place, energy was put into the atom to free the electron, a like amount of energy must be given off when an electron from a higher energy level enters to fill the hole in the shell (satisfying the Law of Conservation of Energy). This energy is emitted as a *characteristic x ray* because its energy is characteristic of the target element and the involved shells. (In fact, there are series of characteristic rays because, for example, the replacing electron leaves a hole which must, in turn, be filled by an electron from a still higher energy level or shell.)

We may now summarize as follows the nature of the *primary radiation*, that is, the x rays emerging from the target:

    a. **General Radiation**—"white" or heterogeneous radiation with a continuous range of energies (and wavelengths) due to deceleration of electrons by strongly positive electric fields of nuclei in target atoms. This is *brems radiation,* constituting about 70 per cent of emitted x rays in the diagnostic range.

b. **Characteristic Radiation**—consists of limited, discrete energies (and wavelengths), constituting about 30 per cent of emitted x rays in the diagnostic range.

Target Material

The target metal must be selected with two aims in mind. First, it must have a very *high melting point* to withstand the extremely high temperature to which it is subjected. Second, it must have a *high atomic number* because (1) the resulting characteristic radiation is of short wavelength (high energy) and therefore very penetrating and (2) there is increased production of brems radiation. Tungsten, a metal with a melting point of 3370 C and an atomic number of 74, satisfies the above requirements in most cases. Hence, it is now generally employed as the target in tubes used in radiology. Rhenium (atomic number 75), coated on molybdenum, is used as the target in advanced types of radiographic tubes for more efficient x-ray production. To facilitate the dissipation of heat from the anode in stationary anode tubes, a tungsten button is imbedded in a block of copper which is a better conductor of heat than tungsten. The electron beam in the tube is focused on a small spot on this tungsten target.

Properties of X Rays

We shall now list some of the properties of x rays that are of importance in the field of roentgenology.

1. They are *highly penetrating, invisible rays,* belonging to the general category of electromagnetic waves.

2. They are *electrically neutral* and cannot be deflected by electrical or magnetic fields.

3. They are *heterogeneous,* having a *wide range of wavelengths,* from about 0.04 Å to more than 1000 Å. Their useful range in radiography is about 0.1 Å to 0.5 Å. Or we may say they have a wide range of *photon energies* (polyenergetic).

4. They liberate *minute* amounts of heat on passing through matter.

5. They emerge from the tube in *straight lines,* diverging from the focus.

6. They travel only at the *same speed as light,* 186,000 miles per sec or 3×10^{10} cm per sec in a vacuum.

7. They are capable of *ionizing* gases indirectly because of their ability to remove orbital electrons from atoms.

8. They cause *fluorescence* of certain crystals, making possible their use in fluoroscopy, and in intensifying screens for radiography.

9. They *cannot be focused* by a lens.

10. They *affect photographic film,* producing a latent image which can be developed chemically.

11. They produce *chemical and biologic changes,* mainly by ionization and excitation. This underlies their use in therapy.

12. They produce *secondary and scattered radiation.* This is discussed in greater detail in a later section.

SPECIFICATIONS OF THE PHYSICAL CHARACTERISTICS OF AN X-RAY BEAM

One may describe a given x-ray beam on the basis of two properties—*intensity* and *quality.* Strictly speaking, the intensity of the radiation at a given point in the beam may be defined as the quantity of radiant energy passing per second through a unit area of the surface perpendicular to the direction of the beam at the designated point. This definition is not suitable for medical radiology and, instead, the concept of *exposure rate* has been adopted. Furthermore, since the main interest in radiotherapy is centered on the interaction of radiation with the patient's tissues, another unit of quantity has been introduced—the *absorbed dose.*

The *quality* of an x-ray beam refers to its penetrating ability. We shall now discuss the concepts of exposure and quality, leaving absorbed dose for a later section.

X-ray Exposure (Quantity)

The best method of measuring x-ray *exposure* in medical radiography is to determine the amount of ionization an x-ray beam produces in air, since the degree of ionization is nearly proportional to the radiation exposure under certain controllable conditions. In other words, exposure is a measure of radiation quantity based on its ability to ionize air. The international unit of radiation exposure has been established on that basis and is called the *roentgen*, symbolized as R. One roentgen is defined as *"that exposure of x or gamma radiation such that the associated corpuscular emission per 0.001293 gram of air, produces in air, ions carrying one electrostatic unit of quantity of electricity of either sign."* It may be simply explained as follows: a roentgen is a definite radiation *exposure* which, in a standard quantity of air (chosen as 0.001293 g of air, which is really the weight of 1 cc of air at a temperature of 0 C and a pressure of 760 mm mercury), produces ionization consisting of a certain number of ions and electrons (called corpuscular emission in the definition). Eventually, all the usable energy of the corpuscles produced by one roentgen is spent in releasing pairs of oppositely charged ions. If these ions are collected separately and measured, it will be found that exactly one electrostatic unit (esu) of electricity of positive or negative sign has been carried by the ions, provided the radiation exposure was exactly one roentgen. In other words, *1 R of x or gamma rays causes ionization in 1 cc of air under standard conditions, such that the ions of only one sign carry 1 esu of electricity.* It should be pointed out that the roentgen is a valid unit only with x or gamma rays *up to 3 million volts.*

The radiation exposure per unit time (eg, R/min) at a given location is the *exposure rate:*

$$\text{exposure rate} = \frac{\text{exposure in R}}{\text{time in min}}$$
$$(R/min)$$

The *total exposure* in R is obtained by multiplying the exposure rate by the exposure time:

$$\underset{(R)}{\text{exposure}} = \underset{(R/min)}{\text{exposure rate}} \times \underset{(min)}{\text{time}}$$

In measuring radiation exposure, we must be careful that our instrument collects and measures all the produced ions; and that more ions are not accidentally lost to the surroundings than enter from the surroundings. The measuring instrument, designed to have a high degree of precision, is the **standard free air ionization chamber.**

For practical calibration of the roentgen output of an x-ray therapy unit, we use a **thimble type ionization chamber** which has been calibrated initially against the more precise type of instrument. The thimble chamber was developed by H. Fricke and O. Glasser, and incorporated in the Victoreen R-meter (see Figure 12.5). This device is convenient and simple to use, and is suffi-

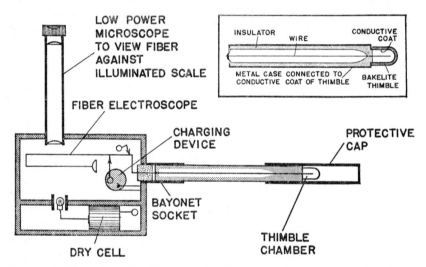

Figure 12.5. Schematic diagram of a Victoreen R-meter. Details of the thimble chamber are shown in the insert. (Simplified from data furnished through the courtesy of the Victoreen Instrument Company.)

ciently accurate for medical radiology, provided it is returned to the manufacturer for calibration at regular intervals. However, its accuracy holds only for a limited range of x-ray or gamma-ray energies, so that outside this range, special caps must be placed over the thimble, or correction factors must be applied.

Radiation exposure as measured in roentgens is not necessarily the same as the total quantity of radiant energy in a beam. The former is of more practical value to the radiologist because the exposure in R is based on the ionizing ability which, after all, is responsible for biologic and radiographic effects.

The **exposure rate** of an x-ray beam may be altered by varying four factors: (1) tube current, (2) tube potential, (3) distance, and (4) filtration.

1. **Tube Current.** As the filament current is increased, more electrons are liberated per second at the filament. Consequently, the *tube current* (mA) increases so more electrons strike the target per sec, increasing the amount of x radiation emitted per sec. The exposure rate is thus directly proportional to the mA, if the other factors remain constant. It should be emphasized that a change in mA does *not* affect the quality of the radiation.

2. **Tube Potential.** As *kilovoltage* (kV) is increased exposure rate increases because the electrons in the tube are speeded up and produce more x-ray photons per sec at the target. The resulting x-ray beam also has more energy than one produced at lower kV. Note that an increase in tube potential increases both the exposure rate and penetrating power of an x-ray beam, but that this relationship is not directly proportional.

3. **Distance.** In determining x-ray exposure rate, we must specify the distance from the target to the measuring device because the x-ray beam leaves the tube focus in a spreading beam shaped like a cone. As the distance from the target increases, the number of rays in each square centimeter of the beam decreases; hence, the exposure rate decreases. Since the measuring device (for example, thimble chamber) has a fixed size, it measures the exposure rate in a constant small area of the beam. *The exposure rate is governed by the inverse square law of radiation,* which is discussed in detail in Chapter 18.

4. **Filtration.** The exposure rate of an x-ray beam may be altered by changing the thickness or type of filter placed in the beam—the thicker the filter and the higher its atomic number, the greater the reduction in the exposure rate beyond the filter. At the same time, the *beam is hardened due to the relatively greater removal of soft (low energy) than of hard (high energy) rays by the filter.*

Quality of X Rays

Definition. The quality of an x-ray beam may be defined as its penetrating power. The term "penetrating power" refers to the ability of a particular x-ray beam to pass through a given part of the body and onto an x-ray film, or to enter the body and deposit energy in a tumor-containing volume.

X-ray Energy and Quality. To understand the relationship between x-ray energy and quality (that is, penetrating power), we must first look into the meaning of x-ray energy. An individual x ray may be regarded as a bit of energy called a *photon* or *quantum*, represented by the equation

$$E = h\nu \qquad (2)$$

where E is the energy of the photon in ergs, h is a constant (Planck's constant), and ν (Greek *nu*) is the frequency of the x ray in hertz (cycles per second). This equation tells us that the *energy of a particular photon is directly proportional to the frequency of its associated wave;* and, therefore, x rays of high frequency have more energy than x rays of low frequency. But since the frequency times the wavelength of an x ray equals a constant (speed of light), as shown in equation (1) on page 150, a wave of high frequency must have a short wavelength; and, conversely, a wave of low frequency must have a long wavelength.

Since the quality of x rays may be defined as their penetrating power which, in turn, is determined by their energy, we may summarize the statements in the preceding paragraph as follows: an x-ray beam of great penetrating power consists of photons which, on the average, have high energy, high frequency, and short wavelength. Conversely, an x-ray beam with poor penetrating power consists of photons which, on the average, have low energy, low frequency, and long wavelength. This relationship is shown in Table 12.1.

TABLE 12.1

INTERRELATIONSHIP OF FACTORS IN RADIATION QUALITY
SPEED OF RADIATION IS CONSTANT.

$c = frequency \times wavelength$

Frequency	Wavelength	Energy	Quality
high	short	high	good penetrating power
low	long	low	poor penetrating power

What influences the *energy* of x rays? We learned earlier that x rays consist of brems and characteristic radiation resulting from the interactions of high speed electrons with target atoms. The larger the kinetic energy of the electrons in the x-ray tube, the higher will be the energy of the resulting x radiation. Since the kinetic energy of the electrons depends on their speed which in turn depends on the kV applied to the tube, it follows that an increase in kV will ultimately produce x rays of higher energy. This discussion may be summarized as follows: the higher the kV, the greater the speed of the electrons, the greater their kinetic energy, the greater the energy of the x rays produced, and the greater the penetrating ability of the x-ray beam. In practice, this is applied directly in that the *penetrating ability of x rays is increased by increasing the kilovoltage; and decreased by decreasing the kilovoltage.* However, this relationship is *not* proportional.

How is the energy of an x-ray beam stated? Since the energy of x rays depends on the speed of the electrons in the tube, which in turn depends on the applied voltage, we customarily use the peak (maximum) voltage to designate the energy of the resulting x-ray beam. For example, if the maximum potential applied across the tube is 100 kV, we then speak of a 100-kV x-ray beam. However, we must bear in mind that the beam really consists of photons whose energies range from a minimum to the peak value, 100 kV. Thus, the beam is *heterogeneous* or *polyenergetic.*

On the other hand, since the energy of monoenergetic radiation such as characteristic x rays and gamma rays is uniform for any given beam, it may be expressed in units of electron volts. One *electron volt* is the energy acquired by an electron when it is accelerated through a potential difference of 1 volt. Thus, the energy of monoenergetic radiation is stated in electron volts (eV) or in some multiple thereof, such as kiloelectron volts (keV), million electron volts (MeV), or billion electron volts (BeV); for example, the gamma rays emitted by technetium 99m all have the same energy—140 thousand electron volts, usually stated as 140 keV.

Insofar as equivalence is concerned, an x-ray beam of given kV has a quality resembling monoenergetic x rays of about ⅓ to ½ the peak energy. Thus, a polyenergetic 120-kV x-ray beam is roughly equivalent to 50 keV monoenergetic x rays.

Heterogeneous Quality of X-ray Beam. As we have already noted, an ordinary x-ray beam is polyenergetic, consisting of a great many photons that differ in energy and penetrating ability. Since photon energy is directly proportional to frequency, and inversely proportional to wavelength, we may say that *an x-ray beam is heterogeneous with respect to wavelength, consisting of many rays of different wavelengths.* Analogy with white light may clarify this. Ordinary white light consists of various colors blended together. Upon being passed through a glass prism, white light separates into its component colors making up the colors of the rainbow, or the *spectrum.* These colors differ in wavelength. For instance, the wavelength of "red" is about 7000 Å, whereas the wavelength of "violet" is about 4000 Å. A light beam made up of a single pure color is called *monochromatic.* Similarly, an x-ray beam consists of different x-ray "colors" or wavelengths, although they are invisible. X rays can be refracted by certain crystals and separated into beams of waves having similar wavelengths, that is, *monochromatic* or, preferably, *monoenergetic* x rays.

Why is an ordinary x-ray beam heterogeneous in wavelength and energy? There are four major reasons:

1. *Fluctuating Kilovoltage.* The applied kV varies according to the wave form of the pulsating current (see Figure 9.4). Since changes in kV are manifested by changes in the energy of the resulting x rays, there must necessarily be a range of photon energies corresponding to the fluctuation of kV.

2. *Processes in X-ray Production.* The x-ray beam leaving the target consists of *general radiation* (brems radiation resulting from deceleration of electrons by target nuclear fields), and *characteristic radiation* (arising from electron transfers within the atoms of the target). The general and characteristic x rays constituting the primary beam have a variety of energies.

3. *Multiple Electron Interactions with Target Atoms.* Electrons traveling from cathode to anode in the x-ray tube undergo varying numbers of encounters with target atoms before being

completely stopped. In these encounters, x-ray photons of various energies are produced.

4. **Off-target Radiation.** Some electrons may strike metal parts of the anode other than the target, producing **stem radiation,** but this has been virtually eliminated by modern tube and collimator design.

X-ray Spectra. A particular x-ray beam can be specified, to a high degree of precision, by sorting out its rays according to wavelength, or its photons according to energy. The meaning of this statement may be clarified by an analogy. Suppose we wish to classify a population of high school senior boys on the basis of height. We would first measure all the boys in the designated population and group them according to height. These data would then be set down as in Table 12.2 and plotted graphi-

TABLE 12.2

HEIGHT DISTRIBUTION OF A CLASS OF
SENIOR HIGH SCHOOL BOYS

Height	Number of Boys
in.	
63	1
64	2
65	4
66	7
67	13
68	14
69	15
70	15
71	14
72	10
73	7
74	1

cally as in Figure 12.6, the number of boys in each group being the dependent variable, and the height the independent variable.

Similarly, we can sort out the rays or photons in a given x-ray beam by means of a special instrument known as an x-ray spectrometer, determining the *spectrum* (that is, relative intensities, a measure of relative numbers) of x rays of various wavelengths. When these data are plotted as in Figure 12.7, a *spectral distribution curve* is obtained, in this case representing x-ray beams produced in a tungsten target by three different applied potentials—30, 40, and 100 kV. A curve is also included for spectral dis-

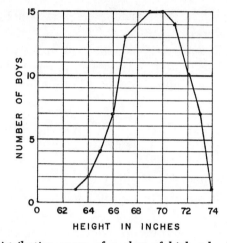

Figure 12.6. Distribution curve of a class of high school senior boys according to their height. Note that the peak incidence is at 68 to 70 inches.

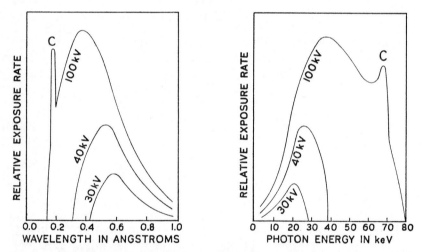

Figure 12.7. Spectral distribution curves of x radiation at 30, 40, and 100 kV. *Characteristic radiation* appears as a peak at *C* (actually, a group of closely spaced peaks) requiring a minimum potential of 69 kV. The remaining curves represent *general radiation*. Note that the curves in the left figure are based on the wavelength of the x rays, whereas those in the right figure are based on photon energy.

tribution according to photon energies, now the preferred method. The peak of the 30 kV curve occurs at about 0.55 Å, indicating that most of the rays in this particular beam have a wavelength of about 0.55 Å. The curve crosses the horizontal axis at 0.41 Å, which is the minimum wavelength (or highest energy) ray occurring in the beam. *The mimimum wavelength depends only on the peak kilovoltage* and is obtained from the equation:

$$\lambda_{min} = \frac{12.4}{kV} \qquad (3)$$

where λ_{min} = minimum wavelength in Å

kV = the peak kilovoltage applied to the tube

If kV = 30, then from equation (3),

$$\lambda_{min} = \frac{12.4}{30} = 0.41 \text{ Å}$$

The curve crosses the horizontal again somewhere above 1 Å, which is the limitation imposed by the glass port and cooling oil of the x-ray tube; any wavelengths longer than this are absorbed by these materials.

A comparison of these curves shows that as the kV is increased, not only is there a relative increase in the number of rays of shorter wavelengths, but also an absolute increase in the number of rays of all wavelengths.

If a sufficiently high kV is applied (at least 69.0 kV), the curves are modified by the appearance of a *sharp peak* at about 0.2 Å, because of the addition of characteristic radiation from the tungsten target.

The same principle may be used in plotting relative intensity as a function of *photon energy,* a more modern way of representing the spectral distribution of an x-ray beam (see Figure 12.7, right). The conversion from wavelength to photon energy is easily made by use of equation (3). For example, the energy of a photon with a wavelength of 0.2 Å is obtained as follows:

$$\lambda = \frac{12.4}{\text{photon energy in keV}} \qquad (4)$$

$$\text{photon energy} = \frac{12.4}{\lambda}$$

$$\text{photon energy} = \frac{12.4}{0.2} = 62 \text{ keV}$$

This method is preferred over that using wavelength because of the more direct relationship of x-ray quality to photon energy. In Figure 12.8 we see that if the kV is kept constant, an in-

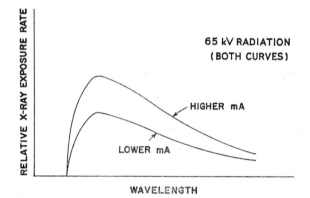

Figure 12.8. Effect of tube mA on relative x-ray exposure rate. With the kV held constant, a higher mA produces a larger exposure rate for all wavelengths. However, the mA has *no* effect on the *minimum* wavelength, which is seen to be the same for both curves.

crease in mA simply increases the exposure rate. However, as anticipated from equation (3), there is no shift in the curves—the minimum wavelength remains the same because it depends *only* on the applied peak kV for a particular target material.

Specification of X-ray Quality—Half Value Layer and Kilovoltage. In view of the polyenergetic nature of the photons in an x-ray beam, the spectral distribution curves described in the preceding section give the most complete representation of the distribution of wavelengths or energies in any given beam. However, this is too cumbersome for radiotherapy and, besides, this degree of precision is unnecessary. The most generally accepted methods of specifying beam quality in radiotherapy include (1) the *half value layer* and (2) the *accelerating potential* (applied voltage).

1. *Half Value Layer.* The half value layer (HVL) is defined *as that thickness of a specified material (usually a metal) which reduces the exposure rate of a beam to one-half its initial value.* The material is used in a thin sheet called a *filter.* By actual measurement, we determine the exposure rate of the x-ray beam

whose quality is to be specified. A series of readings is then taken with increasing thicknesses of filter in the beam. That thickness of the selected metal which reduces the exposure rate 50 per cent is the half value layer. For example, if the exposure rate is initially 90 R/min and it requires 1.5 mm Cu (copper) to decrease the exposure rate by one-half (that is, to 45 R/min), then the HVL = 1.5 mm Cu.

How is the *half value layer of a beam measured?* A brief outline of the procedure will now be presented:

a. Select the x-ray technical factors of kV, mA, and filtration for the *initial beam whose half value layer is being determined.* Collimate the beam to about 10 cm × 10 cm.

b. Place a thimble ionization chamber in the center of the beam at a *fixed distance* from the focus of the x-ray tube. The thimble should be at least several feet from any possible scattering objects such as table, floors, or walls.

c. Determine the exposure rate in R/min (roentgens per minute) without any additional filtration of the beam (other than that which may already be present). Tabulate this exposure rate under *initial beam.*

d. Repeat the procedure after adding a 0.25 mm copper filter, if orthovoltage therapy is used; that is, 150 to 500 kV. All other technical factors remain constant.

e. Repeat the procedure for each of several further additions of copper filters. Assemble the results as in Table 12.3.

TABLE 12.3
EXPOSURE RATES WITH ADDED COPPER FILTERS OF VARIOUS THICKNESSES. kV, mA, AND DISTANCE CONSTANT THROUGHOUT

Added Copper Filtration	Exposure Rate
mm	R/min
0	100
0.25	50
0.50	37
1.0	25
1.5	19
2.0	15

f. Plot the data *graphically* with exposure rate in R/min as the dependent variable and the thickness of added filter as the independent variable, as in Figure 12.9.

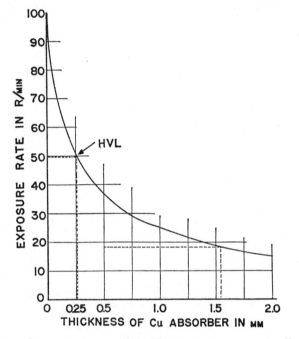

Figure 12.9. Absorption curve of 220-kVp x rays in copper filters of increasing thickness based on data in Table 12.3. Half value layers can be obtained from the curve. If, with no added filter, the exposure rate is 100 R/min, we must add a filter of 0.25 mm Cu to decrease the exposure rate to one-half its initial value, or 50 R/min. The half value layer (HVL) of this beam is therefore 0.25 mm Cu.

If the initial beam is first filtered by 0.5 Cu, we move the vertical axis over to the 0.5 mm line. The corresponding exposure rate is found to be 37 R/min; half of this is 18.5 R/min, corresponding to a reading on the horizontal axis of 1.6 mm Cu. But the initial filter was 0.5 mm Cu. Therefore, the HVL of this particular beam is 1.6 mm − 0.5 mm = 1.1 mm Cu.

As more filtration is added, the beam becomes more homogeneous, that is, the HVL increases and becomes constant.

g. Locate the exposure rate corresponding to one-half the initial value on the vertical axis. Draw a horizontal line from this point to the curve, and then a vertical line from the point of intersection on the curve to the horizontal axis, where the vertical line locates the thickness of copper responsible for the halving of the initial exposure rate. This halving thickness is the half value layer. For an actual example, refer to Table 12.3 and Figure 12.9.

Other metals are used to specify HVL, depending on the kV of the initial beam. Thus, *aluminum* is used for superficial therapy beams, 50 to 150 kV; and *lead* for supervoltage therapy beams, above 500 kV.

2. *Applied Potential (Voltage).* The International Commission on Radiologic Units and Measurements (ICRU) recommends that under certain conditions, the voltage applied to the source of radiation be included in the specification of beam quality. The following data are based on the ICRU report of 1963:

a. *For radiography,* the peak kV is customarily used to designate beam quality, since the HVL does not routinely serve a useful purpose.

b. *For radiotherapy up to 2 million volts (MV), the kV or MV and the HVL should be stated* because in this range of radiation energy, filters cause significant changes in the quality of the beam.

c. *For radiotherapy above 2 MV, the MV only need be stated* because in supervoltage therapy above 2 MV filters have very little effect on beam quality.

Other methods, such as the equivalent constant potential and the photographic, can be used, but their application is limited and will not be described.

"Hard" and "Soft" X Rays

X rays may be roughly classified on the basis of their penetrating power as "hard" or "soft." This is a simple way of expressing the quality of a beam in nonscientific terms.

1. **Hard x rays** are relatively high energy photons with high frequency, short wavelength, high penetrating power, and little absorption in the skin as compared with the interior of the body. Hard x rays may be produced by using:

   a. Higher generating voltage.

   b. Filters such as aluminum or copper which absorb relatively more of the less penetrating rays. The higher the atomic number of the filter and the greater its thickness, the greater is its hardening effect on the x-ray beam.

   c. Tube targets of high atomic number.

   2. **Soft x rays** are relatively low energy photons with low frequency, long wavelength, low penetrating power, and greater absorption in the skin. They may be produced by:

   a. Lower generating voltage.

   b. Lighter filtration.

   c. Targets of low atomic number.

THE INTERACTION OF PENETRATING RADIATION AND MATTER

On passing through a body of matter, an x-ray beam undergoes *attenuation;* that is, its exposure rate gradually decreases. Attenuation consists of two related processes: (1) *absorption* of part of the radiation by various kinds of interactions with the atoms of the body in the path of the beam (see pages 175-181) and (2) emission of radiations comprising *scattered* and *secondary* radiation resulting from interactions in (1). A third factor is the inverse square law.

Scattered radiation refers to those x-ray photons that have undergone a *change in direction* after interacting with atoms.

Secondary radiation is defined as the radiation *emitted* by atoms after having absorbed x rays (for example, characteristic radiation, see below).

It must be mentioned that many of the x rays do not interact with atoms at all, but pass through the body unchanged. This is due to two factors: (1) x rays are electrically neutral so that there is no electric force between them and orbital electrons and (2) atoms contain mainly empty space.

The various types of interactions between x rays and atoms make up one of the most fascinating chapters in x-ray physics. In order to simplify the discussion, as well as to make it more accurate scientifically, we must regard an x-ray beam as consisting of discrete bits of energy traveling with the speed of light. Such a tiny unit of energy is called a *photon* or *quantum* (pl., quanta). This may seem to contradict the electromagnetic wave theory of radiation discussed earlier, but the phenomena of radiation absorption and emission can be explained only by the *apparently contradictory theories* that a given x ray exists at the same time *both as a wave and as a particle of energy.*

This *dual nature* of electromagnetic radiation is now an accepted fact in physics. It tells us how characteristic radiation is produced in an x-ray tube when an electron drops from a higher to a lower energy level in the atom of the target metal (see Figure 12.4). Such a characteristic x ray is a quantum or photon of energy, and its energy and frequency are peculiar to the target element. Thus, an *atom of a given element can emit definite characteristic photons only,* and these are different from those of any other element. The energy of a given photon represents the difference in the energy levels of the shells between which the electron has dropped.

On the basis of this concept, known as the *quantum theory,* an x-ray beam consists of showers of photons (that is, bits of energy) traveling with the speed of light and having no electric charge. This is also known as the *corpuscular theory of radiation,* and applies equally to other forms of electromagnetic radiation such as light and gamma rays.

What determines the amount of energy in a given photon? The following equation, formulated by the German physicist Max Planck, provides the answer:

$$\text{photon energy} = \text{a constant} \times \text{frequency}$$
$$E = h\nu$$

In this equation, h is Planck's constant (does not vary) and ν (Greek "nu") represents the frequency of the associated electromagnetic wave in cycles per sec (hertz = Hz). This equation tells us that the quantum or *photon energy is directly proportional to the frequency,* in agreement with the observation that the penetrating power of x rays, which depends on their energy, increases as the frequency increases. And since the speed of x rays is constant and equals frequency times wavelength, an increase in frequency is accompanied by a decrease in wavelength. Thus, *highly penetrating x-ray photons have high frequency and short wavelength,* while, conversely, poorly penetrating x-ray photons have low frequency and long wavelength.

Before describing the various kinds of interactions of radiation and matter we must know something about the electronic *energy levels* or *shells* within the atom. These are analogous to the mechanical model of the rock on the cliff (see Chapter 3). Since the nucleus carries a positive charge, it exerts an electric force of attraction on the orbital electrons, the force being greater the nearer the electron shell is to the nucleus. Hence, more work would be required to remove an electron from the K shell and out of range of the nuclear electric field, than would be required to remove an electron from one of the outer shells. Since work must be done in moving an electron away from the nucleus in the direction of the outer shells, *the shells are at progressively higher energy levels the farther they are located from the nucleus.* Thus, the K shell represents the lowest energy level, while the Q shell represents the highest.

The energy required to remove an electron from a particular shell and just out of the atom is designated as the *binding energy* of that shell. The binding energy is obviously largest for the K shell because it is closest to the positively charged nucleus, decreasing progressively for successive ones. The binding energy is characteristic of a given element and shell; thus, it is about 70,000 eV for the K shell of tungsten, but only about 500 eV for the K shell of the average atom in the soft tissues of the body.

Since a relatively large amount of energy is needed to remove an electron from an inner shell, such an electron is said to be *bound.* On the other hand, almost no energy is needed to liberate the outermost shell electrons, and they are therefore called *free* or *valence* electrons.

Let us examine in detail the possible sequence of events following the penetration of a body of matter by x-ray photons. Any of the four types of interactions may occur between the photons and atoms lying in their path: (1) photoelectric interaction, (2) coherent or unmodified scattering, (3) Compton interaction with modified scattering, and (4) pair production. Remember that many photons pass completely through the body without experiencing any type of interaction.

1. **Photoelectric Interaction (Photoelectric Effect).** This type of interaction is most likely to occur when the energy—*hv*—of the incident (incoming) photon is *slightly* greater than the binding energy (see page 174) of the electrons in one of the inner shells such as the *K* or *L*. The incident photon gives up all of its energy to the atom; in other words, the photon is *absorbed* and disappears during the interaction. Immediately, the atom responds by ejecting an electron, usually from the *K* or *L* shell, leaving a *hole* in that shell (see Figure 12.10). Now the atom is *ionized positively* (why?) and in an *excited state*. Note that the energy of the incident photon ultimately went to (1) free the electron from its shell and (2) set it in motion as a *photoelectron*.

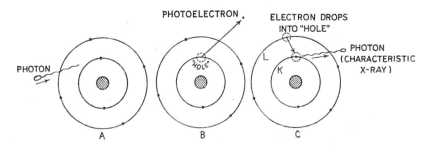

Figure 12.10. Photoelectric interaction with true absorption. A. Incident photon loses all its energy on entering an atom, being *absorbed* in the process. The atom responds by ejecting an *inner shell* electron which becomes a photoelectron. B. Atom is in an *excited state* (also ionized—electron missing). C. Electron from a higher energy level fills vacancy or *hole* in the *K*-shell, a *characteristic* x-ray photon being emitted.

The energy exchange during a photoelectric interaction is beautifully expressed in Einstein's equation:

$$hv_{photon} = W_K + \tfrac{1}{2}\ mv_e^2 \qquad (5)$$

where *hv* is the energy of the incident (entering) photon, W_K is the binding energy of the *K* shell of the element concerned, and $\tfrac{1}{2}mv_e^2$ is the kinetic energy of the photoelectron. (The same equation applies to photoelectric interaction in other shells. However, when photoelectric interaction occurs with electrons in *outer* shells, the incident photons are more likely to be in the ultraviolet or visible light region; as indicated above, inner shell photoelectric interactions are more likely to occur with x- or gamma-ray photons.)

What happens next? Suppose that a *K* electron has been ejected. The hole in the *K* shell is *immediately* filled by an electron moving in from one of the outer shells, usually the *L* shell, because they are at higher energy levels than the *K* shell (see page 174. During the transition of an *L* electron to the hole in the *K* shell the energy difference between these shells is radiated as an x-ray photon known as a *fluorescent* or *characteristic x ray* because its energy (also its frequency and wavelength) is typical of the shells and the element concerned. Thus, the characteristic x rays of copper are different from those of lead.

In pursuing the sequence of events, you may ask, "What happens to the hole in the *L* shell?" It is promptly filled by an electron from a shell still farther out (that is, at a still higher energy level). This continues as each successive hole is filled by electron transition from some higher energy level, until the atom loses its excited state, returning to *normal* or *ground state*. The total energy carried by the characteristic photons in a single photoelectric interaction equals the binding energy of the shell from which the electron was initially expelled—in this case, the *K* shell.

In summary, then, the energy of the incoming photon in the photoelectric interaction involving the *K-shell* has the following fate:

a. Photon enters atom and completely disappears.
b. A *K*-shell electron is ejected, leaving a hole.
c. Atom has excess energy—excited state.
d. A part of photon's energy was used to liberate electron and the rest to give it kinetic energy; ejected electron is a photoelectron.
e. Hole in *K*-shell is filled by electron transition from a shell farther out, accompanied by emission of a characteristic x-ray photon.
f. Holes in successive shells are filled by electron transition

from shells still farther out, each transition being accompanied by a corresponding characteristic photon.

g. Sum of energies of all the characteristic photons equals binding energy of shell from which the photoelectron originated.

Thus, the photoelectric effect gives rise to two kinds of secondary radiation—photoelectrons and characteristic x rays. Since the photoelectrons are negatively charged, they ionize atoms in their path; but they happen to be of low energy—less than 100 keV—so they are absorbed by only about 1 mm of tissue. In fact, this is one of the ways that x rays transfer energy to tissues and produce biologic changes in them. The characteristic x rays are also of low energy—less than 1 keV—and are locally absorbed too. Their energy is less than that of the primary or incident radiation.

Because the absorption of x rays by photoelectric interaction depends strongly on the atomic number of the irradiated material, and on the energy of the radiation, the photoelectric effect is of utmost importance in radiography. The high atomic number of the elements in bone as contrasted with that of the soft tissues, together with the relatively long wavelength x rays used in radiography, contribute to a large difference in photoelectric absorption of these two kinds of structures. This is the most important factor in the production of radiographic contrast—the different degrees of darkness of various areas in a radiograph (see page 181).

2. **Coherent or Unmodified Scattering.** If a *low energy* x-ray photon passes close to a *loosely bound, outer-shell* electron, it may set the electron into vibration. This produces an electromagnetic wave identical in wavelength with that of the incident photon, but *differing in direction* (see Figure 12.11). Thus, in effect, the entering photon has been *scattered without undergoing any change in wavelength, frequency, or energy.* This type of interaction is important with low energy x rays, below the range that is useful in clinical radiology.

3. **Compton Interaction with Modified Scattering.** If an incident (entering) photon of sufficient energy encounters a *loosely*

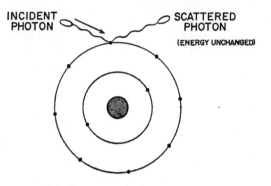

Figure 12.11. Unmodified scattering. The entering photon has undergone a charge of direction only. This is really a special case of the Compton Effect.

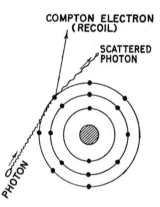

Figure 12.12. Compton interaction with modified scattering. *Part* of the photon's energy has been used up in removing a loosely bound orbital electron. Therefore, the emerging photon has *less energy*, and it has also undergone a *change in direction*.

bound, outer shell electron it may dislodge the electron and proceed in a different direction, as shown in Figure 12.12. The dislodged electron is called a *Compton or recoil electron.* It acquires a certain amount of kinetic energy which must be subtracted from the energy of the entering photon, in accordance with the Law of Conservation of Energy. Consequently, the energy of the incident radiation is decreased, and so when it emerges from the atom it has a lower frequency, longer wavelength, and less energy. This phenomenon was discovered in 1922 by the renowned physicist A. H. Compton and is therefore called the **Compton Effect.** The emerging photon, having undergone a change of direction, is called a *scattered photon.* (The characteristic radiation, arising from electron transitions between outer shells, is of extremely low energy and may be neglected.)

The Compton Effect may be expressed as follows:

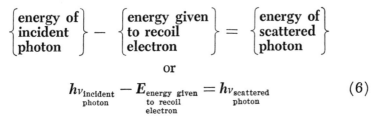

$$\begin{Bmatrix} \text{energy of} \\ \text{incident} \\ \text{photon} \end{Bmatrix} - \begin{Bmatrix} \text{energy given} \\ \text{to recoil} \\ \text{electron} \end{Bmatrix} = \begin{Bmatrix} \text{energy of} \\ \text{scattered} \\ \text{photon} \end{Bmatrix}$$

or

$$h\nu_{\substack{\text{incident} \\ \text{photon}}} - E_{\substack{\text{energy given} \\ \text{to recoil} \\ \text{electron}}} = h\nu_{\substack{\text{scattered} \\ \text{photon}}} \tag{6}$$

Thus, the frequency and energy of the scattered photon are less than those of the incident photon. A scattered photon, in turn, behaves in the same manner as other x rays on interacting with atoms. The recoil electrons cause ionization of atoms in the same way as other high speed electrons.

As the energy of the incident photon *increases,* the probability of occurrence of the Compton interaction *decreases.* But, at the same time, the *scattered photons* (1) have more energy and (2) tend to be scattered more and more in a forward direction, increasing the likelihood of their passing completely through the body and reaching the film. This constitutes the *scattered radiation* which, in radiography, impairs film quality by producing an overall fogging effect.

Despite the decreased chance of occurrence of the Compton Effect at higher photon energies, the probability of the photoelectric effect decreases even more sharply, so as the energy of the incident photon increases above 70 keV (about 180-kV x rays) the Compton Effect becomes the predominant type of interaction.

4. **Pair Production.** A *supervoltage* photon with energy above 1.02 million electron volts (MeV), upon approaching a *nucleus,* may disappear and give birth to a *pair*—a negative and a positive electron (see Figure 12.13). The latter—also called a *positron*—as it comes to rest, combines with any negative electron only to

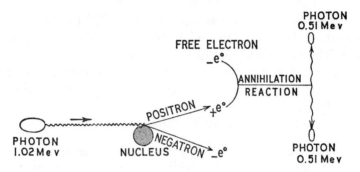

Figure 12.13. Pair production and annihilation reaction.

disappear and give rise to two photons, each having an energy of 0.51 MeV, moving in opposite directions. This is the ***annihilation reaction.*** The two processes exemplify, first, the conversion of energy to matter, and then matter to energy according to Einstein's equation $E = mc^2$. Note that the "magic" number 0.51 MeV is simply the energy equivalent of the mass of an electron at rest. Pair production becomes significant at about 10 MeV, but not until about 24 MeV does it predominate over the Compton Effect insofar as energy absorption is concerned.

Secondary and Scattered Radiation Consists of

1. Primary electrons
    a. Photoelectrons
    b. Compton or recoil electrons
    c. Positron-negatron pairs
2. Secondary and scattered x rays
    a. Characteristic
    b. Coherent or unmodified scattered
    c. Modified scattered (Compton Effect)
    d. Annihilation radiation

The secondary and scattered x ray photons are less energetic (softer) than the incident photons, except for the relatively insignificant factor of unmodified scattering. The production of primary electrons accounts for the absorption of x radiation by matter, since this removes energy from the primary beam. When a *filter* is used to harden a beam of x rays, it does so mainly by photoelectric absorption of *relatively more low energy than high energy photons,* so that the emerging beam retains a relatively greater *percentage* of higher energy, more penetrating photons. If a copper filter is used, it emits its own soft characteristic radiation which can be absorbed by an aluminum secondary filter. The aluminum also emits characteristic rays, but these are so soft that they are absorbed in a few centimeters of air.

Relative Importance of Various Types of Interaction in Radiology

The various types of interaction between radiation and matter will now be discussed as they apply to diagnostic and therapeutic radiology, especially with regard to their relative importance.

1. **Diagnostic Radiology.** In the usual range of 30 to 140 kV, the photoelectric effect predominates insofar as energy absorption from the beam is concerned. Since photoelectric absorption is about four to six times greater in bone than in an equal mass of soft tissue in the radiographic kV range, this type of interaction is responsible for much of the radiographic contrast (difference in film darkness) between these tissues. Differences in density (grams/cc) of various tissues also contribute to radiographic contrast.

On the other hand, the Compton Effect makes its presence known mainly by the *scattered radiation* resulting from interaction with the atoms of the tissues, table tops, etc. In fact, even at 60 kV 90 per cent of the x-ray beam is scattered and only 10 per cent absorbed (latter by photoelectric and Compton processes). As will be shown later, the scattered radiation approaches the film from many directions and degrades radiographic quality, making it necessary to use grids and collimators. As kV is increased the scattered radiation becomes progressively more damaging to radiographic quality because its energy increases and it is therefore more likely to pass completely through the body and reach the film. Note that the characteristic radiation arising in the photoelectric process is also scattered in many directions, but is so soft that it is absorbed locally in the tissues.

2. **Therapeutic Radiology.** When radiation tranverses the body, its interaction with the atoms of the tissues results in the absorp-

tion of energy through the liberation of electrons. The freed electrons then expend their energy in ionizing other atoms, thereby causing tissue damage and concomitant therapeutic effects. This will be discussed more fully below.

The photoelectric effect predominates below about 180 kV. Primary photons in the incident beam are completely absorbed upon interaction with the tissue atoms, setting in motion photoelectrons which ionize atoms of the tissue along their paths.

As the applied voltage is increased from about 180 kV to 3 MV, the probability of the Compton interaction decreases, but the probability of the photoelectric interaction decreases even more, so that the Compton Effect becomes predominant.

Although pair production begins at about 1.02 MeV, it does not contribute significantly to energy absorption until about 10 MeV. Only at 24 MeV and above does ionization by negatrons and positrons becomes the main process of energy absorption.

Detection of X Rays

There are several ways in which the presence of x rays may be detected, though not necessarily measured.

1. **Photographic Effect**—the ability of x rays to affect a photographic emulsion so that it can be developed chemically.

2. **Fluorescent Effect**—the ability of x rays to cause certain materials to glow in the dark. These include zinc cadmium sulfide and calcium tungstate.

3. **Ionizing Effect**—the ability of x rays to ionize gases and discharge certain electrical instruments such as the electroscope.

4. **Physiologic Effect**—the ability of x rays to redden the skin, destroy tissues, and cause genetic damage.

5. **Chemical Effect**—the ability of x rays to change the color of certain dyes.

X-RAY DOSIMETRY

The changes produced in tissues traversed by penetrating radiation are believed to be due to the transfer of energy—*linear energy transfer (LET)*—to the atoms of the tissues by the process of *ionization* and *excitation.* We saw in the sections on the interaction of radiation with matter that ionization accompanies these interactions. With a given type of radiation, such as x rays, the degree of ionization and the resulting tissue effects are related to the amount of energy absorbed at a particular site. This *absorbed dose* of radiation depends on the quantity (exposure in R) and quality (energy) of the radiation, and on the nature of the tissue (muscle, bone, etc). It is thought that more soft than hard x rays are needed for a given tissue effect as, for example, skin reddening (erythema). Such dependence on beam hardness or quality—so-called *wavelength dependence*—may be due to differences in the number and distribution of ions liberated by radiation of various energies; that is, differences in LET.

Exposure—the Roentgen

To specify *exposure,* we make use of the unit described above —the roentgen (R)—a measure of the radiation being delivered to a particular area, based on the ionizing ability of the radiation. In addition, we must state the *quality* of the beam, namely, the peak kV applied to the tube as well as the half value layer for the orthovoltage region, or the energy of the radiation in MV or MeV in the megavoltage region. The mA, time, and distance should also be included where applicable although they have no bearing on the quality of the x-ray beam. The total duration of a therapy course in terms of days or weeks has a distinct bearing on the effect of treatment, since a given number of roentgens administered at one sitting has a much more pronounced tissue effect than does the same dosage fractionated over a period of days or weeks. Thus, we specify x-ray therapy on the basis of the number of roentgens administered, the quality of the beam, and the fractionation of the treatments. (As will be explained later, the size of the treatment field is also important and should be specified.)

Absorbed Dose—the Rad

In the preceding paragraph the discussion was limited to exposure in roentgens. No mention was made of the dose *absorbed* in the tissues. Since only the energy actually transferred to the irradiated volume is responsible for the ensuing biologic changes, *we may define the absorbed dose as the amount of energy that ionizing radiation delivers per gram of irradiated matter.* The unit of absorbed dose is the *rad (radiation absorbed dose),* and is defined as an *energy absorption of 100 ergs per gram.* This applies to any type of radiation (x rays, gamma rays, beta rays, etc) and to any kind of matter, including tissues.

The *roentgen,* it should be recalled, is limited to x rays and gamma rays up to 3 million volts (MV).

What is the relationship of the unit of absorbed dose—the rad— to the unit of exposure—the roentgen? The answer is not a simple one, because it depends on a number of factors. However, in the usual therapy range of 100 kV to 2 MV the absorbed dose in soft tissues exposed to 1 R is approximately 1 rad, with an error of only a few per cent. In any case, this one-to-one conversion is sufficiently accurate in radiotherapy of soft tissues.

On the other hand, absorption in other tissues such as bone depends on the energy of the incident radiation. For example, with an x-ray beam of half value layer 2 mm Cu the absorbed dose per gram of bone is about twice that per gram of soft tissues for the same exposure. Thus, 1 R would produce an absorbed dose of about 1 rad in soft tissue, and about 2 rads in bone. With beams of higher energy—HVL 4 mm Cu and above—the absorbed dose in equal masses of bone and soft tissue is about the same. This is of practical importance in therapy when the beam has to traverse a bony structure; with the softer radiation, one must be careful not to exceed the tolerance of the bone while delivering a therapeutic dose to the soft tissues. With the more energetic beams, having a half value layer of at least 4 mm Cu, the equal absorption in equal masses of bone and soft tissues decreases, but does not entirely prevent the possibility of serious bone damage.

Determination of Tissue Dosage

The practical method of tissue or tumor dose determination will now be discussed. However, we shall introduce this subject by first describing three important entities: (1) in-air exposure, (2) entrance exposure, and (3) tumor exposure.

In-Air Exposure. When the exposure rate is measured at a given point in a beam, *with no scattering material in the vicinity of the measuring device* (other than air), it is called the *in-air exposure rate* and is usually expressed in R per min (abbreviated R/min). The total in-air exposure is obtained by multiplying the in-air exposure rate by the exposure time:

in-air exposure = in-air exposure rate × exposure time
$$R = R/min \times time$$

This is, in fact, the method of calibrating an irradiation therapy machine, the measuring device usually being a Victoreen condenser-R-meter with a thimble chamber appropriate for that particular radiation. The in-air exposure rate is influenced by several factors; thus, it increases with an increase in kV, an increase in mA, a decrease in filtration, and a decrease in the source-skin distance (according to the inverse square law). (See pages 160-161.) Therefore, an x-ray machine should be calibrated for every combination of treatment factors (kV, mA, filtration, and distance) that is to be used. With cobalt-60 gamma rays (1.25 MeV) we are concerned only with distance in measuring in-air exposure rate.

Entrance Exposure and Backscatter. If a radiation measuring device such as a thimble chamber is placed on the surface of the body (or some equivalent material such as Masonite or water) the recorded exposure rate is larger than that in the absence of the body. In other words, the entrance exposure rate in R/min is greater than the in-air exposure rate. This results from *backscatter radiation* which is simply radiation scattered back from matter lying beneath the thimble chamber. Note that this increases the exposure to the skin and must be recognized in planning therapy.

We usually designate the degree of backscatter as the ratio of the measured entrance-R to in-air-R:

$$\text{backscatter factor} = \frac{\text{entrance-R}}{\text{in-air-R}}$$

For example, if the entrance exposure is 125 R and, in the same time interval, the in-air exposure is 100 R, then

backscatter factor = 125/100 = 1.25

Obviously, the larger the backscatter factor, the larger will be the entrance exposure for a particular in-air exposure.

In planning irradiation therapy we take the backscatter factor into account in specifying the entrance exposure, depending upon the energy of the beam. With x rays in the 200-300 kV range the maximum dose is reached virtually at the skin surface; therefore we speak of *skin exposure* (R), obtained by multiplying the in-air exposure by the backscatter factor. On the other hand, with 1-2 million volt radiation maximum dosage occurs about 5 mm below the skin surface (scattered mainly in a forward direction) and that is where the level of the entrance exposure is designated, again obtained by multiplying the in-air dose by the backscatter factor. This is often referred to as the "given R-dose." The differences between these kinds of entrance doses are observed in practice by virtue of the more striking skin effects from lower energy radiation than from an equal dose of higher energy radiation. However, the latter is still capable of causing significant damage to the tissues just beneath the skin.

Published tables give the backscatter factors obtained experimentally under various conditions. The factors affecting backscatter are as follows:

1. *Area of Treatment Field.* As this is increased, the backscatter factor increases due to the larger volume of irradiated tissue and the correspondingly larger source of scattered radiation.

2. *Thickness of Irradiated Part.* As this is increased, the backscatter factor increases for the same reason as in 1. However, beyond a certain critical thickness, further increase in thickness does not influence the backscatter factor.

3. *Energy of Beam.* Above 1 mm Cu, an increase of HVL decreases the backscatter factor because an increasingly greater percentage of the radiation is scattered in a forward direction, and less in a lateral or backward direction. It should be recalled that HVL increases with an increase in kV and filtration. With *megavoltage beams* of at least 1 MV, backscatter is negligible, and, in fact, the exposure rate at a point below the surface is greater than at the surface. Thus, with *cobalt-60* beam therapy

the maximum exposure rate is found at a point about 5 mm below the surface because that is where maximum ionization takes place. But in *orthovoltage therapy,* say 250 kV, maximum ionization produced by the primary beam and backscatter occur so close to the skin surface that, for all practical purposes, they may be considered to lie within the skin.

It must be emphasized that the source-skin distance does *not* affect the backscatter factor.

Tumor Exposure. As a beam of x rays traverses matter the exposure rate gradually *decreases.* There are three factors responsible for the weakening or *attenuation* of the beam.

       1. Absorption
       2. Scattering
       3. Inverse square law

Absorption of radiation is due to the various types of interactions between its photons and the atoms in the irradiated material (or tissues) as described earlier. Photoelectric interaction predominates with radiation energy up to about 140 kV. Compton interaction predominates in the energy region between about 180 kV and 10 MV. Above this limit pair production begins to assume major importance.

Scattering occurs in soft tissues mainly by the Compton process in the 100 kV to the several million volt range. Above 20 MV, scattering results mainly from the annihilation reaction following pair production.

The *inverse square law,* a geometric factor, applies because the beam, on passing through the body, reaches points that are progressively farther from the radiation source. The exposure rate is inversely proportional to the square of the distance between the source of radiation and the point of interest, namely, the tumor.

The *tumor exposure* may be defined as the exposure in R, delivered at a given point in the tumor. Ordinarily, this is determined for a point along the central axis (central ray) of the beam at the center of the tumor. Published tables such as Table 12.4 give the central axis depth doses as a percentage of the entrance

TABLE 12.4
CENTRAL AXIS DEPTH DOSES
(Data adapted from Johns, *The Physics of Radiology*,
by permission of Charles C Thomas, Publisher)

HVL 1.0 mm Cu		SSD 50 cm			Circular Fields
Area (cm²)	0	20	50	100	200
*B.S.F.	1.00	1.20	1.29	1.36	1.42
Depth (cm)					
0	100	100	100	100	100
1	79	94	98	101	103
2	63	83	90	94	97
3	51	72	81	86	91
4	41	62	71	77	82
5	33	52	61	68	73
6	26	44	52	59	62
7	21	37	45	51	55
8	17	31	38	45	48
9	14	26	33	38	42
10	11	22	28	33	36
Cobalt 60		SSD 50 cm			Circular Fields
Area (cm²)	0	20	50	100	200
*B.S.F.	1.00	1.02	1.03	1.04	1.05
Depth (cm)					
0.5	100	100	100	100	100
1	95	96	97	98	98
2	85	89	91	91	92
3	77	82	84	85	86
4	69	76	78	80	81
5	63	70	72	74	75
6	56	64	67	69	70
7	51	58	61	63	65
8	46	53	56	59	61
9	42	49	52	54	56
10	38	45	47	50	52

*Backscatter Factor

dose for various combinations of half value layer, area of treatment port, and source-skin distance.

$$\% \text{ depth dose at } x \text{ cm} = \frac{R/\text{min at } x \text{ cm depth} \times 100}{\text{entrance } R/\text{min}}$$

For example, suppose we wish to treat a malignant tumor whose center lies 8 cm below the body surface. We are to use radiation with HVL 1.0 mm Cu, a treatment distance of 50 cm, and a treatment port that measures 10 cm × 10 cm. We find in Table 12.4 that for an area of 100 sq cm and a depth of 8 cm, the percentage depth dose is 45 per cent. This means that for each 100 R skin exposure the center of the tumor would receive an exposure of $100 \times 45\% = 45$ R. If we are to deliver a total exposure of 5000 R at the center of the tumor, then, to find the skin exposure,

tumor R = surface R × % depth dose

and rearranging,

$$\text{surface } R = \frac{\text{tumor } R}{\% \text{ depth dose}}$$

$$\text{surface } R = \frac{5000}{45\%} = \frac{5000}{0.45} = 11{,}300 \text{ R (approx)}$$

Obviously, this skin exposure is far above the maximum that can be tolerated without serious radiation injury, but if it is divided among a sufficient number of portals of entry to *crossfire* the tumor, no one skin port need receive excessive exposure. Better still, we can use high energy x-ray therapy (for example, 2 MV) or cobalt-60 beam therapy (1.25 MeV), using appropriate percentage depth dose tables and crossfiring ports.

Tumor Absorbed Dose. How we go about determining the *absorbed dose* in a tumor at some depth below the surface? Unfortunately, the published depth dose tables for x radiation up to 400 kV are not in complete agreement, mainly because of the different conditions under which they were obtained. Additional error may be introduced when these tables are used in the radiotherapy department without regard to differences in equipment and operating factors. The ICRU (*Handbook 87*) has devised a procedure for determining depth exposures that minimizes these discrepancies with radiation below 400 kV so that any one of the published central ray depth dose tables may be used.

Now we must convert the exposure in R at the tumor to absorbed dose in rads. The conversion factor, *f*, published in ICRU *Handbook 87* for various photon energies, for water, muscle, and bone, is shown for three qualities of radiation in Table 12.5. To obtain the absorbed dose we simply multiply the exposure by the appropriate value of *f*:

$$\underset{R}{\text{exposure}} \times \underset{rad/R}{f} = \underset{rads}{\text{absorbed dose}}$$

TABLE 12.5

SAMPLE VALUES OF AVERAGE CONVERSION FACTOR f RADS/R*

HVL	Muscle	Bone
mm Cu		
1.0	0.95	1.9
2.0	0.96	1.4
Cobalt 60	0.96	0.92

* Based on ICRU Handbook 87.

Besides the central ray depth exposure, we should know the exposures at other points in the tumor to insure uniformity of irradiation, so as to avoid dangerous over- or under-treatment. Such off-center exposures can be obtained from special charts called *isodose curves* which contain plotted depth dose percentages at various points in the beam along the central axis and elsewhere (see Figure 12.14).

Factors Affecting Percentage Depth Dose

The *percentage depth dose* increases with increasing (1) *half value layer or beam energy*, (2) *treatment field area*, and (3)

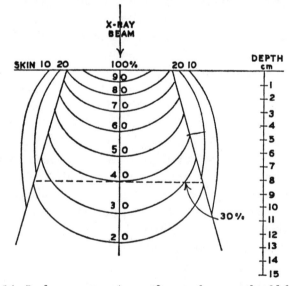

Figure 12.14. Isodose curves. A specific set of curves should be obtained for each combination of treatment factors. The curved lines connect all points receiving identical dosage, expressed as a percentage of surface dose.

source-surface distance. The percentage depth dose decreases with increasing *depth of tumor.* These factors will now be discussed.

1. **Beam Quality.** As already noted, this is specified by the HVL and the kV or MV. The larger the HVL, the greater will be the penetrating ability of the beam and the depth dose percentage. In Table 12.4 contrast the depth doses of the lower energy x rays (HVL 1.0 mm Cu) with the higher energy cobalt-60 gamma rays (1.25 MeV) for equal field areas.

2. **Treatment Field Area.** As the area of the field increases, there is a corresponding increase in the volume of the irradiated tissue with resulting increase in the percentage of scattered radiation. The latter is added to the primary radiation reaching a given point below the surface and therefore contributes to an increased depth dose percentage.

3. **Source-Surface Distance (SSD).** As the distance between the source (x-ray target; ^{60}Co) and surface is increased, the exposure rate at the surface decreases according to the inverse square law. However, due to less divergence of the rays at the greater SSD (see Figure 12.15) a larger fraction of the surface exposure rate reaches a given depth below the surface; hence, the greater the SSD, the larger the depth dose percentage. This is dependent on the inverse square law and can be demonstrated by an example based on the data in Figure 12.15:

Point P is 10 cm below the surface.

At SSD 30 cm

Point P is 10 cm + 30 cm = 40 cm from source F
Exposure rate at P, neglecting absorption in tissue, is

$$\frac{30^2}{40^2} = 0.56 \text{ or } 56\% \text{ of surface exposure rate}$$

At SSD 50 cm

Point P is 10 cm + 50 cm = 60 cm from source F
Exposure rate at P, neglecting absorption in tissue, is now

$$\frac{50^2}{60^2} = 0.69 \text{ or } 69\% \text{ of surface exposure rate}$$

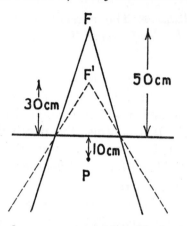

Figure 12.15. Depth dose percentage falls with decreasing source-surface distance, as shown by the greater divergence of the rays. This can be verified by the inverse square law.

4. Depth of Tumor. As the depth of the tumor increases, the exposure rate decreases because of the attenuation of the beam through absorption, scattering, and the inverse square law.

Modification of X-ray Beams by Filters

The use of filters has been mentioned several times, and a more complete discussion of this important subject will now be presented. A filter is simply a sheet of metal placed in the path of a heterogeneous x-ray beam usually with HVL up to about 4 mm Cu, for the purpose of increasing its average penetrating power; that is, *a filter increases the hardness of a beam.* This is brought about through the interaction of the x-ray photons with the atoms of the filter material (see section on interaction of radiation with matter). Recall that these interactions consist of absorption and scattering of photons. As a result, a *greater fraction of low energy* (long wavelength) *than of high energy* (short wavelength) *x rays is removed from the beam by a filter,* although some short wavelength photons are also removed. In other words, a beam of x rays generated at a given kV undergoes an increase in half value layer upon passing through a filter. Needless to say, a filter functions only with a beam of heterogeneous photon energies such as we have under ordinary conditions; in the case of monoenergetic x rays, available only in the experimental laboratory, filtration does not affect the half value layer. Filtration is not used with cobalt-60 therapy beams because of the absence of low energy photons.

Notice that *filtration does not alter the minimum wavelength* of a heterogeneous beam (see Figure 12.16). Minimum wavelength depends solely on the *peak kilovoltage* applied to the tube. However, the maximum wavelength of the beam (that is, the wavelength of the lowest energy radiation) depends on the *inherent filtration* of the tube; that is, the filtration afforded by the glass envelope and the cooling oil layer through which the beam must pass after leaving the target.

The inherent filtration of a tube is usually expressed in the equivalent thickness of aluminum or copper. Any additional filtration one chooses to place in the beam is called *added filtration.* The *total filtration* of an x-ray beam is obviously the sum of the inherent filtration plus the added filtration.

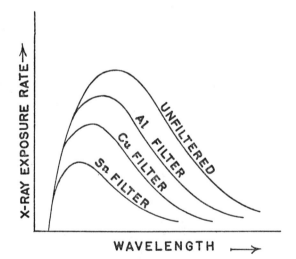

Figure 12.16. Family of curves for a given beam filtered by various types of added filters, as compared with an unfiltered beam. Note that the minimum wavelength is not altered, since it depends only on the peak kV for a given target material. The peak of the curve is shifted toward a shorter wavelength as the atomic number of the filter increases; at the same time, the total exposure rate (area under the curve) decreases.

total filtration = inherent filtration + added filtration

Filtration affects both the *exposure rate* and the *quality* (half value layer) of an x-ray beam:

1. **Exposure rate** is always decreased by filtration because some photons are removed by absorption and scattering. In fact, the thicker the filter and the higher its atomic number, the smaller will be the exposure rate.

2. **Quality** is changed by filtration. As either the atomic number or the thickness of a filter is increased the average penetrating ability of a beam increases; in other words, the half value layer is increased. The type of filter should be appropriate to the energy of the radiation. With radiation produced at 50 to 100 kV, aluminum is most suitable as a filter. Copper is used as the *primary filter* in the 120 to 400 kV range, but an aluminum filter must be placed between the copper and the patient as a *secondary filter* to absorb the characteristic radiation emitted by the copper. The resulting aluminum characteristic radiation is soft, being absorbed in a few centimeters of air.

A special type of filter often used with 250 to 400 kV x rays is known as the *Thoraeus filter* (named after its inventor). This is a variety of *compound filter* consisting of three metals—tin, copper, and aluminum—placed in that order from the tube side to the patient side (see Figure 12.17). The chief advantage of

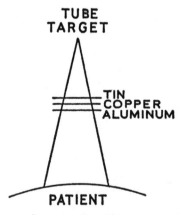

Figure 12.17. Schematic diagram of a Thoraeus compound filter, showing the relation of the component filters.

the Thoraeus filter is that it provides a beam of the same half value layer as a copper and aluminum filter, but with significantly less attenuation of the beam. Thus, the Thoraeus filter is about 25 per cent more efficient than an equivalent copper and aluminum filter, in that it allows the delivery of approximately 25 per cent more radiation of the same half value layer in the same treatment time.

Comparison of Cobalt-60 and 250-kV X-ray Therapy

In recent years irradiation therapy of deep-seated tumors with conventional 200-250 kV x rays has been almost universally replaced by higher energy *megavoltage* radiation such as cobalt-60 gamma rays (1.25 MeV) or 2 MV x rays. There are several outstanding advantages inherent in the use of radiation in this energy range.

1. **Skin-sparing Effect.** As described before, the maximum dose with the higher energy radiation occurs below the skin surface. For example, cobalt-60 gamma rays produce the maximum dose 5 mm (see Table 12.4), and 2 MV x rays 4 mm, below the surface. This shifts the skin reaction, ordinarily visible with 200-kV x rays, to the level of the subcutaneous tissues. While this permits a considerable increase in the total skin dose, nevertheless we must guard against possible injury to the underlying tissues where it is not so readily apparent. In this regard we must avoid placing material in the path of the beam any closer than about 15 cm from the skin, since Compton electrons liberated in such material and reaching the skin can significantly raise the dosage level there.

2. **Bone-sparing Effect.** An important advantage of high energy radiation is the nearly equal absorption occurring in bone and soft tissue, gram for gram. This permits larger doses to be delivered safely when there is bone in the path of the beam. Furthermore, there is no "shading" of the beam by bone, avoiding a decrease in soft tissue dosage beyond the bone. With 200-kV x rays and HVL of 1 mm Cu, absorption in bone is about three times that in soft tissue of the same mass, increasing significantly the likelihood of radiation injury to bone. At the same time, the lower energy radiation is shaded by the bone, so soft tissue beyond may be underdosed.

3. **Greater Percentage Depth Dose.** The higher energy radiation provides a significantly higher percentage depth dose at a given treatment distance. In fact, this may be its greatest advantage over lower energy x rays. For example, according to Table 12.4, at a 10-cm depth, and using a 100 cm² port with an SSD of 50 cm

% Depth Dose with Cobalt 60...............50%
% Depth Dose with X Rays HVL 2 mm Cu..33%

This represents an increase in percentage depth dose of

$$100 \left(\frac{50\text{-}33}{33} \right) = \frac{100 \times 17}{33} = 52\%$$

using the higher energy radiation. At the same time, as we have just seen, the skin-sparing effect further enhances the possibility of delivering an adequate dose to the tumor. It should be mentioned, however, that most authorities agree that the biologic effectiveness of cobalt-60 gamma rays relative to 250-kV x rays is about 85 per cent; therefore, such gamma ray therapy requires a compensating increase in dosage of about 15 per cent.

4. **More-forward Scatter.** Scattered radiation with high energy radiation is predominantly in a forward direction; that is, in the same direction as the primary beam. This results in less side scatter and a better defined beam edge. Furthermore, the isodose curves are flatter, so there is more uniform irradiation of the selected volume.

QUESTIONS

1. What properties of x rays led to their discovery?
2. What are roentgen rays? What are the two apparently contradictory theories as to their nature?
3. How are the frequency and wavelength of x rays related? What is the speed of x rays?
4. How do soft rays differ from hard rays?
5. What are the four essential components of a modern x-ray tube? Discuss the principles of x-ray production.
6. Describe in detail the production of x rays when the electron stream hits the target.

7. What is characteristic radiation and how does it arise?

8. Discuss the differences betwen characteristic and general radiation. What is their relative importance in radiography?

9. Name ten properties of x rays.

10. What is meant by the quality of roentgen radiation? What factors influence it and in what manner? How can it be measured and specified (ICRU recommendations)?

11. Explain fully why a beam of x rays emerging from the target is heterogeneous. What are monoenergetic x rays?

12. What determines the minimum wavelength in an x-ray beam? State the equation that applies.

13. Define and explain half value layer and filters. What purpose does HVL serve?

14. Why is tungsten so widely used as the target material?

15. How is the quantity of x radiation measured?

16. What factors influence x-ray exposure rate?

17. Discuss the four types of interactions of x rays and matter, showing how the corpuscular theory of radiation applies. What is their relative importance in radiography and radiotherapy?

18. Of what do scattered and secondary radiations consist?

19. How are tumor exposure and absorbed dose calculated?

20. Define absorbed dose. Name and define its unit.

21. What is the relationship between exposure rate and exposure? Exemplify this relationship by means of appropriate units.

22. Prove the statement in Figure 12.15 by application of the inverse square law.

23. Describe a Thoraeus filter and discuss briefly its importance in therapy.

24. What are isodose curves? Why are they used?

25. How does a filter influence the minimum wavelength of a beam? The exposure rate?

26. Discuss the advantages of megavoltage therapy over 250-kV therapy. Are there any disadvantages of the former?

CHAPTER *13* X-RAY TUBES AND RECTIFIERS

Thermionic Diode Tubes

In Chapter 12 we learned that x rays are produced whenever a stream of fast-moving electrons undergoes a sudden deceleration or loss of speed. This is achieved by the use of a special type of *thermionic vacuum tube*—the hot filament or Coolidge x-ray tube. Before describing the detailed construction of such a tube, let us review its underlying principles:

1. *A hot metal filament gives off electrons* by a process called *thermionic emission.* The filament is heated by a separate *filament current.* The rate of electron emission increases with an increase in the temperature of the filament which, in turn, is governed by the filament current (measured in amperes).

2. *If no kilovoltage is applied,* the emitted electrons remain near the filament as an electron cloud or *space charge.*

3. *If kilovoltage is applied* between the filament and target so as to place a negative charge on the filament (cathode) and a positive charge on the target (anode), *space charge electrons are driven over to the anode at high speed by the large potential difference.* The maximum and average speeds of the electrons increase as the peak kilovoltage is increased. Electron flow across the gap between the cathode and anode constitutes the *tube current,* measured in milliamperes (mA).

4. *If the applied kilovoltage and resulting electron speed are high enough,* x rays are produced when the electrons strike the target, their *kinetic energy* being converted to heat (about 99.8 per cent) and x rays (about 0.2 per cent).

The most widely used x-ray tubes are basically thermionic

198

diodes, consisting of a tungsten filament cathode, a tungsten target anode, an evacuated glass tube enclosure, and two circuits to heat the filament and to drive the space charge electrons to the anode. However, various modifications of these tubes adapt them to special purposes. The structural and other details of radiographic, therapy, and valve tubes will now be described.

RADIOGRAPHIC TUBES

In this section we shall examine in detail the essential features of radiographic tubes used in medical x-ray diagnosis.

Glass Envelope

The working parts are enclosed in a glass tube or *envelope* containing a high vacuum. In fact, the tube is baked during manufacture to expel air and other gases that may have been trapped in the glass or metal parts, a process called *degassing*. Radiographic tubes are usually cylindrical in shape. Immersion in *insulating oil* in a suitable metal housing prevents high voltage sparkover between the terminals, thereby making possible the design of smaller tubes.

Cathode

The negative terminal or *cathode* consists of (1) the *filament*, (2) its *supporting wires*, and (3) the *focusing cup*. Loosely speaking, the terms "cathode" and "filament" are used interchangeably. The filament, itself, is a small coil of tungsten wire, the same metal that is used in electric light bulb filaments. In most radiographic tubes the filament measures about 0.2 cm in diameter and 1 cm or less in length, and is mounted on two stout wires which support it and also carry electric current. These wires lead through one end of the glass envelope to be connected to the proper electrical source (see Figure 13.1). A low voltage *filament current* is sent through the wires to heat the filament, and one of these wires is also connected to the *high voltage source* which provides the high negative potential needed to drive the electrons toward the anode at great speed. Backing the filament is a negatively charged concave *molybdenum cup* which confines the electrons to a narrow beam and focuses them on a small spot on the tungsten target, known as the *tube focus.*

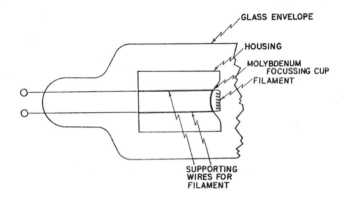

Figure 13.1. Details of the cathode of an x-ray tube (side view).

The filament current, serving to heat the filament and provide a source of electrons within the tube, usually operates at about 10 volts and about 3 to 5 amperes. An increase in the filament current increases the temperature of the filament and the rate of electron emission (see Figure 13.2).

With prolonged use the filament gradually becomes thinner because of evaporation of its metal, especially during short, intense exposures. As a result, the electrical resistance of the filament increases (resistance inversely proportional to cross-section area) and the filament current decreases (Ohm's law), causing a decrease in tube current (mA) and consequent radiographic underexposure. To correct this, the service representative adjusts the filament current upward just enough to obtain the proper electron emission for the required tube current. It will be found that *as the tube ages, a progressively lower filament current setting is required for a desired tube milliamperage.*

Some tubes, such as those used in older mobile equipment, have a single filament. However, most modern radiographic tubes, in both stationary and mobile equipment, are provided with two filaments mounted side by side. One is usually smaller than the other, so focal spots of two different sizes appear on the target. Such tubes are called *double focus* tubes, although

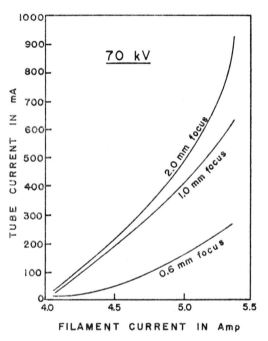

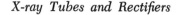

Figure 13.2. Relation of tube current (mA) to filament current. Note that a small increase in filament current produces a large increase in mA. Furthermore, mA is larger with a larger tube focus for a given filament current.

you can see that they are really double filament tubes. Note that only one filament is activated for a given exposure, being se-

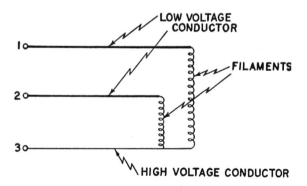

Figure 13.3. Simplified diagram of the connection of two filaments of an x-ray tube (double focal spot tube) to three wires, one of which is common to both filaments. The filaments are mounted side by side, facing the target.

lected by a switch at the control panel. Supporting the two filaments are three stout wires, one of which is connected in common with both filaments. In Figure 13.3, if terminals 1 and 3 are connected to the low voltage source, the large filament lights up. If, instead, terminals 2 and 3 are connected to the low voltage source, the small filament lights up. The high voltage is applied through wire 3 which is common to both filaments, providing either one with a high negative potential.

Anode

Two types of anode are available: (1) stationary and (2) rotating.

1. **Stationary Anode.** A gap of a few centimeters separates the cathode from the anode which consists of a block of copper in which is imbedded the target, a small tungsten button. These metals are selected for definite reasons:

a. *Tungsten*—so placed that it forms the actual target of the x-ray tube, the high speed electrons striking it directly. Since more than 99 per cent of the kinetic energy of the electrons is converted to heat at the target, the *high melting point* of tungsten—3370 C—makes it especially suitable as a target metal. Furthermore, its high atomic number favors the production of photons of sufficiently high energy (short wavelength) for radiography. (A special mammographic tube is available with a molybdenum target—see pages 319-320.)

b. *Copper*—a much better conductor of heat than tungsten, carrying the heat away from the target more rapidly and thereby protecting it from overheating, within certain limits. Even greater heat-loading capacity can be achieved by circulating air, water, or oil around the anode, permitting larger exposures than would otherwise be possible.

The electron stream bombards a limited area on the target called the *focus* or *focal spot*. As will be shown later, using the smallest practicable focus enhances the sharpness of the radiographic image. Therefore, radiographic tubes are provided with a small filament and a molybdenum focusing cup. However, as the focus size decreases it experiences a greater concentration of heat with the attendant danger of overloading and melting the target. Tubes are so designed that the focus is small enough to give satisfactory radiographic detail, but not so small as to restrict their practical usefulness. Most stationary anodes have focal spot widths of 2 mm and 4 mm.

Anodes of radiographic tubes are constructed on the *line-focus principle* to provide a focus which, when projected toward the film, is smaller than is its actual area on the target, thereby *permitting heavier exposure while producing a relatively fine focus.* The line-focus principle can best be explained by a diagram, as shown in Figure 13.4. Here, the actual focus is a rec-

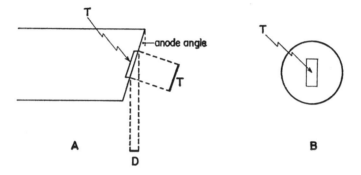

Figure 13.4. Line-focus principle. In *A* the target is shown in side view, and *D* is the projected or *effective focus.* In *B* the target is shown in face view with the actual rectangular focus, *T*, at its center. The same principle is used in rotating anodes.

tangle as in *B.* Assume this retctangle, *T,* to measure about 2 mm × 4 mm. In *A,* the side view of the tube, note that the target is inclined, making an angle of 17° with the vertical. Since the film will be placed at some distance directly below the target, the *effective focal area* from the standpoint of the film is only *D* which is about 2 mm × 2 mm. Perhaps it would become clearer if you were to imagine yourself lying supine in the position of the film and looking upward at the target—the focus would resemble a small square rather than a rectangle. In other words the focus is foreshortened, just as a pencil appears shorter when it is held obliquely in front of the eyes, than it is when held parallel to the eyes.

2. **Rotating Anode.** In 1936 there became available a radically different kind of anode of ingenious design—the *rotating anode*. As its name indicates, this anode, attached to the shaft of a small induction motor, *rotates during x-ray exposure*. Rotating anodes were heretofore constructed of solid tungsten in the form of a disc. More recently, an improved type of rotating anode has been introduced, consisting of a *molybdenum disc* coated with a *tungsten-rhenium alloy*. The addition of rhenium diminishes roughening of the focal area, thereby maintaining a high x-ray output (emissivity). The usual diameter of rotating anodes is about 3 in., although 5-in. models have been introduced for extra-heavy duty. Speed of rotation is ordinarily 3300 to 3600 rpm (revolutions per min), but special high-speed models capable of 10,000 rpm are available for special procedures. A diagram of a rotating anode tube is shown in Figure 13.5.

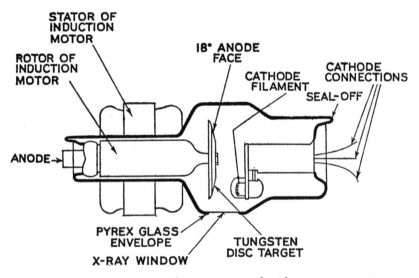

Figure 13.5. Rotating anode tube.

As already mentioned, rotating anodes utilize the *line-focus principle*; they therefore have a beveled edge as shown in Figure 13.6. The cathode is so placed that the electron beam strikes the beveled edge of the anode, as shown in Figure 13.5, the degree of beveling being called the *target angle*. Anodes are available

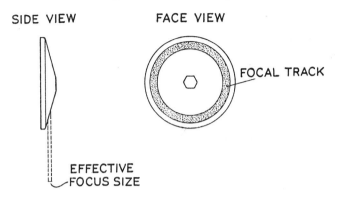

SIDE VIEW   FACE VIEW

FOCAL TRACK

EFFECTIVE
FOCUS SIZE

Figure 13.6. Line-focus principle with a rotating anode. The *focal track* is the actual area of impact of the electrons, or the actual focal area. Note how much larger the focal track is than the effective focus.

with targets of various angles such as 10°, 12°, and 17°. These influence the size of the exposure that can be used in a very short exposure time without damaging the tube (see Table 13.1).

TABLE 13.1

EFFECT OF TARGET ANGLE OF A ROTATING ANODE
ON SHORT-EXPOSURE RATING RELATIVE TO THAT OF A
STANDARD TARGET ANGLE OF 17°

Target Angle	Approximate Increase in Short Exposure Rating
12°	30%
10°	60%

Outside the tube is the stator of the induction motor, the rotor being inside the tube, connected to the anode. Self-lubricating bearings coated with **metallic barium** or **silver** reduce friction, thereby prolonging tube life.

How does such an anode operate? During exposure the target rotates rapidly—3300 to 3600 rpm—while being bombarded by the electron stream. As it rotates, it is constantly "turning a new face" to the electron beam, so that the heating effect of the beam does not concentrate at one point, as in the stationary anode, but spreads over a large area called the *focal track* on the beveled face of the anode. However, the effective focus remains constant in position relative to the film because of the extremely smooth motion of the anode. As a result, it is possible to have an effective focus of 1 mm or even smaller—the 0.3 mm fractional focus tube. Furthermore, it becomes feasible to pass high milliamperage through the tube. For instance, with a stationary anode, the upper limit for a 4 mm focus is 200 mA, whereas with a rotating anode, the upper limit for a 2 mm focus is 500 mA, the maximum exposure time being approximately the same in both cases. Rotating anode tubes usually have two filaments to provide 1 and 2 mm focuses; the smaller one is used to obtain fine radiographic definition, whereas the larger one can tolerate heavier loading in shorter exposure time to minimize motion in special radiographic procedures.

Special rotating anode tubes with 0.3 mm focal spots, and even smaller, have been designed for *magnification radiography.* Called *fractional focus tubes,* they permit placing the film at a distance from the part to permit direct magnification of the radiographic image. This is described on page 325.

You can see from the preceding discussion that the modern rotating anode tube has successfully combined the extremely small focus with reasonably large heat-loading capacity. This has made possible the superb sharpness of detail in modern radiography.

Space Charge Compensation

The low-voltage current supplied to the x-ray tube filament heats it to incandescence, causing the emission of electrons (thermionic emission). The rate of electron emission, or the number emitted per second, depends on the temperature of the filament which, in turn, is governed by the current supplied to it. In fact, *a small increase in filament current produces a large increase in the rate of electron emission and resulting space charge.*

If *no* high voltage is applied across the tube, the electrons remain in the vicinity of the hot filament as a *space charge.* This gives rise to the *space charge effect* which may be simply explained as follows: since the space charge consists of electrons, it is negative and therefore tends to hold back the further emission of electrons (like charges repel). In fact, some electrons are actually repelled from the space charge back into the filament.

At any particular value of the filament current, an equilibrium is reached between the rate of electron emission by the filament and the rate of return to the filament from the space charge. Thus, *the space charge has a definite size for a given value of the filament current in a given tube,* and as just indicated, a small change in filament current produces a large change in the size of the space charge.

If a sufficiently high voltage (kV) is now applied across the tube, imparting a large negative charge to the filament and an equally large positive charge to the anode, the space charge electrons are driven toward the anode at a speed that depends on the kV (but not directly proportional). When tube current (mA) is plotted as a function of tube voltage (kV) a family of *characteristic curves* is obtained as shown in Figure 13.7. As expected, there is a different curve for each value of filament current (amp). Furthermore, the larger the current, the greater will be the eventual mA.

Let us now examine the characteristic curves. Note that the lowest curve (representing a filament current of 4.2 amp) has an initial curved portion and then flattens out at about 100 kV, after which a further increase in kV causes no additional increase in mA. We can explain this as follows: as the kV is progressively increased, more and more of the space charge electrons per second are driven to the anode; that is, there is an increase in mA. This is called the *space-charge limited region* of tube operation. Above 100 kV all of the space charge electrons are driven to the anode just as they are emitted from the filament, so that further increase in kV cannot increase mA; hence, this portion of the curve is flat, representing *saturation current* or the *temperature-limited region.* Here the mA can be increased only by increasing the filament current and temperature.

The remaining curves in Figure 13.7 do not have a flat portion below 150 kV, so at the corresponding filament currents the tube operates only in the space-charge limited region. Operation of an x-ray tube in the space-charge limited region is undesirable in radiography because kV cannot be changed independently of mA—when the technologist changes the kV setting there is an unavoidable change in mA. To correct this situation a *space*

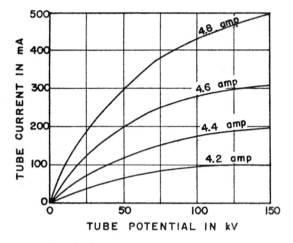

Figure 13.7. Family of characteristic curves of a typical rotating anode tube (*Machlett Dynamax "60"*; *Dunlee*). Note the progressive rise in mA as kV is increased. A space charge compensator is used to flatten the curves so that kV can be changed without a change in mA as explained in the text.

charge compensator is introduced into the filament circuit. Its function is to lower automatically the filament current just the right amount, as the kV is raised (and conversely), to keep the mA constant over the useful kV range. Now the technologist has full control of tube operation, being able to change kV without producing any appreciable effect on mA.

In modern automatic equipment with monitor-type mA selection, the space charger compensator has an additional function; it also selects the correct filament current to provide the desired space charge and resulting mA.

Tube Rating Charts

The high milliamperages needed for radiography place a heavy load on the tube. Most radiologic tube failures are due to "burning" the *filament* from application of excessive amperage which raises filament temperature beyond the safe limit; or it can result from prolonged heating of the filament at normal amperage, just as an electric light bulb filament burns out after a certain period of normal use. In modern installations, a special *booster circuit* in the filament circuit limits the filament current to a low value until the x-ray exposure switch is closed. At that instant, the booster circuit automatically raises the filament current sufficiently to provide the correct filament temperature. This helps significantly in prolonging tube life with high mA technics.

While relatively few tube failures are due to damage to the target, the rating of tubes for maximum operating technics is based on the ability of the *anode* to accumulate heat without melting. As a general rule, *an increase in kV produces a smaller load on the anode than does an increase in mA and time (mAs) for the same radiographic density* (film darkening), provided the safe exposure limit of the tube has not been exceeded.

The safe limits for tube loading can easily be found by referring to the *tube-rating chart* provided by the manufacturer for a particular tube. Such a chart indicates the maximum safe kV time, and mA for a single exposure. The chart in Figure 13.8 is

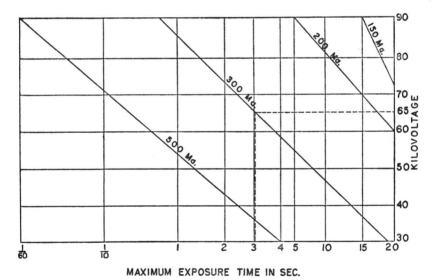

MAXIMUM EXPOSURE TIME IN SEC.

Figure 13.8. Tube-rating chart, arbitrary values. The chart for any particular tube must be obtained from the manufacturer.

fictitious and is shown only by way of an example. It includes a series of lines representing various mA values of tube current. The vertical axis of the graph represents kV, and the horizontal axis represents maximum safe exposure time in sec. The chart is quite simple to use. Assume that it applies to a particular tube of *given focus size,* and with a *given type of rectification.* Suppose we are to make an exposure of 300 mA and 65 kV, and we wish to determine the maximum allowable exposure time. Starting at 65 on the vertical axis, follow horizontally from this point to the point where it meets the 300 mA curve. Then from this point of intersection, follow vertically downward to the horizontal line, crossing it at 3 sec, the maximum exposure time with these factors. With longer exposures, the tube would most probably be damaged.

It must be emphasized that the correct chart must be selected, depending on the following conditions:

1. **Type of Rectification.** The maximum safe exposure is less with self or single-valve-tube than with full-wave rectification.

2. **Size of Tube Focus.** The maximum safe exposure is less with a small focus than with a large one.

3. **Tube Design.** This affects the maximum safe exposure; therefore the appropriate chart supplied by the manufacturer should be used.

4. **Cold vs. Hot Tube.** The charts apply only to a relatively cold tube, one that has not been subjected to heavy loading just prior to use.

5. **Type of Power Supply.** The rating will be different for single phase and three phase.

The tube rating chart for the particular tube in use should be mounted in the control booth, especially if the equipment is not provided with automatic exposure limiters. These devices prevent overloading the x-ray tube by not permitting excessive exposures to be made.

When numerous exposures are made, especially in a short period of time, we must take into account the *ability of the anode to store heat.* This is measured in *heat units,* defined by the following equation for single phase operation:

$$\text{Heat Units (H.U.)} = kV \times mA \times sec \qquad (1)$$

For 3-phase operation, the equation must be multiplied by 1.35.

Note that heat units are not fictitious, but really represent energy:

$$1 \text{ kV} \times 1 \text{ mA} \times 1 \text{ sec} = 1 \text{ watt-sec}$$

which is a unit of energy. The *heat storage capacity* of the anodes of various diagnostic tubes ranges from about 100,000 H.U. to about 250,000 H.U., depending on the size of the anode and the cooling method. The cooling characteristics of anodes are specified by the *heat dissipation rate,* which must be determined from cooling curves supplied by the manufacturer of that particular tube (see Figure 13.9). If the anode has been heated to its full

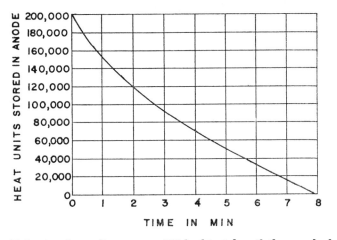

Figure 13.9. Anode cooling curve. With this tube, if the anode has the maximum 200,000 heat units already stored, then at the end of 2 min the stored heat will have decreased to 120,000 H.U. It would then be safe to load the anode with additional 80,000 H.U. for a total maximum of 200,000 H.U. For complete evaluation, a housing cooling curve should also be consulted. (Based on data furnished by courtesy of Machlett Laboratories, Inc.)

storage capacity by a large number of exposures, the safety of further exposures depends on the heat dissipation rate. Suppose a number of chest radiographs are to be made in rapid succession, requiring on the average an exposure of 70 kV, 100 mA, and 0.1 sec. The heat units developed per exposure would be obtained by the above equation as follows:

$$\text{H.U.} = 70 \times 100 \times 0.1$$
$$\text{H.U.} = 700$$

If the anode cooling curve shows that it can discharge heat at rate of 40,000 H.U. per minute, divide 40,000 by 700 = about 54 exposures per minute as the safe maximum. Obviously, this number is unlikely in ordinary roentgenography but it could assume major importance in special procedures requiring repeated frequent, heavy exposures. In brief, for ordinary radiography the customary tube-rating chart has the greatest practical value, whereas in special procedures and in extremely busy x-ray departments the heat storage capacity and the cooling curves of both the anode and the tube housing must be taken into account.

The equation defining heat unit storage also demonstrates that an x-ray tube operates more efficiently at higher kilovoltages. For example, a given radiographic technic calling for an exposure of 60 kV, 100 mA and 1 sec would develop the following number of heat units in the anode:

$$\text{H.U.} = 60 \times 100 \times 1$$
$$\text{H.U.} = 6000$$

If the kV is increased by 10 and the time reduced one-half, the exposure would remain practically the same, but now let us see how many units would be developed in the anode:

$$\text{H.U.} = 70 \times 100 \times \frac{1}{2}$$
$$\text{H.U.} = 3500$$

Thus, with an increase in kV the heat units in the anode have been decreased almost one half, even though the exposure has remained almost constant.

The *short-exposure rating* (that is, the ability of an anode to withstand a large heat unit input in a small fraction of a second) does not necessarily follow the heat storage capacity of the anode. For example, as described on page 246, the short-exposure rating is higher with three-phase than with single-phase current for a tube with the same heat storage capacity. Furthermore, as shown in Table 13.1, the short-exposure rating of a rotating anode tube increases significantly as the target angle decreases. Still a third factor in this problem is the speed of rotation of the anode; thus, with high speed anodes—10,000 rpm—the short-exposure rating is increased about 70 per cent over that of the conventional rotating anode at 3600 rpm.

With minor degrees of overloading of the tube, tiny depressions appear in the target, a condition known as *pitting*. Excessive pitting may reduce the x-ray output of the tube. The use of rhenium-molybdenum targets instead of tungsten minimizes pitting.

After prolonged use of an x-ray tube, the glass aperture through which the x-ray beam passes becomes purple due to a chemical change induced by the x rays. This does not affect the output of the tube.

In the ordinary oil-immersed tube, adequate *cooling* results from the transfer of heat to the oil, tube, housing, and surrounding air by simple convection and conduction. For heavier loads, as in chest surveys, special procedures, and high voltage radiography, cooling is improved by mounting a small fan outside the tube housing. Some specially designed tubes are cooled by a forced oil circulating system, the oil flowing into the hollow anode and then between the tube and housing; this provides a very high rate of heat dissipation. Heat storage capacity may be enhanced by the use of rotating anodes with surface areas of 100 per cent greater than usual. The oil incidentally provides efficient insulation, while the metal housing helps make the tube rayproof by shielding out unwanted radiation.

X-RAY THERAPY TUBES

These are x-ray tubes designed exclusively for use in clinical radiation therapy. Such tubes, operating at relatively low mA values of tube current, are manufactured in various models for the desired kV range.

In x-ray therapy, four main kV energy ranges are available for treatment at different depths in the body. Some authorities would avoid special terminology, preferring simply to state the energy range.

1. **Low Voltage.** This includes a range of about 50 to 120 kV and is used to treat lesions in the skin. It is also known as *superficial therapy.*

2. **Intermediate Voltage.** This is about 130 to 150 kV and is used mainly in treating lesions a few centimeters below the skin.

3. **Orthovoltage.** The usual range of 160 to 500 kV provides x rays of relatively high penetrating power, applicable in deep x-ray therapy for the treatment of lesions below the surface; now largely replaced by megavoltage.

4. **Megavoltage (Supervoltage).** Operating at potentials above 1 MV, the specially designed tubes for this type of equipment will not be described here.

The construction of orthovoltage therapy tubes resembles radiographic tubes, but in the therapy tube the filament is usually larger, the anode is always stationary, and the focal spot is larger because a relatively large beam is desirable. The anode consists of a tungsten mass imbedded in a large block of copper, its face usually being inclined at an angle of about 30° to generate a beam of reasonably uniform intensity. As is true of radiographic tubes, therapy tubes have a separate filament current to heat the filament and produce a space charge; and a high voltage to drive the electrons to the target at high speed.

Since there may be a tremendous amount of heat liberated in the anode, particularly in the orthovoltage region, special means of cooling the tube must be provided. Superficial therapy tubes are usually cooled in the same manner as radiographic tubes, depending on the anode heat load. On the other hand, intermediate and orthovoltage therapy tubes most often have hollow anodes cooled by circulating oil between the tube and housing. A pump then transfers the hot oil to a special tank where it is cooled by water circulating in coils.

The oil serves another purpose. Because it is highly refined and moisture free, it has superb insulating ability. Such insulating oils permit a reduction in the size of the tube housing, thereby making it more easily maneuverable. In some therapy units—special *self-contained units*—the tube and transformer are mounted in the same casing, suspended over the patient.

VALVE TUBES (VACUUM DIODE RECTIFIERS)

The use of valve tubes in rectification has already been described. They are *thermionic diode tubes* having the same general construction as x-ray tubes, but differing from them in certain details because of their specific function. We shall now describe their structural components and operation.

Cathode

The filament of a valve tube consists of a coil of tungsten wire which is longer and thicker than the filament of an x-ray tube. It lies in the *longitudinal axis* of the tube, supported by a large molybdenum spiral as shown in Figure 13.10. Modern valve tube

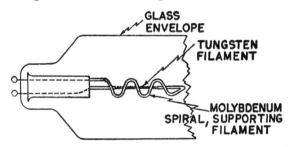

Figure 13.10. Cathode of a valve tube (vacuum rectifier).

filaments are *thoriated* for more efficient release of electrons, and for greater durability. The filament of a valve tube is heated through its own filament circuit, and is also provided with a connection to the high voltage circuit. Thus, the valve tube as well as the x-ray tube has two circuits.

Anode

As shown in Figure 13.11, the valve tube anode, having a *large*

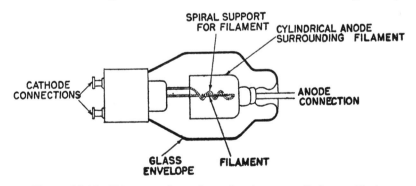

Figure 13.11. Diagram of a valve tube (vacuum diode rectifier).

surface and a *cylindrical shape* (resembling a metal can), surrounds the filament. This makes it possible for a current of large cross-sectional area to pass from the cathode to the anode.

Principles of Operation

A valve tube normally operates *below saturation;* that is, in the space-charge limited region (see page 207). Because of the large size of the filament, it emits a generous space charge when heated by the filament current so that enough electrons are available for any reasonable combination of kV and mA. The superabundance of space-charge electrons, together with the large cylindrical anode, is responsible for the *low electrical resistance* and consequent *small voltage drop* across the valve tube; for example, with a 500 mA current in the x-ray tube, the voltage drop in the valve tube should not exceed 3 kV. Since the speed of the electrons in the tube varies with the kV drop across the tube, the electrons do not gain significant speed with a drop of only 3 kV, and therefore lack sufficient energy to produce x rays during normal operation. However, in a defective valve tube, as thermionic emission decreases, the resistance may increase until the voltage drop becomes large enough to accelerate the electrons to the point of x-ray production. At the same time, a large fall in kV across the valve tube would decrease the kV available for the operation of the x-ray tube.

Bear in mind that the mA in a valve tube must be equal to that in the x-ray tube because they are connected in series and *all parts of a series circuit carry an identical current.* Since current is defined as the rate of flow of electricity, one may say that an electric current represents the flow of a certain *number of electrons per second.* During operation, the same number of electrons per second pass from cathode to anode in the x-ray tube as in the valve tube. But in the valve tube, the individual electrons move less rapidly and in a larger group, while in an x-ray tube the same number of electrons move in smaller groups, but at a very high speed. For example, suppose that in the valve tube 1 trillion electrons were passing to the anode in one second; each electron would move slowly but at the end of the second, 1 trillion would have reached the anode. Now consider 1 trillion electrons in an x-ray tube. Under the large kV drop, each electron moves at a very high speed, taking only a tiny fraction of a second to reach the anode, but at the end of a second, 1 trillion have reached the anode, the same number per sec as in the valve tube.

The valve tubes are placed within the oil in the transformer casing for the sake of convenience. Incidentally, this affords protection from x radiation that might be emitted in the event of valve tube failure as described above (page 216). Furthermore, this arrangement decreases the bulk of the x-ray apparatus, at the same time making the valve tube assembly shockproof.

SOLID STATE RECTIFIERS

In recent years, intensive investigation of a class of materials known as semiconductors has led to the development of an entirely different type of high voltage rectifier—the *solid state diode rectifier*. To understand its basic principles we must look into the conduction of an electric current in solids from a slightly more advanced point of view than in Chapter 6.

One of the ways in which we can classify matter is on the basis of its *conductivity*, defined as the ease with which electrons move about within it. Accordingly, there are three classes of solids: (1) *conductors* (excellent conductivity—electrons can be made to drift readily from point to point) (2) *nonconductors* or *insulators* (very poor conductivity—high resistance to movement of electrons) and (3) *semiconductors* (intermediate conductivity).

What accounts for the difference in conductivity of these three classes of matter? Energy-state diagrams come to our aid. An *energy state* is the energy of a particular electron, corresponding to its energy level in the atom. In Figure 13.12 we see the energy

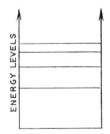

Figure 13.12. Possible energy states of an electron in a single atom. The higher energy states are in the direction of the arrows.

states that an electron may have in a single atom. Since a piece of matter is an aggregation of a fantastic number of atoms usually arranged in a crystal lattice (that is, a regular pattern), their energy levels merge into a series of *energy bands.*

There are three types of energy bands—*valence, forbidden,* and *conduction.* No electron can remain in the forbidden band; this band represents the energy needed to raise an electron from the valence band to the conduction band where the electron can move freely. The various bands are shown in Figure 13.13.

In a *metallic conductor,* shown in Figure 13.13A, the outermost or *valence* electrons are very loosely bound; in fact, the valence band actually overlaps the conduction band. Thus, an extremely weak electric field (applied emf) causes the electrons in the valence band to drift in a particular direction as an electric current. Because electrons can be made to flow in a metal with the application of so little energy, electric current in a conductor is not easily controllable (as it is in a semiconductor).

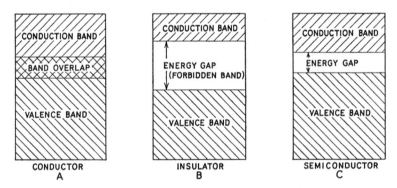

Figure 13.13. Energy level diagrams of three types of matter. In a conductor (*A*) electrons move freely from the valence band to the conduction band. In an insulator (*B*) there is a large energy gap between the valence and conduction bands preventing free movement of electrons. In a semiconductor (*C*) there is a small energy gap between the valence and conduction bands making it possible to raise electrons to the conducting band, thereby controlling resistance.

In an *insulator,* shown in Figure 13.13B, all electron-containing bands are completely filled. Since a very large energy gap exists between the valence band and the conduction band in insulators, an extremely large electric field would be needed to raise an electron to the conduction band, so that, for all practical purposes, an insulator does not conduct an electric current (we may say that it has virtually infinite resistance).

Now we come to the third class of solids—*semiconductors.* These are intermediate between conductors and insulators in their ability to conduct an electric current. In Figure 13.13C you can see that a small energy gap exists between the valence band and the conduction band. Thus, only a small amount of energy is needed in semiconductors to raise the valence electrons to the conduction band. Note that this energy gap is much smaller in semiconductors than in insulators (compare Figure 13.13B with 13.13C). Because the valence and conduction bands do not overlap in semiconductors, current flow can be controlled in them; for example, the addition of carefully measured amounts of certain impurities to a semiconductor modifies its conductivity. In fact, its conductivity can be regulated by the kind and amount of impurity.

Of the available semiconductors, we shall limit our discussion to *silicon* (an element in ordinary sand which is silicon dioxide) because it is now being used in the manufacture of solid state rectifiers for x-ray equipment.

Basic to an understanding of semiconductors is the rule that, in combining chemically, most atoms share their valence electrons in such a way that the outer shell of each contains 8 electrons. This is known as the *octet rule.*

The silicon atom has 4 valence electrons (in its outermost shell) as shown in Figure 13.14. Silicon atoms aggregate into

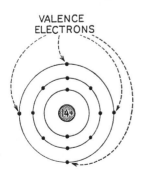

VALENCE
ELECTRONS

• (Si) •

SYMBOL FOR SILICON
WITH 4 VALENCE ELECTRONS

Figure 13.14. The silicon atom has 4 valence (outer shell) electrons as indicated by the broken lines.

crystals by sharing these valence electrons with each other—so-called *covalent bonding*—as shown in Figure 13.15. Such bonds, in effect, saturate the outer or valence shells, thereby satisfying the octet rule and enhancing the stability of the crystal structure. Pure silicon is an example of an *intrinsic* semiconductor.

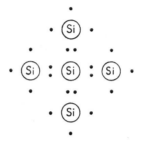

Figure 13.15. Diagram of the crystal lattice of silicon atoms which share valence electrons in pairs by means of *covalent bonds*. Each silicon atom is surrounded by 4 others (actually, in 3 dimensions). Note that the central Si atom, by electron sharing, now has an octet—8 valence electrons. In a real crystal there are tremendous numbers of Si atoms forming covalent bonds.

If a small amount of *arsenic* is added to silicon, it contributes to the crystal lattice an *extra* electron, not needed for bonding silicon and arsenic atoms (see Figure 13.16). This is a loosely bound electron which can easily be raised to the conduction band

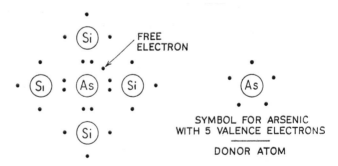

Figure 13.16. N-type silicon. Addition of a minute amount of arsenic with its 5 valence electrons introduces a free electron into the crystal. Under the influence of an applied emf, this free electron can be raised to the conduction band and made to drift through the crystal as an electric current.

by a small applied electric field. Arsenic is known as a *donor* atom because it furnishes electrons. Silicon "doped" with arsenic is called *n-type silicon* (*n* stands for negative). Its conductivity can be controlled by the amount of arsenic doping—the more arsenic, the more electrons available and the greater the conductivity (lower resistance).

On the other hand, if *gallium* with its 3 valence electrons is added to the silicon as shown in Figure 13.17, there is a shortage

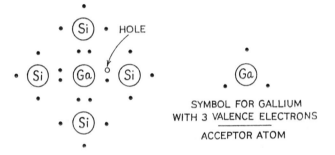

Figure 13.17. P-type silicon. Addition of a minute amount of gallium with its 3 valence electrons results in a deficit of 1 shared electron in the crystal lattice. An absent electron is equivalent to a positive charge, here called a "hole"; this can accept an electron.

of one electron in the crystal structure called a "hole." Such a hole behaves like a positive charge, gallium-doped silicon being known as *p-type silicon* (*p* stands for positive). Since the hole can accept an electron, gallium is an *acceptor* atom.

How do these two types of semiconductors behave? In Figure 13.18A, representing an n-type semiconductor, electrons drift in an applied electric field just as they do in a conductor. But in a p-type semiconductor, shown in Figure 13.18B, an applied electric field causes electrons to move toward holes, at the same time leaving new holes which are then, in turn, filled by other electrons. Thus, the holes, in effect, drift in the opposite direction from that of the electrons as though the holes were positive charges.

Suppose, now, we were to combine n-type and p-type silicon at a *junction* as in Figure 13.19. Some electrons from the n-type

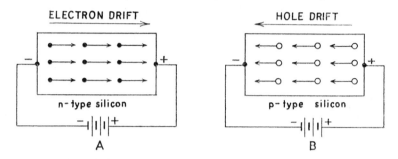

ELECTRON DRIFT

HOLE DRIFT

n-type silicon

p-type silicon

A

B

Figure 13.18. A, behavior on *n*-type silicon when an electric field (emf) is applied. Free electrons are raised to the conduction band and drift through the crystal just as in a conductor. B, in *p*-type silicon, electrons enter from the negative terminal of the battery and combine with holes which may be regarded as drifting toward the negative terminal.

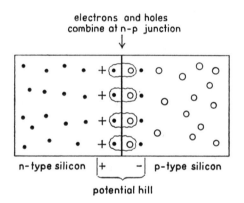

electrons and holes
combine at n-p junction

n-type silicon   p-type silicon

potential hill

Figure 13.19. Behavior of an *n-p* junction. When *n*-type and *p*-type silicon are brought into contact, holes and electrons combine near the junction leaving, respectively, negatively and positively charged ions. A potential "hill" (potential difference) is built up until further combination of holes and electrons ceases.

silicon combine with holes of the p-type near the junction, leaving positive and negative ions, respectively, as shown in the figure. A potential difference exists between these oppositely charged ions, called the *barrier voltage* or *potential hill.*

Let us see how an n-p junction rectifies—that is, *conducts current in only one direction.* Suppose we apply a current across this junction in the direction shown in Figure 13.20A, with the negative terminal of the source at the n-end, and the positive terminal at the p-end. Electrons in the n-type silicon move toward the junction and, at the same time, holes in the p-type silicon also move toward the junction where the electrons and holes combine. As a result, the n-type silicon accepts electrons from the negative terminal of the battery, while the p-type silicon passes electrons to the positive terminal. Thus, the diode (that is, the n-p junction) continues to conduct current as long as the battery terminals are connected in the manner shown. This is called a *forward biased diode.*

Now if the battery terminals are reversed, as in Figure 13.20B, electrons leave the n-type silicon, and holes leave the p-type sili-

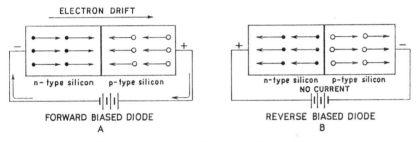

Figure 13.20. Effect of an applied electric field (emf) on an *n-p* junction. In *A*, with the negative terminal connected to the *n*-type silicon, electrons drift toward the junction where they combine with holes. The net effect is that the electrons enter the *n*-type silicon and leave the *p*-type silicon. In *B* the battery terminals have been reversed; now electrons are withdrawn at the positive end, and holes at the negative end. Both are immediately depleted and no current flows. When ac is applied, current flows only during the forward-biased half of the cycle; in other words, the *n-p* junction behaves as a *rectifier.*

con (that is, electrons enter the latter). This causes an increase in the number of positive ions on the n-side of the junction, and an increase in the number of negative ions on the p-side of the junction, thereby increasing the barrier voltage. As a result, the barrier voltage instantaneously becomes equal and opposite to the applied voltage and no current flows across the diode. This is now a *reverse biased diode.* Thus, if an alternating current is applied to an n-p junction it undergoes rectification, being conducted across the junction only when the negative half of the ac cycle is directed from n to p. The symbol for a solid state diode rectifier is shown in Figure 13.21.

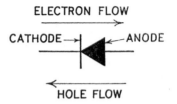

Figure 13.21. Symbol of a solid state diode rectifier.

A practical silicon rectifier for x-ray equipment consists of a stack of individual diodes, each containing an n-p junction. These diodes, called *modules,* are connected in series. The complete assembly is enclosed in a sealed ceramic tube containing an insulating liquid. Each module is capable of withstanding a maximum of 1000 volts, so that for a 150-kV generator, 150 modules are required. The rectifier stack is of the *controlled avalanche type* to minimize breakdown of individual modules. A silicon rectifier for x-ray equipment is shown in Figure 13.22.

Figure 13.22. Silicon rectifier for x-ray equipment. Various sizes are available, from about 8 to 10 inches in length.

What are the advantages of the silicon rectifier over the valve tube? They may be listed as follows:

1. Compact size.

2. No filaments, hence no filament transformers needed for solid state rectifiers. This is especially important in three-phase equipment where 12 rectifiers are ordinarily used.

3. Low forward voltage drop—less than 300 volts at maximum current rating of 100 mA.

4. Low reverse current—0.002 mA at maximum voltage.

5. Long life due to rugged construction.

Because of these advantages, solid state rectifiers are gradually replacing valve tubes in modern radiographic units.

RECTIFIER FAILURE

The technologist should form the habit of checking for satisfactory operation of the rectifier diodes during each exposure. This can be done by simply watching the milliammeter, which indicates *average mA*. With four diode (full-wave) rectification, the *peak mA* is actually about one and one-half times the reading of the milliammeter. Whenever one rectifier diode in such a system fails, the system operates as a two-diode (half-wave) rectifier, this being reflected in a drop in the milliammeter reading to about one-half the normal value, and producing radiographic underexposure. An even more serious situation would exist if, with an x-ray unit having manual control of filament current, the filament current were to be increased to obtain the desired mA; the peak mA would now be about three times the average, with resulting danger of overloading the tube.

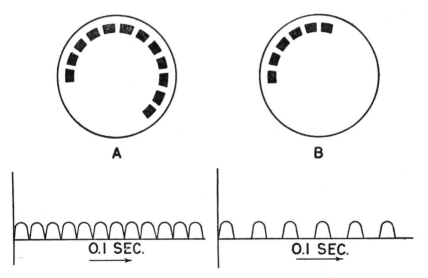

Figure 13.23. Spinning top test. In *A* the radiograph of the spinning top shows 12 spots in $\frac{1}{10}$ sec, representing 120 pulsations per sec; this is full-wave rectification. In *B* there are 6 spots in $\frac{1}{10}$ sec, representing 60 pulsations per sec; this is half-wave rectification.

The simplest device for testing the competence of any type of rectifier diode is the *spinning top*. This is a flat metal top which has a small hole punched near one edge. It is placed on a film exposure holder containing an x-ray film, and made to spin during an x-ray exposure of $\frac{1}{10}$ sec. If all four diodes are operating, the circuit is fully rectified and there should be 120 pulsations per sec with 120 corresponding peaks of x-ray output (on 60 cycle current. Therefore, in $\frac{1}{10}$ sec there will be twelve peaks and the image of the spinning top on the x-ray film will show twelve spots, as in Figure 13.23A. If only six dark spots appear in the radiograph of the top, there were only six pulsations in $\frac{1}{10}$ sec or sixty pulsations per sec, indicating that the circuit is half-wave rectified, provided the exposure timer is known to be operating correctly. The spinning top may be used in a similar way to test the accuracy of the x-ray exposure timer if all four rectifier diodes are known to be functioning.

QUESTIONS

1. Describe the main features of a thermionic tube.
2. Compare the structural details of radiographic tubes, therapy tubes, and valve tubes.
3. Describe with the aid of a diagram, a double focus, stationary anode tube.
4. What effect does evaporation of the filament have on tube operation? How is it corrected?
5. What are the advantages of a rotating anode tube?
6. Describe the line focus principle and discuss its importance in roentgenography. In what type of tubes is it used?
7. What is meant by saturation current?
8. How does the "space charge effect" influence the operation of a tube? How does a change in filament current affect the space charge?
9. Using the chart in Figure 13.8, determine the maximum safe exposure time at 70 kV and 500 mA.
10. Discuss the importance of heat storage and heat dissipation rates. Define heat unit.
11. Why does an x-ray tube have a greater capacity with full-wave rectification than with half-wave rectification?

12. Compare the values of peak milliamperage and milliammeter readings in a full-wave and half-wave rectified circuit.

13. Show by a labeled diagram the construction of a valve tube. Why must it operate below saturation conditions?

14. Compare the current in an x-ray tube and valve tube connected in the same circuit. Compare the kV drop in each under normal conditions.

15. How are radiographic and deep therapy tubes cooled?

16. Describe the spinning top method and state its purpose.

17. What are the main causes of x-ray tube failure? How can this be minimized?

18. What are semiconductors? How is their conductivity controlled?

19. Describe the principle of the n-p junction.

20. What intrinsic semiconductor is most often used in solid state rectifiers in medical radiography?

21. Discuss solid state rectifiers on the basis of principle and construction. What are their advantages and disadvantages relative to valve tubes?

22. Explain why $kV \times mA = watts$.

23. Show how heat units are related to energy according to equation (1).

CHAPTER 14   X-RAY CIRCUITS

THE MAJOR ITEMS of an x-ray unit have been covered in some detail, but they have been discussed individually without relation to each other. There now remains the task of combining them to create a functioning x-ray machine. A number of auxiliary devices are necessary for the proper operation of x-ray equipment, and these will also be described.

Source of Electricity

The electric power company supplies a high voltage alternating current to the *pole transformer* outside the building. The transformer, which is mounted atop a pole, is a step-down transformer. It reduces the voltage coming from the power house to 115 or 230 volts, depending on the type of equipment to be served. The voltage supplied to the radiology department is called the *line voltage.*

A *three-wire system* conducts the current into the building, the two outer wires being "hot" and the middle one "grounded" (neutral). At the instant that one outside wire is 115 volts above ground the other is 115 volts below ground, thereby producing a potential difference of 230 volts. Thus, connection to the two outside wires provides a 230-volt source, whereas connection to one outside wire and the middle wire supplies 115 volts. This relation is shown in Figure 14.1.

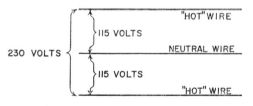

Figure 14.1. Three-wire system which brings electric power into the building. By connecting the electrical appliance or equipment to the appropriate wires as indicated, one may obtain 115 volts or 230 volts.

Most x-ray units now operate on 230 volts, although some of the less advanced equipment, and especially the older mobile units, require only 115 volts.

Note that the current delivered to the Radiology Department is *alternating current* (ac) because it is necessary for the operation of the x-ray transformer. Under unusual circumstances when only direct current (dc) is available, it must first be converted to ac so that it can be stepped up by the transformer.

The Main X-ray Circuits

To simplify the discussion of the electrical connections of an x-ray unit, we may regard the circuit as being divided at the x-ray transformer. The part of the circuit connected to the primary coil of the transformer is the *primary* or *low voltage* circuit, whereas the part connected to the secondary coil is the *secondary* or *high voltage* circuit. This scheme is perfectly natural, since the two coils of the transformer are electrically insulated from each other, and since the profound difference in voltage in the two circuits requires separate consideration. However, some parts of the equipment are connected to both the low and high voltage sides.

1. **Primary Circuit.** Included in this portion of the circuit is all the equipment that is connected between the electrical source and the primary coil of the x-ray transformer. Each essential component will be described in proper sequence.

a. *Main Switch.* This is usually a double-blade, single-throw switch, illustrated in Figure 14.2.

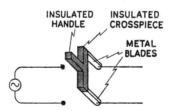

Figure 14.2. Double-blade single-throw switch. When the switch is closed toward the left, the metal blades contact the corresponding wire and the circuit is completed to the x-ray machine.

b. *Fuses.* Each wire leading from the main switch is provided with a *fuse*—an insulated cylinder with a metal cap at each end, in the center of which lies a strip of easily melted metal joining the caps. Overloading the circuit with excessive amperage causes the metal strip within the fuse to melt because of the heating effect of the current, thereby breaking the circuit. Figure 14.3

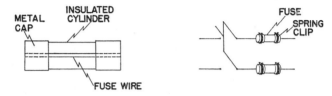

Figure 14.3. Diagram of a fuse (left), and its connection in a circuit. When the circuit is overloaded (excessive amperage) the wire in the center of the fuse melts, breaking the circuit.

shows the construction of such a fuse and its electrical connection. Fuses protect the equipment and reduce fire hazard. A blown fuse is easily replaced by spring clamps at each end.

c. *Autotransformer.* Discussed in detail in Chapter 10, the autotransformer controls the voltage supplied to the primary of the transformer, thereby providing various kilovoltages for the x-ray tube.

d. *Prereading Kilovoltmeter.* This is an ac *voltmeter* connected in parallel with the autotransformer. When used as a prereading meter, a voltmeter must first have been calibrated (standardized) by the manufacturer against peak kilovoltages by means of spark gap measurements. These depend on the fact that a given kV will cause a spark to jump between two metal spheres separated by a particular distance or *air gap* (at standard temperature, atmospheric pressure, and relative humidity). The kilovoltages obtained by such spark gap measurements are recorded directly on the scale of the prereading kilovoltmeter for each corresponding autotransformer setting. In other words, the prereading kilovoltmeter, since it is in the primary circuit, does not measure kV directly, but rather indicates what the kV will be in the secondary circuit because it has been initially calibrated to do so at the factory. The main advantage of the prereading kilovoltmeter is that *slow* fluctuations in line voltage can easily be corrected. If, for instance, the line voltage should drop, the autotransformer controls are adjusted manually until the correct kV shows on the prereading kilovoltmeter.

Some types of x-ray equipment do not employ a prereading kilovoltmeter. Instead, the desired kV is obtained by means of kV-labeled *pushbuttons* which select the appropriate transformer settings. The calibration of these settings is similar to that of a prereading kilovoltmeter. But these fixed settings indicate kV accurately only if the line voltage is the same as it was during the actual calibration. Since the line voltage may fluctuate slowly, depending on the other electrical equipment in use, a means of detecting such fluctuation is provided by a *compensator voltmeter* connected in parallel across a portion of the primary side of the autotransformer. A line on the compensator voltmeter scale indicates the correct line voltage. If the pointer is above or below this mark on the meter, a *line voltage compensator* permits adjustment of the line voltage to the correct value. The compensator utilizes the primary side of the autotransformer, varying the number of turns on the primary side until the compensator meter needle indicates that the correct line voltage has been established; only then will the correct kV be obtained.

e. *Timer and X-ray Exposure Switches.* These switches control the primary circuit to the primary coil of the transformer, serving to complete the x-ray exposure. The switches themselves are modifications of an ordinary pushbutton, as shown in Figure 14.4,

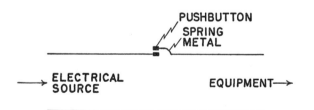

Figure 14.4. Pushbutton switch. When the button is depressed, it makes contact with the wire beneath it, closing the circuit.

and can be used for either hand or foot operation. But these are not designed to withstand the high amperage in the primary circuit (as high as 90 amp with a 200 mA unit), or to prevent electrical shock. Therefore, these switches operate a *remote control switch*, which in turn closes the primary circuit.

f. *Remote Control Switch.* Operated by either hand or foot switch, its basic design can best be appreciated by studying Figure 14.5.

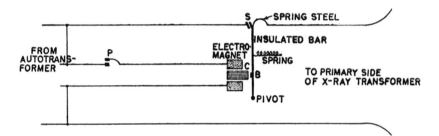

Figure 14.5. Remote control switch. The pushbutton, *P*, when depressed, completes the circuit through the coil of the remote control switch. This magnetizes the core, *C*, which attracts the metal button, *B*, on the insulated bar. This closes the primary circuit by bringing together the contacts at *S*. When the pushbutton is released, the reverse occurs, breaking the circuit.

g. *Timer.* Included in the exposure switch circuit is a timer which can be set to start and stop an exposure of preselected duration. There are four main types of manual exposure timers in use today. They include (1) the *synchronous timer*, (2) the *motor-driven impulse timer*, (3) the *simple electronic timer*, and (4) the *electronic impulse timer*.

1) *Synchronous Timer*—consists of a small synchronous motor rotating at a fixed speed, turning a disc at a slower speed through a system of reduction gears. The disc operates "make" and "break" contacts, and the farther the disc is set to rotate, the longer the time between the closing and opening of the timer circuit and, hence, the longer the exposure. Synchronous timers are reasonably accurate for exposures of $\frac{1}{20}$ to 20 sec, but have been largely replaced by timers that can provide shorter exposures to a high degree of precision.

2) *Motor-driven Impulse Timer*—also activated by a synchronous motor, but operates at a higher speed to provide exposures of $\frac{1}{120}$ or $\frac{1}{60}$ sec to $\frac{1}{5}$ sec. It is so designed that the contactors

open and close the timing circuit at or very near a zero point of the ac cycle, so that it actually measures the time in impulses; hence its great accuracy.

3) *Electronic Timer (Simple Type)*—provides a range of $\frac{1}{30}$ to 20 sec in less advanced equipment. However, it opens and closes the timing circuit within $\frac{1}{4}$ cycle from the zero point. While this is sufficiently precise for exposures longer than $\frac{1}{10}$ sec, it introduces an unpredictable and significant error in exposures of $\frac{1}{20}$ sec or less.

4) *Electronic Impulse Timer*—provides accurate exposures of $\frac{1}{120}$ sec (1 impulse), and longer exposures in multiples of one impulse. But, in contrast to the motor-driven impulse timer, the electronic impulse timer utilizes special electronic tubes—*thyratrons*—to maintain the exposure for a particular number of impulses. Thus, each pulse being $\frac{1}{120}$ sec in duration, an exposure of $\frac{1}{20}$ sec would require six pulses. When operating properly, this type of timer opens and closes the timing circuit precisely at zero points in the cycle.

Automatic control of radiographic exposure became possible with the advent of *phototiming*, originated by Morgan and Hodges. With this system, a film is exposed to the x-ray beam and as soon as it has received the correct amount of radiation for the desired degree of film darkening, the exposure terminates automatically. This requires a special type of diode known as a *phototube.* Its cathode is coated with an alkali metal such as potassium or cesium which has the peculiar property of *giving off electrons when struck by light.* If the anode of the phototube has, in the meanwhile, been given a positive charge from an outside source, the electrons emitted by the cathode are attracted to the anode, constituting a current in the phototube. By placing a fluorescent screen between the phototube and the x-ray source, we can make the phototube indirectly sensitive to x-rays, the brightness of the screen depending on the intensity of the x radiation. As shown in Figure 14.6, the phototube-fluorescent screen combination is placed behind the cassette whose back is radiotransparent (x-ray transmitting). When a predetermined quantity of radiation has reached the fluorescent screen, depending on the part being radiographed, the resulting current in the phototube operates a capacitor, thyratron, and relay circuit which activates a contactor to terminate the exposure automatically. The complete unit, known as a *phototimer* (see Figure 14.6),

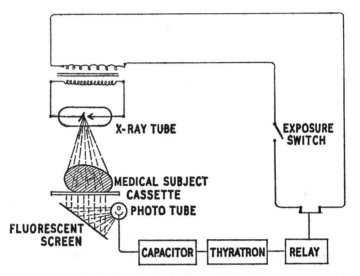

Figure 14.6. Simplified phototimer circuit. When the exposure switch is closed, x rays passing through the patient and reaching the fluorescent screen of the phototimer assembly cause the screen to emit light which activates the phototube. Current then flows in the phototube and charges the capacitor while the exposure is in progress. When the predetermined voltage is built up on the capacitor, the thyratron becomes conductive, activating the relay which opens the exposure circuit and terminates the exposure even while the manual exposure switch is still closed.

makes possible extremely accurate reproduction of radiographic density, provided the anatomic part is carefully centered. The most useful application of the phototimer at the present time is in spotfilm and chest radiography, and in photofluorography, although it is also used to a limited extent in general radiography.

h. *Circuit Breaker.* Additional protection against overloading the circuit is provided by a circuit breaker which can easily be reset, while a blown fuse has to be replaced. It is usually connected in series with the exposure switch, timer, and remote control switch. Figure 14.7 illustrates schematically the operation

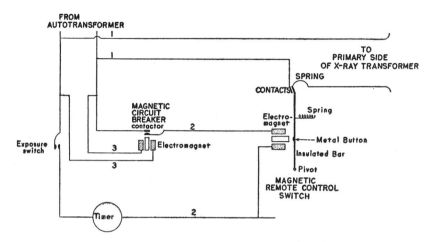

Figure 14.7. Magnetic circuit breaker together with the magnetic remote control switch, as they are connected in the primary circuit.

of a magnetic circuit breaker, shown in the exposure circuit, 2. When the exposure switch is closed, circuit 2 is completed through the timer, the electromagnet of the remote control switch, and the circuit breaker contacts which are in the closed position. This activates the electromagnet of the remote control switch closing the primary circuit, 1, that leads to the primary side of the x-ray transformer. If there should be a momentary surge in the primary current, this will, through circuit 3, increase the strength of the electromagnet in the circuit breaker to the point where it will open circuit 2, thereby interrupting the current to the electromagnet of the remote control switch, opening the remote control switch, and breaking the primary circuit, 1. The circuit breaker must then be reset manually before another exposure can be made.

i. *Filament Circuit of X-ray Tube.* The primary circuit supplies the *heating* current for the filament of the x-ray tube, but this current must first be reduced to 3 to 5 amp and 4 to 12 volts by a *rheostat* (variable resistor); in accordance with Ohm's law, the greater the resistance, the smaller the current (amp). Another type of filament current regulator is an electromagnetic device called a *saturable reactor* (see page 133). In series with

this is an *oil-immersed step-down transformer*, which further reduces the voltage to the required value. This is shown in Figure
14.8. The transformer is oil-immersed for added insulation be-

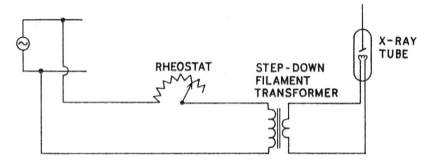

Figure 14.8. Filament circuit. The rheostat varies the filament current (and
voltage), thereby controlling filament temperature, electron emission, and
tube current (mA). A saturable reactor is being used instead of the rheostat
in advanced design equipment. The step-down transformer further helps
reduce the voltage and also insulates the primary circuit from the high
voltage in the secondary circuit.

cause its secondary coil is on the high voltage side of the x-ray
unit, and the primary circuit must be protected from the high
voltage in the secondary.

There is often included in the filament circuit an x-ray tube
filament stabilizer because relatively small changes in the filament voltage or current produce a large change in electron emission and consequent tube current (mA). The filament stabilizer
corrects for instantaneous fluctuations in line voltage and current
that may be caused by momentary demand elsewhere on the line,
such as in the starting of an elevator or air conditioner. It may
be so effective that a variation in line voltage of 10 per cent will
cause no greater change than ½ per cent in the filament voltage.
The stabilizer consists of a capacitor and a small, modified split
transformer which are so arranged that they compensate for a
rise or fall in line voltage and provide a more uniform filament
voltage, maintaining the filament current more nearly constant.

j. *Filament Ammeter.* In order to measure the filament current, and hence the amount of heat developed in the filament, an
ammeter is connected in series in the *filament circuit.* By previous calibration, we can establish the readings of this meter that

correspond to desired mA in the x-ray tube at a given kV. A space charge compensator provides automatic adjustment of filament current to maintain constant mA over a wide range of kV values.

k. *Primary Coil of X-ray Transformer.* This is the last connection of the primary circuit. Both the primary and the secondary coils are immersed in special transformer oil to provide adequate insulation.

In general, the wires in the primary circuit must be relatively large, because of the high amperage. Circuit breakers and fuses should be conveniently located for easy resetting or replacement.

2. **Secondary Circuit.** This includes the secondary coil of the transformer and all devices to which it is connected electrically.

a. *Secondary Coil of X-ray Transformer.* As we have indicated earlier, this consists of many turns of electrically insulated wire that is thinner than the wire in the primary coil because of the very small current in the secondary circuit. The transformer steps up the primary voltage to provide the high voltage required to operate the x-ray tube, the step-up ratio being about 500:1.

b. *Milliammeter.* In order to measure the mA in the x-ray tube, a *milliammeter* is connected in series in the high voltage circuit. The milliammeter is grounded together with the midpoint of the secondary coil of the x-ray transformer. Since it is grounded, it is at the same potential as the person manipulating the controls, and can therefore be safely mounted in the control panel.

Attention must be called to the fact that the *tube current is measured in milliamperes by the milliammeter* placed in the high voltage circuit, while *the filament current in measured by an ammeter placed in the low voltage filament circuit.* The milliameter measures average values and gives no indication of the peak values of tube current.

c. *Ballistic Milliampere-Second Meter.* In modern equipment operating at 200 mA or more, there is added in series with the regular milliammeter, a *ballistic milliampere-second meter.* Due to the large mass of the rotating mechanism, it turns slowly so that the attached indicator needle registers the product of mA and time, that is, it indicates *milliampere-seconds* (mAs). This device is needed because the ordinary milliammeter does not have time to register the true mA at exposures of less than $\frac{1}{10}$

sec. The ballistic meter is therefore needed to measure mA at very short exposure times; and because of its great sensitivity, it must be used in conjunction with an impulse timer or an electronic timer. At exposure times longer than $\frac{1}{10}$ sec, the ordinary milliammeter will register the correct average mA.

d. *Rectifier.* In all but self-rectified units, a system of rectification is included to change ac, supplied by the transformer, to dc. As we have already noted, rectification increases the heating capacity of the x-ray tube, safely permitting larger exposures.

e. *Cables.* These conduct the high voltage current from the rectifier to the x-ray tube. The midpoint of the secondary coil of the transformer is grounded. However, the total kV across the ends of the transformer secondary remains unchanged because one-half is below, and the other half above, ground potential. For example, if there is 90 kV across the transformer and the center is not grounded, the kV will fluctuate between 90 kV above ground and 90 kV below ground according to the ac sine wave. If the center of the secondary coil is grounded, the peak kV will still be 90, but in this case one half or 45 kV will be above ground and one half will be 45 kV below. The difference between 45 above zero and 45 below zero is still 90. Under these conditions each cable has to be insulated for only 45 kV (rather than 90 kV), representing a significant saving in weight and cost.

The cables are so constructed that there is no danger of shock through intact insulation. Ordinarily, a very bulky thickness of insulation would be required to prevent the high voltage from sparking over to the patient or other objects near the cable. However, the grounded woven wire sheath surrounding the cable eliminates the danger of sparkover. Figure 14.9 shows a shock-

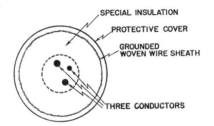

SPECIAL INSULATION

PROTECTIVE COVER

GROUNDED
WOVEN WIRE SHEATH

THREE CONDUCTORS

Figure 14.9. Cross section of a shockproof cable.

proof cable in cross section. The cable, as illustrated in the figure, has three conductors which, when connected to the *cathode*, make contact with the leads to the two filaments of a double focus x-ray tube (see Figure 13.3).When such a three-conductor cable is connected to the *anode*, only one of the conductors actually carries current.

f. *X-ray Tube.* The ultimate goal of the x-ray equipment is the operation of an x-ray tube, and this is the last piece of equipment connected in the high voltage circuit.

Completed Wiring Diagram

In order to visualize clearly the relationship of the various parts of the x-ray unit, we must connect them correctly in the circuit. Figure 14.10 is a schematic representation of the wiring of a full-wave rectified x-ray unit employing four rectifier diodes (these may be valve tubes or solid state diodes). Most of the auxiliary items, though desirable or even necessary for proper operation, have been purposely omitted to avoid complicating the diagram. The hook-up of these accessory devices has been indicated in the preceding sections. We can readily modify this basic diagram; for self-rectification, we simply omit the rectifier and connect the transformer to the tube terminals.

The X-ray Control Panel

In order to operate x-ray equipment, we must be able conveniently to adjust kV, mA, and time; and we must also have readily accessible meters to check the operation of the equipment. The controls and meters are mounted compactly in a *control panel*—a separate unit which is connected electrically to the x-ray equipment and is comparable in function to the dashboard of an automobile.

X-ray control panels vary greatly in complexity, depending on the design of the individual x-ray machine. If we realize that all control panels, regardless of their intricacy, are similar in their basic design, the problem of operating unfamiliar equipment becomes a relatively simple matter. The more complicated units are variations of the basic pattern.

In Figure 14.11 is shown a simple, nonautomatic control panel

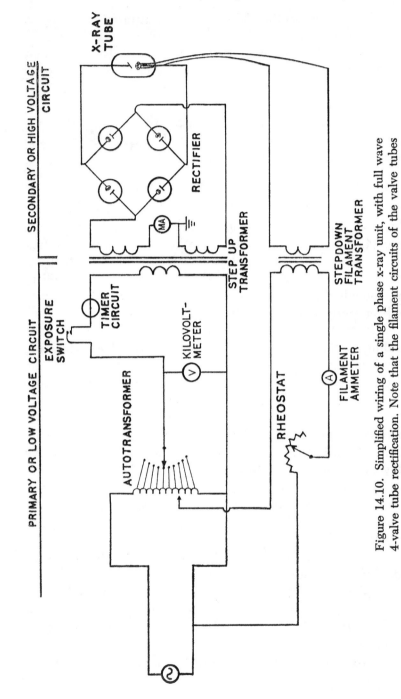

Figure 14.10. Simplified wiring of a single phase x-ray unit, with full wave 4-valve tube rectification. Note that the filament circuits of the valve tubes have been deliberately omitted to avoid complicating the diagram.

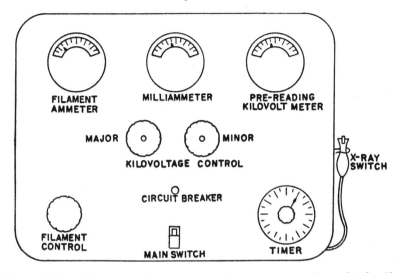

Figure 14.11. The essential components of an x-ray control panel. The filament control knob regulates the rheostat in the filament circuit, varying the filament current and consequently controlling the tube current (mA). The filament ammeter measures the current in the filament circuit. The kV control operates the autotransformer in steps of 10 kV (major) and 1 kV (minor). Actually, the autotransformer varies the primary voltage which is registered on the pre-reading kilovoltmeter in "kilovolts." The milliammeter indicates the average tube current in mA.

comprising all the essential items; familiarity with this diagram will make it much easier to understand the operation of the more complex types of equipment. The three meters shown in the figure include the milliammeter, filament ammeter, and kilovoltmeter. The milliammeter registers in mA, the current in the secondary circuit, including the x-ray tube.

A filament ammeter indicates filament current, which is the current that heats the tube filament, and which ultimately determines mA. The filament current can be varied by turning the control knob in the lower left hand corner of the diagram, labeled "filament control." This control really operates the rheostat in the filament circuit in order to vary the resistance of the circuit.

A kilovoltmeter does not actually measure kV directly, but is connected across the primary circuit and therefore measures volts. However, these readings have been calibrated beforehand against known tube kilovoltages and the meter scale has been marked accordingly in kV (see page 230). A kilovoltmeter serves also as a compensating voltmeter, because any drop in the line voltage produces a lower kV reading on the meter. The technologist then adjusts this manually by turning the "major" and "minor" kV controls which operate the autotransformer, a device that varies the voltage input to the primary of the x-ray transformer. The major control varies the kV in steps of ten, while the minor varies it in steps of one.

An important modification of this basic plan is the type of control panel wherein the kilovoltmeter is omitted and, instead, a range of kilovoltages can be selected either by *kilovoltage selector* control knobs or by *pushbuttons* operating through the autotransformer. These kV settings have been arrived at by previous calibration at the factory. In other words, when the settings read "56," it means that if the kV were actually measured, it would be 56 kV. With this system a line voltage meter and a line voltage compensator must be included to correct for variations in line voltage.

Modern radiographic equipment includes a *milliamperage selector.* Instead of a filament control knob, there is an array of pushbuttons which are used to select the mA and tube focus automatically. Thus, if the appropriate pushbutton of the mA selector is set for 100 mA and large focus, the milliammeter should register 100 and the large focus should be activated when the x-ray exposure is made.

The timer, indicated in the lower righthand corner of the diagram, serves to time the x-ray exposure which is initiated by the pushbutton x-ray switch. In addition, there is a main switch; when this is in the "on" position, the primary voltage is applied to the autotransformer, and also to the kilovoltmeter which is connected in parallel across the primary circuit. At the same time, the current also flows through the *filament only* of the x-ray tube and registers on the filament ammeter. Thus, *when the main switch is closed, only the filament ammeter and the kilovoltmeter should normally be activated.* Now, when the x-ray switch is closed, the primary current is applied to the primary side of the x-ray transformer, the voltage is stepped up by the transformer, and the current finally passes through the x-ray tube. This is indicated on the milliammeter. Thus, *the milliammeter does not register until the x-ray switch is closed.*

The construction and operation of the circuit breaker have already been discussed.

Diagnostic equipment is often supplied with a special radiographic-fluoroscopic *changeover switch,* enabling the radiologist to do spotfilm work automatically. When the cassette in the spotfilm device is shifted into the radiographic position, the control settings in the control panel switch over to the radiographic position. When the spotfilm is withdrawn from the radiographic position, the settings automatically return to fluoroscopy. Phototiming is especially desirable in timing spotfilm exposures.

You will find it advantageous to correlate the sections on the various items of x-ray equipment with the operation of the basic control panel, thereby developing a much clearer concept of the function of the various parts.

THREE-PHASE GENERATION OF X RAYS

Available for a number of years in Europe, three-phase generators are being imported into the United States in ever increasing numbers. This trend is due to the rapidly growing field of angiography with its need of large tube currents and very short exposure times.

The circuitry of three-phase equipment is necessarily more complex than that of single-phase. At present, there are three main types of three-phase circuits: 6-pulse-6-rectifier, 6-pulse-12-rectifier, and 12-pulse-12-rectifier. All use *solid state diode rectifiers.*

A three-phase generator is designed to operate on three-phase current, which consists of three single-phase currents out of step with each other by ⅓ cycle or 120°. Therefore, the transformer primary and secondary coils must each have three windings. These are arranged in either a *delta* or a *star* configuration (see Figure 14.12). The above mentioned circuits all have delta-wound primary coils, but vary in the form of the secondary wind-

Figure 14.12. Types of windings used in transformers for 3-phase equipment.

ings. In Figure 14.13 is shown a wiring diagram of a 12-pulse-12-rectifier generator having a balanced circuit and a voltage ripple of only 3.5 per cent.

With three-phase equipment, three autotransformers are needed for kV selection, one for each phase. Furthermore, because of the nearly constant potential voltage curve, the high voltage

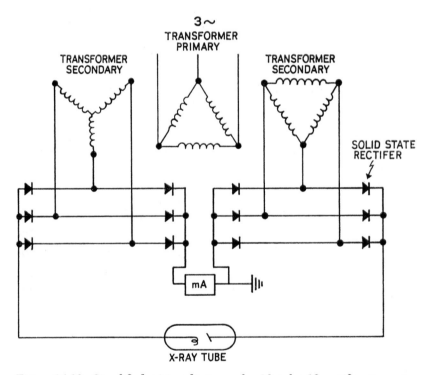

Figure 14.13. Simplified wiring diagram of a 12-pulse-12-rectifier generator in a balanced circuit. Note the delta winding of the transformer primary, and the delta and star windings of the secondaries.

circuit cannot be closed and opened at or near zero potential as in the conventional single-phase unit. This introduces the danger of power surge which can damage the equipment. Therefore, special timer circuits have been designed to make the contactors open and close sequentially instead of simultaneously. Power surge is further suppressed by the solid state rectifiers themselves.

At the present time, three-phase generators are available with ratings up to 1600 mA at 150 kV, and with exposure time as short as $\frac{1}{1000}$ sec. Furthermore, as shown in Figure 14.14, the voltage

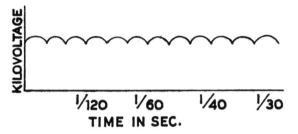

Figure 14.14. Rectified voltage curve with three-phase x-ray generator. The voltage remains nearly constant throughout the alternating current cycle, the slight variation being called *ripple*.

is the resultant of three out-of-phase single-phase voltages and never reaches zero. In fact, the voltage shows a small fluctuation or *ripple* and may be regarded as nearly **constant potential.** Thus, the effective kV with three-phase generators is about 95 per cent of the peak kV, as contrasted with about 70 per cent in single-phase full-wave rectification.

Comparison of radiographic technics with the two systems shows that with the same mAs, 85 kV three-phase is equivalent to 100 kV single-phase generated x rays. On the other hand, if one uses 100 kV with both types of equipment, the mAs for three-phase will be about 0.4 that of single-phase for equal radiographic density. However, in this case the contrast will be less with the three-phase generator because of the greater half value layer of the x-ray beam. When technics are adjusted to obtain the same density and contrast, the **radiation exposure of the patient is the same with both types of equipment.** Hence, from the standpoints of radiographic quality and patient exposure, there is no appreciable improvement with the three-phase generator.

Probably the single most important advantage of the three-phase system is the higher tube rating at *very short exposures.* Figure 14.15 shows tube rating curves comparing the maximum

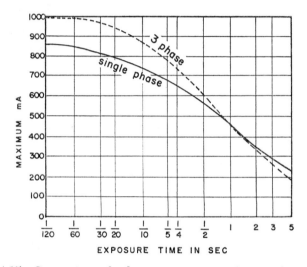

Figure 14.15. Comparison of tube rating curves with typical single phase and 3-phase x-ray generators. Note the higher rating with 3-phase for short exposures, and the slightly higher rating with single phase for long exposures.

mA that can be used with three-phase and single-phase equipment at various exposure times (Dynamax "61" tube, 1 mm focus, 100 kV). Note that at $\frac{1}{120}$ sec the maximum permissible values are 1000 mA with three-phase, and 860 mA with single-phase. On the other hand, at 5 sec the situation is reversed; now the maximum permissible values are 190 mA with three-phase, and 230 mA with single-phase. The reason for this is that there is a larger power input and resulting anode heat production in the x-ray tube during long exposures with three-phase equipment.

In summary, then, the advantages of the three-phase generation of x rays are as follows:

1. High mA at very short exposures, especially useful in angiography, spotfilm radiography, and tomography.

2. Nearly constant potential characteristics.
3. Higher effective kV.
4. Less strain on shockproof cables because of small ripple.

RECENT ADVANCES IN MOBILE X-RAY EQUIPMENT

Battery Powered Mobile X-ray Units

Ordinary mobile x-ray units, operating on 115 or 230 volts, require heavy cables to avoid excessive voltage drop, especially if they are long. Furthermore, in the event of a power blackout such equipment may be useless.

Self-contained mobile apparatus powered by storage batteries eliminates the need of an external power supply except to recharge the batteries. Depending on the particular model, these batteries may be either of the wet cell or the nickel-cadmium type.

One type of heavy-duty mobile unit* operating on a rechargeable nickel-cadmium battery pack provides a reasonably high mA for a rotating anode tube with 1 and 2 mm focal spots. If the rotor does not run for more than 10 sec, exposures totaling *10,000 mAs* are obtainable at 100 kV! Output is nearly at constant potential and all exposures are at 100 mA. X-ray exposure is selected in 24 steps of mAs ranging from 1.0 to 300 mAs. A separate timer is not necessary, since the exposure time is simply obtained by dividing the mAs by 100 mA. Thus, exposure times range from 0.01 sec at 100 mAs, to 3 sec at 300 mAs. Maximum ratings are as follows:

kV	mAs
110	200
100	250
50-90	300

In operation, the battery direct current is converted to high frequency alternating current, and the voltage is stepped up by a transformer and rectified by 4 silicon rectifiers. A meter indicates

---

* General Electric CMX-110 Cordless Mobile X-ray Unit.

battery condition at any instant. Recharging the battery requires an ordinary 105 to 125 volt, 50 to 60 cycle, 5 amp current, and stops automatically when the battery pack is completely recharged.

Capacitor (Condenser)-Discharge Mobile X-ray Units

One of the early methods of generating x rays utilized the principle of storing a quantity of electricity in a capacitor (condenser) and then discharging it through an x-ray tube. At this point review the capacitor on pages 80-81. For our present purpose, we may say that the *capacitor stores mAs.*

The voltage and charge on a capacitor are related by the following equation:

$$Q = CV \qquad (1)$$

where Q = charge in coulombs
V = potential in volts
C = capacitance in farads (constant for a
particular capacitor)

If the appropriate units are selected, equation (1) is readily adapted for x-ray equipment; thus, with a capacitance of 1 microfarad (μF),

$$mAs = \mu F \times kV$$

In other words, if the capacitor is designed to have a capacitance of $1\mu F$, then 1 kV of potential is acquired for every mAs quantity of electric charge stored in the capacitor. For example, at a potential of 75 kV, it will have a charge of 75 mAs (that is, a total of 75 mAs will be available to the x-ray tube). To charge the capacitor in 10 sec requires only 75 mA/10 sec = 7.5 mA, a small current, indeed.

The time it takes to discharge a capacitor is called the time constant, T (Greek letter, *tau*), which is determined by the equation

$$T = RC$$

where R and C are the resistance and capacitance of the circuit, respectively. By selecting the proper capacitance for a particular circuit we have a means of controlling the time for complete discharge.

In practice, we should avoid too great a drop in kV during exposure, a situation that would exist if the capacitor were allowed to discharge completely. Instead, some type of interval timer must be used to control the duration of the exposure. This is called *wave-tail cutoff*, and consists of stopping the discharge of the capacitor at some preselected point on the discharge curve (see Figure 14.16). The first portion of the diagram shows the

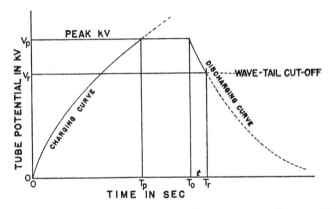

Figure 14.16. Charging and discharging characteristics of a capacitor discharge radiographic unit. V_p is peak kV; V_r is residual kV after capacitors have discharged during exposure time $(T_r - T_o)$; and T_p is time required to charge capacitors to peak kV.

charging curve as the capacitor is charged to a desired peak kV, V_t, in time T_p. At time T_o the exposure is initiated and the capacitor discharges along the discharging curve which, in this case, is cut off at time T_r. The tube current decreases along a similar curve. Wave-tail cutoff is represented by the dotted portion of the discharging curve.

If we were to start with 100 kV and a tube current of 500 mA, and used 30 mAs for an exposure, the exposure time, t, would be obtained from

$$mAt = mAs$$
$$500t = 30$$
$$t = 30/500 = 0.06 \text{ sec}$$

During this exposure time the tube potential will have dropped from 100 kV to 70 kV (a loss of 30 kV accompanying a loss of charge of 30 mAs), when the total capacitance is 1 μF.

How do we start and stop the discharge of the capacitors, thereby activating and deactivating the x-ray tube? This requires a *grid-controlled x-ray tube.* An ordinary x-ray tube is diode, having two electrodes. On the other hand, a grid-controlled tube

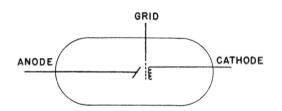

Figure 14.17. Simplified diagram of a grid-controlled x-ray tube. Exposure is started by removing the negative charge on the grid (nearly to zero); it is stopped by restoring a negative charge to the grid.

has a third electrode or *grid,* and this type of tube is therefore a *triode.* The grid consists of several crisscrossed wires bridging the cathode's focusing cup which is completely insulated from the filament. When a relatively small negative potential is placed on the grid, it prevents electrons from passing to the anode. The negative potential on the grid is called *grid bias.* A grid bias of −2 kV with the equipment under discussion will prevent the capacitors from discharging through the x-ray tube; thus, the grid bias acts as a sort of switch that interrupts the circuit through the tube. When the exposure switch is activated, grid bias is instantaneously reduced to almost zero, permitting the applied kV to drive the space charge electrons to the anode as in an ordinary diode x-ray tube. When the preselected mAs value has been reached in the exposure (determined either by an mAs control device or a phototimer) a grid bias of −2 kV is automatically applied and wave-tail cutoff achieved, thereby stopping the exposure.

Figure 14.18 shows a simple block diagram of a capacitor-discharge unit. For those who may be interested in greater detail, Figure 14.19 contains a wiring diagram of such equipment. As

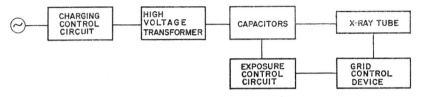

Figure 14.18. Block diagram of a capacitor discharge radiographic unit.

shown, one capacitor is charged during each half cycle of the alternating current to one-half the desired kV. The two capacitors in series discharge almost continuously through the tube during exposure, with a relatively small ripple in kV. Note that the kV of the two capacitors in series is added.

In actual radiography with capacitor discharge apparatus, after the main switch is turned on, the line voltage is adjusted and the kV selected. Next, the capacitors are charged by depressing the charger button. When the preselected kV has been acquired by the capacitors, charging automatically stops. The mAs is selected by another control, although this step is omitted with phototiming (optional). Finally, the exposure button is activated, causing the capacitors to discharge through the x-ray tube.

Note that after completion of the exposure, kV may not return immediately to zero because of a residual charge on the capacitors. Since this is a shock hazard, the high voltage cables cannot

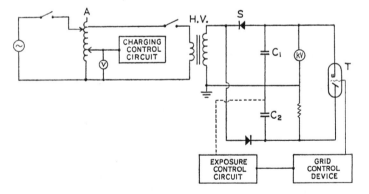

Figure 14.19. Simplified wiring diagram of a capacitor discharge radiographic unit. *A* is autotransformer. *H.V.* is high voltage transformer. *S* is solid state diode (rectifier). C_1 and C_2 are capacitors.

be disconnected while any charge remains. Another warning should be sounded here—if a charge remains on the capacitors, leakage may cause generation of x rays without activation of the exposure switch.

Field Emission X-ray Tubes*

Electronics deals mainly with the use of electrons that have been given off by metals. There is supposed to be an electric *potential barrier* or *hill* behaving as a sort of electric "fence" on metallic surfaces. Electrons can therefore be emitted only if given enough energy to penetrate the potential barrier.

We learned earlier that liberating energy can be imparted to electrons in three ways: (1) *heat*—thermionic emission as in an ordinary x-ray tube, (2) *light*—photoelectric emission as in a photocell, and (3) *bombardment with other radiation* such as electrons, other particles, x rays, and gamma rays.

In recent years physicists have discovered a fourth way of bringing about electron emission—*field emission.* This involves the application of very strong electric fields (that is, high voltage) powerful enough to make electrons leave the surface of a *cold* metal (Schotkey effect). For a particular metallic element, the probability of an electron's escaping the potential barrier depends very strongly on the applied field or voltage. For example, an increase in field strength of only 1 per cent causes a 10 per cent increase in current density (amp per sq cm of emitter surface).

In general, the emitted electrons have nearly uniform energy, for only electrons at the highest energy level in the emitter atoms (that is, valence electrons) are likely to escape the potential barrier.

At this point, be reminded that field emission occurs *without* heating the metal; in other words, here is the possibility of designing a *cold cathode* vacuum diode x-ray tube. To dispel any doubts about this, you should know that field emission of electrons has been accomplished at a temperature of 4 K (−269 C)!

One of the exciting prospects of field emission, aside from the elimination of the hot cathode, is the extremely high current

---

* Manufactured by *Field Emission Corporation.*

densities obtainable. Since there are about 10^{22} free electrons per cc of metal (swarming around in the lattice structure), a tremendous number will strike the potential barrier when a high voltage is applied. In fact, about 1 per cent escape, yielding a current density of about 1 million amp per sq cm of emitter surface steadily, and about 10 times this value in pulses. Nowhere near this current density is obtainable with pure thermionic emission in a conventional x-ray tube.

Unfortunately, there is a pronounced space charge effect that limits the ultimately achievable current density. However, in practice the emitter can function well at current densities below those causing a strong space charge effect.

Let us proceed now to the design of a field emission diode. The emitter or cathode is, in its *simplest* form, a tungsten needle measuring about 1 to 0.01 micron (1/thousandth to 1/100 thousandth of a millimeter) in diameter because this tiny dimension— many times smaller than a pin point—provides a very intense current at a reasonable applied voltage. The hemispheric needle tip consists of a single crystal of tungsten. A cone-shaped, divergent beam of electrons is produced, arising from only a small area of the tiny tip.

Our interest naturally centers around the application of the field emission principle to a functional x-ray tube. But first recall that in an ordinary x-ray tube the information carried by the x-ray beam in "bits" per sec is proportional to the number of electrons striking the target per second; or, stated differently, proportional to the current density in the cathode ray beam in, say, mA per sq. cm. Generally, we use a 1 or 2 mm focal spot and a tube current ranging up to a few hundred mA.

On the other hand, in a field emission x-ray tube the cathode or emitter consists of several rows of electron-emitting needles resembling combs, surrounding a *conical anode* (see Figure 14.20). While the anode presents a large surface to the emitted electrons to improve heat dissipation, it gives off x rays from a very small area. It need not be explained further that x rays are produced when the fast electrons strike the anode, in the same way as in an ordinary x-ray tube.

At present, field emission is being used in some types of por-

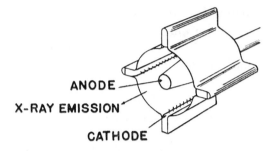

Figure 14.20. Diagram of a field emission x-ray tube (from data furnished by courtesy of the Field Emission Corporation).

table x-ray apparatus, especially for the *operating room* and *nursery*. In addition, the principle is being applied to industrial and biologic research, such as in high speed radiography and photography, and in radiobiology. It is anticipated that field emission units, because of their compactness, extremely large radiation output, and use of ordinary house current will assume an ever-increasing role in radiology as newer designs become available.

Some Advantages of Newer Mobile Type Apparatus

Unlike ordinary mobile x-ray units, battery-powered, capacitor-discharge, and field emission equipment is not subject to troublesome line voltage fluctuations. Another advantage is their nearly constant potential output (with low mAs in the capacitor-discharge type). Radiographs are equivalent in quality to those produced at high kV; for example, 85 kV at constant potential is equivalent to about 100 kV with conventional full wave rectification.

The decision as to which type of mobile equipment to purchase depends on several factors, such as use, cost, and availability of service, just as in the case of other equipment.

QUESTIONS

1. Describe with the aid of a diagram the overload circuit breaker.
2. What is the function of a fuse and how does it work?
3. What is the principle of the prereading kilovoltmeter?

4. Why is a remote control switch necessary? How is it constructed?
5. What are the four main types of exposure timers? Describe phototiming.
6. What is the advantage of a circuit breaker over a fuse?
7. Show by diagram the filament circuit and its important components.
8. Why can a milliammeter be safely mounted on the control panel? What is a ballistic mAs meter and under what conditions is it an essential part of the control apparatus?
9. Explain why the center of the transformer secondary coil is grounded.
10. Explain how the kilovoltage control knob varies the kilovoltage. Which device does it operate in the primary x-ray circuit?
11. Which device in the filament circuit is operated when the filament control knob is manipulated? What function does it serve?
12. Describe a shockproof cable with the aid of a diagram.
13. Prepare a simple wiring diagram of an x-ray machine, including four-valve-tube rectification.
14. Of what does a three-phase current consist?
15. How does the design of a three-phase transformer differ from that of a single-phase? What type of rectifiers is used with three-phase equipment?
16. What is meant by ripple? What is its importance in radiography?
17. What are the advantages and disadvantages of the three-phase over the single-phase generator?
18. Name three types of mobile apparatus other than the conventional kind.
19. Discuss briefly the principle of the capacitor discharge unit. How is exposure time controlled?
20. What is meant by field emission? State some of its advantages over other methods of generating x rays.

CHAPTER *15* X-RAY FILM, FILM HOLDERS, AND INTENSIFYING SCREENS

In the early days of roentgenography, glass photographic plates coated with an emulsion sensitive to light were used to record the x-ray image. The disadvantages of plates included the danger of breakage, the hazard of cutting one's hands, the difficulty in processing, and the inconvenience in filing these plates for future reference. With the introduction of modern films, these disadvantages were eliminated.

Composition of X-ray Film

There are two essential components: the *base* and the *emulsion*.

1. **The Base.** Modern safety film has a base about 0.008 in. thick, consisting of a transparent sheet of *polyester plastic* which is usually tinted blue. It is called "safety" because it is non-explosive, being no more inflammable than the same thickness of paper.

2. **The Emulsion.** This consists of microscopic crystals of *silver bromide* suspended in *gelatin.* The gelatin, similar to that used in cooking, is obtained from cattle skins and is treated with mustard oil to improve the sensitivity of the emulsion by providing *traces of sulfur.* To the gelatin dissolved in hot water are added, in total darkness, silver nitrate and potassium bromide to produce *silver bromide.* Heating this mixture to a temperature of 50 to 80 C—a process called *digestion*—further improves the sensitivity of the emulsion. After cooling, it is shredded, washed, reheated, and mixed with additional gelatin. Finally, the emulsion is spread on both sides of the polyester base in a layer about 0.001

inch in thickness and allowed to cool under controlled conditions. Film coated on both sides is said to be *duplitized*. Single-coated film is also available for special purposes. It must be emphasized that from the moment the ingredients are brought together in the emulsion until the finished sheet film is packed in boxes, the entire process must be carried out in *total darkness*. The composition of the emulsion has been designed to be sensitive both to blue-fluorescent light and to x rays. Furthermore, it must be capable of rendering the radiographic image of a variety of tissues with optimum density, contrast, and sharpness, qualities that will be discussed later.

To be suitable in radiography, an emulsion should have two important characteristics:

1. *Speed* or *Sensitivity*—the relative ability of an emulsion to respond to radiation such as light and x rays. An emulsion is said to be fast or have high speed if a small exposure produces a radiograph of adequate density. Ordinary x-ray film is rated as being *fast*.

2. *Latitude*—the ability of an emulsion to record a radiographic image with a reasonably long range of tones, from white, through various shades of gray, to black; that is, *long scale contrast,* which will be discussed in detail later. Another aspect of latitude is the margin of error permissible with any given technic. Obviously, an emulsion should have sufficient latitude to allow a reasonable degree of error in exposure without serious impairment of radiographic quality. Both aspects of latitude—*long tonal range* and *permissible margin of error in exposure*—are closely related. In general, excessive latitude should be avoided because it makes it more difficult to see small detail.

Types of Films

There are three main types of x-ray films in use at the present time in medical radiography.

1. **Screen Film** is the type most often used in radiography. Although sensitive *mainly* to the blue light emitted by intensifying screens, this film also responds to the direct action of x rays. Screen film is used chiefly in cassettes with intensifying screens,

a combination that has made possible the high speed and excellent quality of modern radiography. Film manufacturers produce screen films having a variety of speeds and contrasts to suit the individual radiologist's preference, both for general radiography and for special examinations.

2. **Nonscreen** or **Direct-exposure Film** has a *thicker emulsion* than screen film and is not so readily penetrated by light. Therefore, non-screen film depends *mainly* on the *direct action of x-rays* during exposure and should *not* be used with intensifying screens. Non-screen film is about *four times as fast* as screen film in cardboard holders, and therefore requires only about *one-fourth the exposure* of screen film for equal blackening. If reduced speed is acceptable, screen type film may be used with cardboard holders in the radiography of small parts, but non-screen film exhibits greater contrast and latitude as well as greater speed. On the other hand, the usual combination of screen film with intensifying screens has greater speed and contrast (but less latitude) than non-screen film in cardboard holders.

3. **Mammography Film** is a modified single-emulsion industrial film. It is characterized by slow speed, high contrast, and very fine grain to provide the utmost detail in radiography of the breast. Such film is now available for automatic processing.

Practical Suggestions in Handling Unexposed Film

1. Films deteriorate with age; therefore, the expiration date stamped on the box should be observed, the older films being used first.

2. Since moisture and heat hasten deterioration, films should be stored in a cool, dry place.

3. Films are sensitive to light and must be protected from it until processing has been completed.

4. Films are sensitive to x rays and other ionizing radiations, and should be protected by distance and by shielding with protective materials such as lead.

5. Films are marred by finger prints, scratches, and dirty intensifying screens; and by crink marks due to sharp bending.

6. Rough handling causes black marks due to static electricity. These appear as jagged lines, black spots, or tree-like images after development.

FILM EXPOSURE HOLDERS AND INTENSIFYING SCREENS

Each x-ray film must be carried to the radiographic room in a suitable container which not only protects it from outside light but also allows it to be exposed in radiography. Film holders are available in standard film sizes.

There are two types of film holders used in general radiography: *cardboard holders and cassettes.*

1. **Cardboard Film Holder**—an enclosed light-proof envelope into which the film is loaded **in the darkroom.** The film *with its paper wrapping* (if it is provided with such a wrapper) is placed in the folder and the long flap of the envelope is folded over it, followed by the two shorter side and end flaps. After being closed and secured by the hinged clip, the holder is ready to be taken to the radiographic room for exposure. During exposure, the front of the film holder must face the tube, since the back cardboard is lined with lead foil to prevent fogging of the film by x rays scattered back from the table. Figure 15.1 shows a cardboard film holder in cross section.

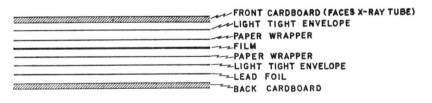

FRONT CARDBOARD (FACES X-RAY TUBE)
LIGHT TIGHT ENVELOPE
PAPER WRAPPER
FILM
PAPER WRAPPER
LIGHT TIGHT ENVELOPE
LEAD FOIL
BACK CARDBOARD

Figure 15.1. Cross section of a cardboard film holder.

2. **Cassette**—a case measuring about one-half inch in thickness and having an aluminum, stainless steel, or plastic frame, a hinged lid with one or more flat springs, and a bakelite or light metal front. One of a pair of intensifying screens is mounted on the inside of the cassette front, and the second screen is mounted on the inside of the lid. The front of the cassette faces the x-ray tube during exposure. We load the cassette by raising the hinged lid **in the darkroom,** slipping a film, with its **wrapping paper removed,** gently into the cassette of the same size, and closing the lid by means of the flat springs. The film is thus sandwiched

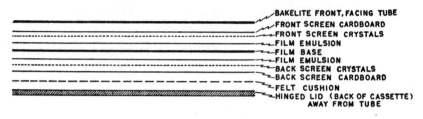

Figure 15.2. Cross section of a cassette with intensifying screens and film.

between two screens and in close contact with them. Figure 15.2 shows a cassette in cross section.

Intensifying Screens

We have already mentioned the use of intensifying screens and shall now describe their structure and function.

1. **Composition.** An intensifying screen consists of microscopic crystals of a *phosphor* (fluorescent material), incorporated in a binding material, and coated on one side of the white reflecting surface of a sheet of high grade cardboard or plastic. This coating is called the *active layer.* It is, in turn, coated with a thin, smooth, abrasion-resistant material, its edges being sealed against moisture. The most widely used phosphor is *calcium tungstate* because it meets the following requirements:

a. Emits blue and violet light to which x-ray film is particularly sensitive.

b. Responds well in the kV range ordinarily used in radiography.

c. Does not deteriorate appreciably with use or with age.

d. Does not have significant afterglow or lag (see page 268).

e. Can be used to manufacture screens of uniform quality.

2. **Principle.** Phosphors have the peculiar property of *fluorescing* or giving off visible light when struck by x rays. This process involves the raising of valence electrons of the phosphor atoms to the conduction band (see page 218) of the crystal lattice by the absorbed x-ray photons. Return of these electrons to the valence band in less than 10^{-10} sec is accompanied by the

emission of characteristic radiation in the form of visible light. Thus, in fluorescence we have the absorption of energy in one form, x rays, and its conversion to another form, visible light, which is emitted in all directions by the tiny phosphor crystals. The total amount of light emitted in a particular region of the active layer depends on the quantity of x rays striking that region, and is a summation of the light given off by the innumerable, closely packed crystals.

In the object being radiographed, those parts which are readily penetrated by x rays will appear on the *screen* as brighter areas than those parts which are poorly penetrated. Thus, the screen image is made up of zones of various degrees of brightness. These, in turn, ultimately produce corresponding differences in darkening of the x-ray film. Since the film emulsion is particularly sensitive to the *blue light* emitted by the screens, this *photographic effect* on the film is of major importance. In fact, *about 98 per cent of the recorded density (blackening) on the film exposed with intensifying screens is photographic in origin,* that is, due to light emitted by the screens. Only about 2 per cent results from direct x-ray exposure. Thus, screens intensify about 10 to 60 times the effect of x rays on the film emulsion, thereby reducing the exposure needed to obtain a particular degree of film blackening. Intensifying screens have made possible the use of grid radiography of thick anatomic parts such as the abdomen, skull, and spine. Furthermore, screens have contributed to the development of high speed radiography with significant reduction in the radiation exposure of the patient.

To summarize this discussion, we may say that the *exit* or *remnant radiation* (radiation that has passed through the patient) passes through the front of the cassette and impinges on the front intensifying screen. This emits light which varies in brightness depending on the amount of radiation reaching any particular area of the screen. In fact, there is a light and dark pattern on the screen corresponding to the transmission of x rays through the various body structures in the path of the beam. The light emitted by the screen affects the film emulsion with which it is contact, reproducing the light and dark pattern, although this is reversed on the finished radiograph. X rays also pass on directly to the film causing a variable, though small, degree of darkening. Some x rays reach the back screen which fluoresces and affects the film emulsion nearest it in the same manner as the front screen. Thus, a film is coated on both sides with a sensitive emulsion so that one side receives light from the front screen, and the other from the back screen, as shown in Figure 15.3.

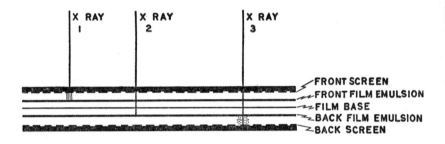

Figure 15.3. Schematic representation of the principle of the intensifying screen. X-ray photon *1* strikes the front screen causing it to fluoresce at that point. Another x-ray photon such as *2* may pass directly to the film and affect it at that point. Or a photon such as x ray *3* may pass through the front screen and the film, striking a crystal in the back screen and exciting fluorescence in it. Actually the screens and film are in close contact.

3. **Screen Speed.** A screen is said to be fast or to have high speed when a relatively small x-ray exposure produces a given output of light and causes a certain degree of blackening of a film. Conversely, a screen is said to be slow when a relatively large exposure is required for a given amount of film blackening. *The intensifying factor of a pair of intensifying screens may be defined as the ratio of the exposure required without screens to the exposure required with screens to get the same degree of blackening of x-ray films.* Another name for intensifying factor is *speed factor.*

$$\text{intensifying factor} = \frac{\text{exposure without screens}}{\text{exposure with screens}}$$

Since the denominator is always less than the numerator, the intensifying factor of a pair of screens always exceeds unity, which means simply that the exposure with screens is less than that without screens for the same amount of film blackening. However, the intensifying factor varies for a given screen because of its dependence on kV and temperature, as described below. The intensifying factors of various types of screens range from about 10 to 60. In other words, the exposure with screens varies from about $\frac{1}{10}$ to $\frac{1}{60}$ that without screens for equal film blackening. Screens are available in four speeds, in decreasing order: high speed, par (medium) speed, detail, and ultra detail.

The *speed* of intensifying screens depends on two main classes of factors—intrinsic and extrinsic.

a. **Intrinsic Factors** are those inherent in the composition of the screens and include the phosphor, the thickness of the active layer, the size of the phosphor crystals, and the reflectance of the backing.

(1) *Phosphor.* Calcium tungstate is the main phosphor now being used in screens. Special activators are added to increase the speed of this phosphor.

(2) *Thickness of Active Layer.* For a given phosphor, the greater the thickness of the active layer the greater will be the speed, up to a limit.

(3) *Size of Phosphor Crystals.* The larger the crystals, the greater will be the speed. However, in practice crystal size is essentially the same for all screens.

(4) *Reflectance of Backing.* The greater the amount of light reflected back to the active layer by the cardboard or plastic backing, the greater will be the speed. However, this may be undesirable with thick active layers because of the detrimental effect on image sharpness.

Thus, a faster screen would probably be composed of a more luminescent phosphor such as especially activated calcium tungstate in a thicker layer. But note that *as the thickness of the active layer is increased the radiographic image becomes more blurred (that is, less sharp)* because the average distance the emitted light has to travel to the film emulsion is increased and there is a greater spread of the light rays (see Figure 15.4).

Figure 15.4. Effect of active layer thickness of an intensifying screen on image definition. Note that the light originating at points deep within the thinner active layer (right figure) spreads less before reaching the film, thereby producing a sharper image.

Consequently, there is a practicable limit to the thickness of the active layer.

Theoretically, we would expect the size of the crystals to influence image sharpness because a larger crystal would produce a larger spot of light than a smaller crystal, but this is not important in actual practice. The crystals are extremely minute, averaging about 5 micra (0.005 mm) with a range of 4 micra (slow or "detail" screens) to 8 micra (fast screens). This range of crystal size has no significant effect on image sharpness.

In any case, regardless of the type of screen, *image sharpness is superior with direct x-ray exposure of films in cardboard holders,* provided the patient can be suitably immobilized.

b. **Extrinsic Factors** include the conditions under which the screens are used, namely temperature and kV.

(1) *Temperature.* As room temperature rises, screen speed decreases and film speed increases. These two tendencies nearly cancel each other, so that no temperature correction is necessary for screen speed under normal conditions. However, at the high temperatures prevailing in the tropics, basic mAs values have to be increased about 20 per cent at 110 F. On the other hand, in extreme cold such as may be encountered in industrial radiography, a decrease of about 25 per cent in the basic mAs is required at 15 F.

(2) *Kilovoltage.* An increase in kV increases screen speed. Thus, at 40 kV the intensifying factor with medium speed screens is about 30, whereas at 80 kV it increases to about 75.

4. **Screen Contact.** The film must be sandwiched uniformly between the two screens, with perfect contact throughout. Even the smallest conceivable space between the film and screen at any point will permit the light rays emerging from the screen at that point to spread over a wider area, producing a blurred image of that particular point (see Figure 15.5). You can easily

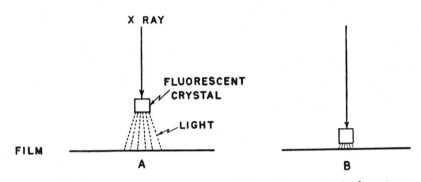

Figure 15.5. Poor screen contact causes a blurred image. In *A*, there is an appreciable space between the fluorescent crystal and the film (when screen contact is poor), so that a tiny but spreading bundle of light rays strikes the film, producing a blurred image instead of a fine point. In *B*, the crystal is in close contact with the film so that the image is about the same size as the crystal. (The crystal is magnified many times in the diagram.)

ascertain the uniformity of contact of the screens with the film by placing a piece of hardware cloth with ⅛ in. mesh over the front of the cassette and making a film exposure—40 kV, 10 mAs, and 40 in. focus-film distance, with a 3 mm Al filter. With satisfactory screen contact, the image of the wire mesh is sharp over the entire radiograph; but in areas of poor contact the image appears blurred and patchy (see Figure 15.6).

Screen contact is usually, though not universally, good in smaller cassettes. However, larger cassettes, especially 11 x 14 and 14 x 17, are often subject to poor contact even when new. Therefore, a layer of contact felt, fiberglass, or foam rubber should be cemented (preferably by an expert) as a cushion between the screen and the cassette. When the screen is closed, the padding tends to equalize the pressure applied to the screens, squeezing them tightly against the film.

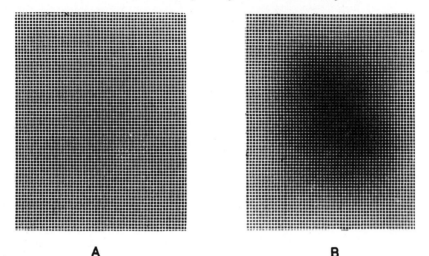

A   B

Figure 15.6. Tests for screen contact. In *A*, with excellent screen contact, the image of the wire screen is uniform. In *B*, with poor contact at the center, the image of the wire is blurred, producing a typical blotchy area in the radiograph.

A new type of cassette, invented by Amplatz, is used mainly in angiography to obtain the ultimate in film-screen contact. This is achieved by producing a vacuum inside the cassette after it has been loaded with film. Known as a *vacuum cassette*, it is virtually indispensable in angiography with radiographic magnification.

An improved cassette (Eastman Kodak) is made of high-impact plastic. It is light in weight and can be dropped without damage. The front is deliberately warped inward to assure excellent film screen contact.

It is most important that all new cassettes and, periodically, old ones as well, be checked for contact as described above. Whenever a radiograph appears sharper in one area than in another (for example, edges sharper than the center), poor screen contact should be suspected and the cassette should be tested without delay.

The *causes of poor screen contact* include (1) warped cassette front, (2) cracked or twisted cassette frame, (3) loose hinges or spring latches, and (4) local elevation of a screen by a foreign body underneath. Rough handling and dropping of cassettes are mainly responsible for damage leading to poor screen contact. Another important cause is the placement of a cassette without a tunnel under a patient in bed often resulting in a warped cassette.

5. **Image Sharpness with Screens.** In previous sections we described the relationship of intensifying screens to radiographic image sharpness. These data may be summarized to good advantage under a single heading. In general, a radiographic image is never so sharp with screens as with direct radiographic exposure of x-ray film in a cardboard holder, provided the patient can be adequately immobilized. In actual practice, image sharpness with screens becomes worse as the thickness of the active layer increases. As noted before, a thicker active layer causes greater diffusion of light and blurring of the image. But remember that as the active layer thickness is decreased to improve image sharpness (as in detail screens) the screen becomes slower. Another fact to bear in mind is that regardless of the type of screen, image sharpness suffers when film-screen contact is poor.

6. **Care of Screens.** Intensifying screens should be mounted with rubber cement or special adhesive tape. Water-soluble paste should never be used. The screens ordinarily come in pairs, one screen being thicker than the other. The thin screen is mounted inside the front of the cassette, whereas the thicker screen is mounted inside the lid.

Screens should be kept scrupulously clean, since dust and other foreign materials absorb light from the screen and cast a white shadow on the film. Screens may be washed with cotton and bland soap, rinsed with cotton moistened with water, and then dried with cotton. Excessive amounts of water should be avoided. The screens are further dried by standing the half-opened cassette on its side for about one-half hour in a dust-free room. Special screen cleaners are also available, but should be used only as recommended by the manufacturer.

Care must be taken not to nick, scratch, or chip the screens. This can occur accidentally by carelessly digging the film out of the cassette with the fingernail, or by scratching the screen with a corner of the film. In loading the cassette, one should carefully slip the film into the cassette with the lid elevated about 2 inches. The lower leaf of the folded paper covering the film (if present) is allowed to hang down over the edge of the cassette, and after the film is in place, the paper is slipped away. This manipulation should be done gently to avoid static marks. To remove the film after exposure, one must carefully raise a corner of the film, being certain that the fingernails do not slip across the surface of the screen.

The cassette must always be kept closed, except when it is being loaded or unloaded, to prevent accidental damage to the screen surface and help keep out dust. Scratch marks and dust block the light from the screen, leaving a *white spot* on the finished radiograph.

Fluoroscopic Screens

A fluoroscopic screen is similar in principle to an intensifying screen, but certain practical differences exist. *Zinc cadmium sulfide* is still the best phosphor for fluoroscopic screens, emitting green light to which the eyes are most sensitive at the low intensity of the fluorescent light. Crystal size is 30 to 40 microns (0.03 to 0.04 mm). The screen is backed by lead glass to protect the fluoroscopist.

Fluoroscopic screens are characterized by two kinds of light emission:

1. *Fluorescence*—the emission of light while the screens are being exposed to x rays.

2. *Phosphorescence*—the persistent emission of light after the x rays have been turned off. It is relatively strong in fluoroscopic screens, but weak in intensifying screens, and is called *afterglow* or *lag.*

QUESTIONS

1. Show by cross-section diagram the structure of x-ray film.
2. What is meant by safety film? Nonscreen film?
3. What are two differences between screen and nonscreen film? What is mammography film?
4. Describe the structure of an intensifying screen. How does it differ from a fluoroscopic screen?

5. Define the intensifying factor of an intensifying screen and discuss the effects of the important intrinsic and extrinsic factors.

6. What are three differences in the composition of a slow screen and a fast screen? How do they differ in their effect on radiographic sharpness?

7. How does an intensifying screen intensify the x-ray image? Approximately what fraction of the film density results from light emitted by the screens?

8. Why is an x-ray film duplitized (coated on both sides with sensitive emulsion)?

9. Why must the paper wrapper be removed before a film is placed in a cassette? Why may it be left on with a cardboard holder?

10. What is screen lag? Phosphorescence? Fluorescence?

11. How does active layer thickness affect image sharpness? Explain. What other factors cause impairment of image sharpness by screens? Explain.

12. Why should we avoid scratching or otherwise marring a screen surface?

13. What is the effect of poor screen contact on radiographic sharpness? Why?

14. Describe a test for film-screen contact.

CHAPTER 16 THE DARKROOM

THE IMPORTANCE of the darkroom in radiography cannot be exaggerated. Radiography unquestionably begins and ends in the darkroom, where the films are loaded into suitable light-proof holders in preparation for exposure, and where they are returned for processing into a finished radiograph. In general, a darkroom is a place where the necessary handling and processing of film can be carried out safely and efficiently, without the hazard of producing *film fog* by accidental exposure to light or x rays to which the film is so sensitive.

The term "darkroom" is not entirely accurate, since complete blackout is unnecessary. In fact, as will be shown later, considerable safe illumination can be provided to facilitate darkroom procedures. Although the term "processing room" is more precise, it is not widely used. We shall therefore continue to refer to it as the "darkroom," bearing in mind that it is dark only insofar as *it must exclude all outside light.*

This chapter will deal only with the darkroom designed for the *manual processing of films.* Much of the material applies also to darkrooms for automatic processing, which will be discussed on pages 290-296.

Location of the Darkroom

Since location is of fundamental importance, it should be determined during the planning stage of the Radiology Department. The darkroom should be centrally located with relation to the radiographic rooms in order to save time and eliminate unnecessary steps. When the darkroom is adjacent to the radiographic rooms, the intervening walls should be shielded with the correct thickness of lead all the way to the ceiling; $\frac{1}{16}$ in. lead is usually adequate. On the other hand, the darkroom should be located as remotely as possible from stored radium and other radioactive materials because even as little as 5 mR total exposure from radium causes detectable fog; in any case, the shielding of the walls would be no greater than that required for protection of personnel, unless films were stored for unusually long periods of time.

Outside windows should be avoided because they are extremely difficult to render lightproof. Furthermore, they serve no useful purpose, since much more satisfactory methods of ventilation are available.

The darkroom should be readily accessible to plumbing and electrical service.

Building Essentials

Walls between the darkroom and adjoining x-ray rooms should contain enough lead thickness, or its equivalent in other building materials, for adequate protection of the films. This is especially important because efficiency demands that the darkroom be close to the radiographic rooms (see above).

In a busy department, the efficiency of work flow can be improved by having *passboxes* built into the walls at appropriate locations. These are light-tight and x-ray proof, being provided with two interlocking doors so arranged that both cannot be opened at the same time. The cassettes, after radiographic exposure, are placed in the passbox through the outside door, and then removed through the inside door by the darkroom technician. The most suitable location for the passbox is obviously near the film-loading bench although this is not always possible.

Darkroom walls should be covered with chemical-resistant materials, particularly near the processing tanks. Such materials include special paint, varnish, or lacquer. Ceramic tile or plastic wall covering provides a more durable finish.

Floor covering should consist of chemical-resistant and stain-proof material such as asphalt tile. Porcelain or clay tile may be used, but a non-skid abrasive should be incorporated to minimize the danger of slipping. Ordinary linoleum and concrete are unsuitable because they are readily attacked by the processing solutions.

Entrance

The entrance should be conveniently located with relation to the darkroom equipment. The simplest type of entrance is a single door which must be made absolutely light-tight by weatherstripping. Such an entrance is now generally used in offices and hospitals, but it should be provided with an inside lock to prevent opening while films are being processed.

Another type of protective entrance is a small hall with two electrically interlocked doors, so designed that one door cannot be opened until the other is completely closed, thereby preventing entrance of light (see Figure 16.1). A separate door should

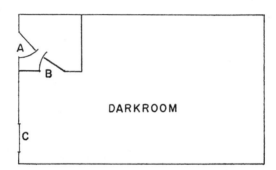

Figure 16.1. Darkroom floor plan with interlocked doors. If door *A* is opened, an electrical relay prevents anyone from opening door *B*, and vice versa. *C* is an emergency door, which is usually kept closed.

be provided for emergency use and for moving equipment into and out of the darkroom.

A more elaborate type of entrance is the *maze,* shown in Figure 16.2. Note that a complete turn must be executed in going through the three doorways. Serving as a light trap, the maze requires no doors, especially if the walls are painted black. Because of the high cost of the required floor area, mazes are rarely used today.

Size

Darkroom size will vary, of course, with the size of the department, but it must be large enough to house conveniently all the

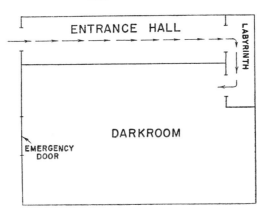

Figure 16.2. Darkroom with maze or labyrinth entrance.

necessary benches for loading and unloading films, a film storage bin, cupboards, processing tanks, refrigerator, and dryer. On the other hand, the darkroom should not be too large, since the excessive distances between various units of equipment result in wasted time and steps. The loading bench should be located across the room from the tanks in order to avoid accidental splashing of water and solutions on the films, benches, cassettes, or other equipment that may come in contact with the films. However, this is not a factor in the darkroom with an automatic processor.

Ventilation

Essential to efficient work and to the technologist's health are the adequate removal of stale, humid air, and the supply of fresh air. Air conditioning is definitely the preferred methd of ventilating the darkroom. However, in a small office or department in areas where climate permits, an exhaust fan may provide adequate air circulation. The system should be absolutely light proof and should include a means of filtering out dust at the point of entry of the fresh air.

Lighting

A properly designed darkroom should have three types of illumination: safelight, general, and radiographic.

1. **Safelight.** We must have a source of light which will not fog films and still provide adequate illumination under processing conditions. For this purpose, there are available *safelight lamps* with filters of the proper color, such as the Wratten Series 6B filter for ordinary x-ray film. It is necessary to maintain a working distance of at least three feet from the safelight and use a bulb with the wattage indicated on the lamp housing. If a brighter bulb is used, light transmitted even through the correct filter may cause film fog. Safelights are designed for either indirect ceiling illumination or for direct lighting.

The safety of darkroom lighting may be tested as follows: subject a film in a cassette to a very small x-ray exposure, just enough to cause slight graying; screen film is more sensitive to fogging by light after initial exposure to the fluorescence of intensifying screens. Then remove the film from the cassette in the darkroom, cover one-half the film with black paper, and leave it exposed under conditions simulating as closely as possible those normally existing when a film is being loaded and unloaded. Process the film as usual. If the uncovered portion appears darker than the covered, you may conclude that the darkroom lighting is unsafe, and must then make every effort to eliminate the source of light responsible for the fogging. This may be due to a cracked or an incorrect filter, a light leak in the safelight housing, or a leak around the entrance door to the darkroom.

Note that the darkroom walls do not have to be painted black. Any color which enhances the safelight illumination is desirable, since this improves illumination without danger of film fogging.

2. **General Illumination.** A source of overhead lighting is needed for general purposes such as cleaning, changing solutions, and carrying out other procedures that do not require safelight illumination.

3. **Radiographic Illumination.** A fluorescent illuminator for viewing wet radiographs should be mounted over the washing compartment to avoid contamination of processing solutions. In more elaborate darkrooms for manual processing the wash tank is installed partly through the wall into the light area to permit viewing the wet radiographs outside the darkroom, thereby minimizing traffic in the darkroom.

Apparatus and Equipment

Processing chemicals are bought ready-mixed as dry powders or as liquid solutions and made up to the proper volume by the addition of water according to the instructions printed on the label. Two stirring paddles are needed, one for the developer and one for the fixer. A thermometer is used to determine the temperature of the developing solution and to check the temperature of water used to prepare the solutions. All apparatus coming in contact with the processing solutions must be made of materials such as plastic, hard rubber, enamelware, or stainless steel of the proper composition, to avoid corrosion and resulting contamination of the solutions.

Comprising the *major equipment* are the *processing tanks*. The simplest type consists of a three-compartment tank, one end-compartment being used for developing and the opposite one for fixing. The middle compartment serves both to rinse and wash the films and should be supplied with running water. Thermostatic temperature control is necessary for optimal results.

Stainless steel of the proper alloy composition is the best material for processing tanks because not only does it resist corrosion but it also permits rapid equalization of the temperature of the processing solutions with that of the water in the middle compartment (stainless steel is an excellent conductor of heat in contrast to rubber or stone). Insulation of the outside walls of the tank prevents condensation of moisture and helps maintain the proper temperature within the compartments.

A more satisfactory arrangement is that shown in Figure 16.3,

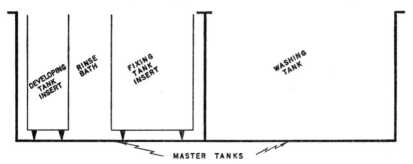

Figure 16.3. A popular type of processing tank arrangement. The developing and fixing tanks are separate units placed in the master tank on the left. Note the relative sizes of the tanks.

consisting of a large, insulated, stainless steel, double-compartment master tank. Two stainless steel insert tanks are placed in one of the compartments, the first insert being the developing tank and the other, the fixing tank. Water between the inserts in this compartment serves both to rinse the films and to control the temperature of the solutions. A fixing tank should have about twice the volume of a developing tank since the time required for fixing films is approximately twice that for development. The other main compartment serves as the washing tank and should be about twice the size of the fixing tank, since washing requires about twice as long as fixation.

Inserts should be measured when first installed to determine their correct volume. If the level of the solution is to be one inch from the top, then

$$\text{volume in gallons} = \frac{(\text{height}-1)\underset{\text{inches}}{} \times \underset{\text{inches}}{\text{length}} \times \underset{\text{inches}}{\text{breadth}}}{231}$$

The wash tank should be deep enough for the water to pass completely over the top of the hangers for thorough washing.

A conveniently located water inlet should be provided for washing the tanks and for making up the solutions. Thermostatically controlled warm water and refrigerated water are essential and should be circulated around the processing tank inserts to maintain optimal temperature.

Various types of *drying devices* are now available to speed up the drying of films. Most of these consist essentially of a rack to hold the films and a fan to circulate air around them. A source of heat is usually included to hasten drying. Such driers are commercially available as enamel or stainless steel cabinets. The latter are by far the most desirable since enameled metal must be repainted at frequent intervals due to chipping and corrosion. The drier should be vented to the outside to minimize heat and humidity in the darkroom.

A *lightproof storage bin* for unexposed film should be placed under the loading bench. Vertical partitions subdivide the bin to accommodate film boxes of different sizes. Counterweighting the drawer of the bin makes it close automatically when released. A warning light must be posted on the front of the bin to prevent its being opened in the presence of white light. *Film hangers* of proper size and in sufficient number should be available to hold the films during processing, but these are unnecessary with automatic processing.

Finally, a mechanical *corner cutter* facilitates the trimming of the film corners. Films with rough corners are more difficult to handle because they cannot be readily slid over each other. Besides, the rough corners increase film thickness, wasting filling space. With automatic processing, film corners do not have to be trimmed.

QUESTIONS AND PROBLEMS

1. What is the best location for the darkroom? How should the walls be shielded?
2. Discuss the various kinds of darkroom entrances, including the advantages and disadvantages of each.
3. What are interlocking doors? Passboxes?
4. What materials are preferred in the construction of processing tanks? What is the best material for tanks and dryers?
5. Describe a darkroom safelight lamp. What precautions are necessary to assure that the safelight is really safe?
6. How do you test a darkroom for safe illumination?
7. What colors should be used in painting the walls of the darkroom? Of the maze?
8. Make a diagram of a convenient arrangement of the processing tanks, and indicate their comparative sizes.
9. A developing tank measures 8 in. wide x 16 in. long x 19 in. high. How many gallons of developer does it hold when filled to a level one inch below the top?
10. How can you prevent x rays from entering the darkroom, and why?
11. Describe briefly a film drier.
12. Why is running water essential in the darkroom?

CHAPTER *17*   CHEMISTRY OF RADIOGRAPHY AND FILM PROCESSING

Introduction

THE BASIC PRINCIPLES of radiographic photography have been well established for many years, and are applicable both to manual and to automatic processing. Nevertheless, as will be described later, certain changes have had to be introduced to adapt conventional film processing to automation, these changes being perhaps more *physical* than chemical. Therefore, a thorough understanding of radiographic image formation and photographic chemistry should precede the study of both manual and automatic processing.

Radiographic Photography

Fundamental to an understanding of the production of an image on an x-ray film, is the concept of the *latent image*. The film emulsion contains silver bromide in the form of minute crystals which are invisible to the naked eye. When struck by radiant energy, such as light or x rays, these crystals undergo an electrochemical change which increases their susceptibility to the action of certain chemicals. Such exposed silver bromide crystals make up the *latent image, defined as that invisible image, produced in the film emulsion by light or x rays, which is converted to a visible or manifest image upon development.*

Let us examine in greater detail what happens when light or x rays strike a silver bromide crystal. Although the process is incompletely understood, there is sufficient evidence for a reason-

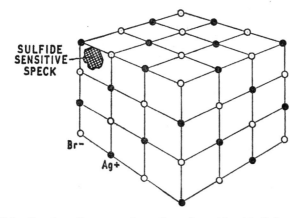

Figure 17.1. Lattice diagram of a silver bromide (AgBr) crystal. The straight lines joining the Ag+ and Br− ions represent the electrovalence forces holding the ions together in the crystal. The sensitive speck (sulfide) renders the crystal highly sensitive to light and x rays.

able explanation. Figure 17.1 contains a diagram of the *lattice* structure of a silver bromide crystal. In it, the positive silver ions alternate with the negative bromide ions, all being held together by ionic bonds (see page 49). Incorporated in many of the crystals during the ripening phase of emulsion manufacture are *silver sulfide* particles serving as *sensitive specks* or *development centers* on the surface of the crystals. Only those crystals having sensitive specks can be affected by exposure to light or x rays. A photon entering such a sensitized crystal may interact with a bromine ion liberating a loosely bound (valence) electron. This valence electron then drifts toward a sensitive speck to which it imparts a negative charge. A positive silver ion is now attracted to the negatively charged speck, picks up the electron, and is neutralized to a silver atom. Another electron soon attaches itself to the speck, and another silver ion drifts to it and is neutralized. This process, initiated by the photons entering the crystal, may be repeated many times in an extremely small intreval of time. The number of susceptible crystals so affected depends on the number of photons falling on a given area of the film. Crystals in which the sensitive specks have acquired silver atoms during exposure are invisible and constitute the *latent image.*

Radiographic Chemistry

When the exposed crystal (latent image) is subjected to the action of the reducing agents in the developer, the process initiated by photon action is greatly speeded up, *the sensitive speck acting as a development center for the entire crystal.* The speck rapidly picks up electrons from the reducing agents, and more silver ions migrate to the speck to become neutralized to silver atoms, growing into thread-like particles of *metallic silver.* Thus, the *dark areas in a radiograph consist of metallic silver in a very fine state of subdivision,* the amount of silver deposited in a given area increasing with the amount of radiation received by that area. The basic concepts of photographic chemistry are shown in Figure 17.2.

A. EXPOSURE

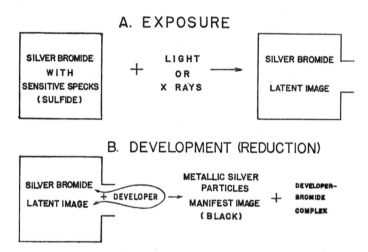

B. DEVELOPMENT (REDUCTION)

Figure 17.2. Scheme of the basic theory of photographic chemistry. In *A*, radiant energy in the form of light or x rays converts a silver bromide crystal containing a sensitive speck (silver sulfide) into a *latent image.* In *B*, the developer enters this altered crystal and reduces it to metallic silver, which constitutes the *manifest image.*

During development the bromine ions diffuse out of the developed crystals and into the solution. This, in addition to the gradual exhaustion of the reducing agents, eventually causes deterioration of the developer to the point where it must be discarded.

What happens to the portions of the film emulsion that are not affected by light or x-ray photons? Since the silver salts in these areas have not been altered, they are relatively unaffected by the developer. However, they must be removed in order to render the unexposed film areas *transparent,* and also to prevent *fogging* by subsequent exposure to light. The unexposed and undeveloped silver bromide is eliminated from the emulsion by immersion in a *fixing agent,* sodium thiosulfate ("hypo"), a process called *fixation.* As a result, the areas from which the silver salts have been removed become clear, while the black areas remain black since metallic silver is not dissolved by the fixing agent in the ordinary course of processing. However, prolonged immersion in the fixing solution will cause bleaching of the image; this may be appreciable even in twenty-four hours, the effect being more rapid with solutions prepared from liquid than from powdered fixer.

In a finished radiograph, the degree of blackening of a particular area depends not only on the total direct x-ray exposure, but even more on the amount of fluorescent light emitted by the screens in contact with this area. The x rays responsible for both these effects represent the *exit* or *remnant radiation;* that is, the radiation remaining in the beam after it has traversed the various thicknesses and densities of tissue interposed between the tube and the film. For example, in a radiograph of the hand the light areas represent the bones. The latter, because of their high calcium content, absorb a large fraction of the incident x rays so that very few x-ray photons pass through them to the silver bromide crystals in the film emulsion underlying them. Therefore, the reducing action of the developer on these crystals produces almost no darkening. The soft tissues, on the contrary, absorb only a relatively small fraction of the incident x rays and therefore the areas of film emulsion beneath the soft tissues receive a relatively large amount of remnant radiation. As a result, more silver bromide crystals are affected in these regions and the developer causes considerably greater blackening. Notice that the finished radiograph is really a *negative* (corresponding to a negative in ordinary photography), and represents the shadows cast by tissues of different thicknesses and densities, due to differences in their absorption of x rays, mainly by the photoelectric effect. As already noted, the radiographic effect of the exit radiation is heightened with intensifying screens.

MANUAL PROCESSING

We shall now turn to the practical aspects of *film processing* which consist of (1) conversion of the latent image to a visible or manifest image and (2) the preservation of that image. This includes primarily development, rinsing, fixing, washing, and drying. Each of these important steps will be described in detail. Optional steps such as fixer neutralization and detergent rinse will also be mentioned. It must be strongly emphasized at the outset that *cleanliness is of paramount importance*. The hands should be rinsed thoroughly in the wash tank immediately after immersing films in the processing solutions, and dried on a clean towel. This not only prevents contamination of solutions, but also protects the hands from chemical irritation. Tanks and mixing utensils must also be kept scrupulously clean.

Development

The function of development is to convert the latent image to a visible image by means of a *developing solution* which contains four essential ingredients.

Developing Solution. Composition:

1. *Organic Reducing Agents*—usually include a mixture of *hydroquinone* and *metol* (or elon, a synonym for metol). The metol acts during the early stage of development, producing the basic gray image, whereas the hydroquinone acts more slowly, building up density and contrast to the desired level. Two such reducing agents in proper combination assure optimum development of the radiographic image.

2. *Preservative—sodium sulfite*—protects the organic reducing agents (hydroquinone and metol) from oxidation by the air, thereby prolonging the effective life of the developer.

3. *Accelerator—sodium carbonate* or *sodium hydroxide*—swells the emulsion slightly, making it easier for the developing agents to enter the emulsion, and the products of the reaction

to diffuse out. Furthermore, hydroquinone can act only in an alkaline medium. Thus, gel swell and alkalinization speed up development.

4. *Restrainer—potassium bromide*—preferentially holds back any action of the developer on the unexposed silver bromide grains without preventing the action of the developer on the exposed grains. Thus, it inhibits fogging of the lighter areas without interfering with the development of the radiographic image.

Practical Factors in Development. The two most important factors in development are (1) the *temperature* of the developing solution and (2) the total *time* of development. A correctly exposed film will be completely and properly developed at a certain combination of temperature and time. But incorrect exposure is almost impossible to correct by varying the temperature and the time of development. A simple analogy comes to mind: if a poor batter is mixed, a poor cake will result no matter how long or at what temperature it is baked.

The optimum temperature for development is 68 to 72 F. With cold developer, that is, below 60 F, the *action of hydroquinone ceases* and the resulting radiograph lacks contrast and density. However, even at 60 F quality is not impaired if developing time is sufficiently long. Developer that is too warm, that is, above 75 F, may soften the emulsion and produce chemical fog. At a given, appropriate temperature, there is a definite, correct developing time, this being best determined from the data furnished by the manufacturer of the films and solutions being used. However, various manufacturers recommend nearly the same developing times so that for practical purposes one may use the graph shown in Figure 17.3. This graph is a composite of the data supplied by the three leading film companies (Eastman, du Pont, and GAF), for *medium speed screen film* and *rapid developer.* From this graph, we can readily determine the correct developing times at various temperatures, based on a normal time of 3 min at 68 F. At 60 F the developing time is 5¼ min and at 75 F it is 2 min. With nonscreen film, the developing time is increased about 50 per cent because of the greater thickness and silver content of the emulsion. Notice that an increase in temperature of the solution increases the speed of chemical action, so that developing

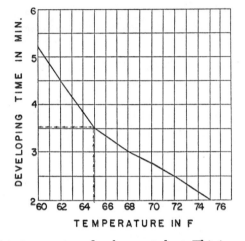

Figure 17.3. Time-temperature development chart. This is a composite based on the data furnished by the three leading American manufacturers for medium speed film and rapid developer.

time must be decreased according to the graph. Processing of films on the above basis is called *time-temperature development.*

With manual development radiographic exposure is based on *five-minute development at 68 F.* This produces radiographs of superior contrast with about 5 kV less exposure than is required for 3-min development.

Only the time-temperature method assures optimum film processing. Except for rare emergencies, the inspection method of estimating the completeness of development should never be used, because of its inaccuracy and inconsistency. Furthermore, the repeated removal of films from the developer during "inspection" may result in slight fogging by oxidation in air.

When films are immersed in the developer, they should be *agitated gently* at first, and about once every minute thereafter to insure uniform development and prevent streaking. During development, bromides are freed from the emulsion and if these are not removed by agitation, they tend to inhibit development wherever they cling to the film. This is most frequently manifested by light streaks below the letters of the identifying labels on the film. Agitation of the film removes these waste products and allows fresh solution always to be in contact with the film emulsion, thereby assuring uniform development.

Replenishment. With continued use of the developing solution, its activity diminishes as the hydroquinone and metol are gradually exhausted during the reduction of exposed silver bromide to metallic silver. Furthermore, the level of developer in the tank drops progressively because some is carried out with each film. Because of the loss of strength and volume of the developer, a special solution—*replenisher*—is added periodically to maintain constant activity and volume and assure uniform results. The composition of the replenisher differs from ordinary developer in having (1) no bromide, because this ion is released from the film emulsion and accumulates in the developer, and (2) a higher concentration of hydroquinone, metol, and alkali, because these ingredients are disproportionately exhausted with continued use of the developer.

To achieve optimum replenishment with this method, the films must be taken out of the developer *rapidly*, without being allowed to drip back into the tank. This removes approximately 2¾ oz of developer with each 14 X 17 film, and correspondingly less with smaller films; amounts that are completely replaced by periodic replenishment to the correct level of the solution in the tank. The solution must be *stirred thoroughly* after each addition of replenisher.

With careful replenishment, the developer should be usable for two or three months, and then the tank should be emptied, thoroughly scrubbed, rinsed, and filled with fresh developer. For optimum results, the developer should be discarded after it has been replenished with about four times its volume of replenisher.

Rinsing

After development, the film must be rinsed for about 30 sec in running water to remove most of the adhering chemicals, thereby diminishing contamination of the next solution, the fixer. Rinsing can be made more efficient by immersing the films in a tank containing a dilute solution of acetic acid (about 1 per cent), which neutralizes the alkali carried over from the developer. Otherwise, the alkali gradually neutralizes the acid in the fixing solution, decreasing its effectiveness as well as its useful life. *Without thorough rinsing, the fixer does not act evenly and the radiograph becomes streaked.*

Fixation

The purpose of this step is (1) to *remove the unexposed and undeveloped silver salts,* (2) to *preserve the film image,* and (3) to *harden the emulsion* so that it will not be easily damaged.

Fixing Solution. This consists of four essential ingredients:

1. *Fixing Agent—hypo (sodium thiosulfate* in powdered fixer; *ammonium thiosulfate* in liquid fixer)—*clears* the film by dissolving out the unexposed, undeveloped silver bromide, leaving the metallic silver in the exposed and developed areas of the film more readily discernible. Without fixation, the undeveloped silver bromide would eventually turn black, fogging the radiographic image.

2. *Preservative—sodium sulfite—*protects the sodium thiosulfate from decomposition and helps clear the film.

3. *Hardener—chrome alum or potassium alum—*"tans" or hardens the gelatin in the emulsion, thereby protecting it against scratches.

4. *Acid—sulfuric acid or acetic acid—*serves two purposes: it neutralizes the alkali still remaining on the film, and provides an optimum medium for the fixer and hardener.

Fixing time depends on the age of the fixer and the number of films processed. A satisfactory fixer requires about 1 to 4 min to *clear* a film—that is, to remove all the unexposed silver salts—but it takes two or three times the clearing time to *harden* the emulsion. Non-screen film, because of its thicker emulsion, requires longer fixation. The solution should be discarded when it has become so exhausted that the fixing time is prolonged beyond 10 min. When the films are immersed in the fixing bath, they should be *agitated* and *separated* to allow uniform action of the chemicals and prevent streaking. To avoid fog, white light should not be admitted until the films have been fixed for at least 1 min. Prolonged fixation is to be condemned because the *hypo becomes so firmly bound to the emulsion that it cannot be removed in washing,* thereby causing eventual brown discoloration of the radiograph. Besides, prolonged fixation may cause *bleaching* of the image. The optimum temperature of the fixing bath is 65 to 75 F; at very low temperatures, the action of the chemicals is retarded, while at high temperatures the emulsion may be softened and easily damaged.

Replenishment requires the periodic addition of fresh fixer after an equal volume of older fixer has been discarded. Not only does this prolong the useful life of the fixing solution, but it also maintains its full strength. The level of solution in the tank remains constant (except for slight loss due to evaporation), because as much liquid is carried in per film as is carried out.

Washing

An important step in film processing is thorough *washing*. Before the films reach the final wash, they have been subjected to the action of a multitude of chemicals which must be completely removed to prevent later discoloration of the radiograph and impairment of its value as a permanent record. Washing requires running water which changes rapidly enough to insure virtually complete removal of the processing chemicals from the film. Ordinarily, if the volume of water flowing per hour is eight times the capacity of the tank, washing is complete in 20 min at 68 F. (Nonscreen film, with thicker emulsion, requires 40 min). This is the minimum wash period regardless of how rapidly the water changes. A slower rate of circulation makes for longer washing, but prolonged washing tends to soften the emulsion. The time must obviously be based on the last film placed in the washing tank, which should be so designed that even the tops of the hangers are completely immersed.

Excessively high temperature should be avoided because of softening of the emulsion. Extremely cold water retards the washing process.

To reduce washing time, a *fixer neutralizer* bath is strongly recommended. After the films have been removed from the fixer, they are rinsed briefly in the rinse tank, then immersed for 2 min in a tank containing a special solution which removes the fixer from the emulsion. Washing will then be complete (in running water) in 5 min.

Drying

Films can be dried by various methods. If the number of films is small, they can be dried in wall-mounted racks. For maximum efficiency, special film driers should be provided, as described in Chapter 16.

The time of drying can be reduced about 50 per cent by the use of a *wetting agent*. After washing is complete, the films are immersed for 2 min in a tank containing household liquid detergent at the rate of about 1 teaspoon per gal of water, after which they are placed in the drier.

Standard Manual Processing Technic

By summarizing a standard technic, we may obtain a clearer idea of the successive steps in processing a group of films.

1. Check temperature of developer with immersed thermometer.

2. Stack cassettes containing exposed films on loading bench.

3. Set time clock at proper developing time, but do not start it.

4. Shut all doors and exclude all white light. The darkroom is now "dark," illuminated only by safelights.

5. Unload cassettes one at a time and clip each film into hanger of proper size, clipping first into bottom clips, then turning hanger with its handle up and inserting film into clips mounted on tension springs. Tension should be equal at four corners so that film will not bulge.

6. Immerse films in developer, agitate gently, and separate by placing fingers of each hand between successive hangers.

7. Rinse hands in water and dry quickly but thoroughly.

8. Start timer. Agitate films gently, about every minute.

9. Load cassettes with fresh film and stand them against wall, or place them in some satisfactory receptacle.

10. When time clock rings, lift films from developer *rapidly* and immerse and agitate them in rinse bath for 20 sec, then drain well and place in fixing bath, agitating them for 20 sec.

11. After fixation is complete (2 to 3 times clearing time, usually about 5 to 10 min) wash films in running water about 20 to 30 min.

12. Optional modification—after fixation, rinse films and immerse for 1 min in hypo neutralizer solution, and then wash for 5 min.

13. Immerse films in solution containing wetting agent (household detergent, 1 teaspoonful per gal) for 2 min.
14. Dry films.
15. Remove films from hangers and trim corners.

Darkroom Errors

It may be of some help to indicate the causes of the more common film defects.

1. **Fog.** There are many causes of film fogging; that is, a generalized darkening of the film.

a. *Exposure to Light.* This may occur when the darkroom is not light proof; the safelight contains too large a bulb; the safelight housing or filter is cracked; the safelight filter series is incorrect; or the exposure of the film to the safelight is prolonged, especially at short distances.

b. *Exposure to X Rays or Radionuclides.* Films should be shielded from these sources of radiation by distance and sufficient thicknesses of lead.

c. *Chemical Fog.* The many causes include overdevelopment or development at excessively high temperatures; oxidized, deteriorated developer, which may also stain the film (oxidized developer is brown); prolonged or repeated inspection of films during development; and contamination from corroded tanks.

d. *Age Fog.* Either mottled or uniform fogging due to outdated films, or films stored under conditions of high temperature and excessive humidity.

2. **Stain.** Various types of discolorations may appear on films at different intervals after processing. These can generally be avoided by the use of reasonably fresh solutions and correct processing.

a. *Brown.* Oxidized developer.

b. *Variegated Color Pattern.* Inadequate rinsing.

c. *Grayish-yellow or Brown.* Excessive fixation, or use of exhausted fixer.

d. *Grayish-white Scum.* Incomplete washing.

3. **Marks and Defects.** There are several different kinds of characteristic markings which appear when films are not handled gently.

a. *Crinkle Marks* are curved black or white lines about 1 cm in length which result from bending the film acutely over the end of the finger.

b. *Static Marks* are lightning or tree-like black marks on the film, caused by static electricity due to friction between the film and other objects such as intensifying screens and loading bench. To avoid this, films should always be handled gently. In addition, the loading bench should be grounded in order to prevent the buildup of static electricity.

c. *Water Marks.* Droplets of water on the film surface may leave round dark spots of various sizes because of migration of silver particles.

d. *Cassette Marks.* Dust particles, fragments of paper, hair, defects in the screens, etc, will leave a corresponding white spot on the film.

e. *Reticulation Marks* are a network of fine grooves in the film surface caused by marked differences in the temperatures of the processing solutions.

f. *Streaking* is caused by a variety of technical errors and is one of the most troublesome types of film defects. It usually results from: (1) failure to agitate the films in the developer; (2) failure to rinse the films adequately; (3) failure to agitate the films when first immersed in the fixer; and (4) failure to stir the processing solutions thoroughly after replenishment.

AUTOMATIC PROCESSING

One of the most significant improvements in radiography has been the perfection of *automatic processing of films*—a completely new system. Equally important has been the significant reduction in the cost of such equipment, making it available even to the department with a limited budget.

There are many advantages of automation in the darkroom. First, it *shortens total processing time to as little as 1½ minutes,* in contrast to about one hour for the manual method. (Even with hypo neutralizer and wetting agent, manual processing can be reduced, at best, to about 40 min.) Second, it *improves quality control* due to more accurate temperature regulation and more precise replenishment in the automatic processor. Third, it in-

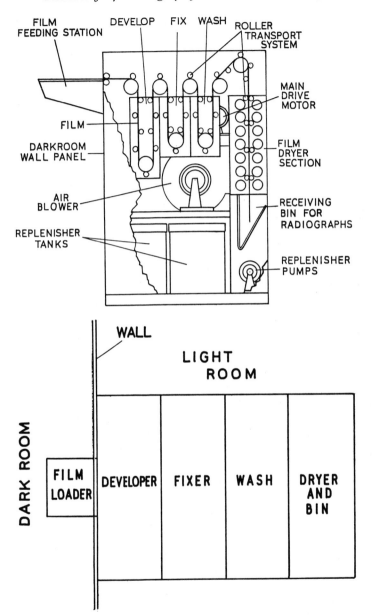

Figure 17.4. Automatic processing. In the upper part is shown a cutaway diagram of the processor (side view). In the lower part is shown a diagrammatic top view. (Courtesy of Eastman Kodak Company.)

creases the capacity of the Radiology Department or office by expediting work flow.

The manufacturers of the first automatic processors were beset by four main problems: transport mechanism, processing chemicals, temperature control, and film characteristics. A discussion of these problems and their solution is essential to the understanding of the principles of automatic processing, especially at their points of departure from manual processing.

1. **Transport Mechanism.** The basic mechanism of the automatic processor is a series of rollers which transport the films from the loader through the various sections—developing, fixing, washing, and drying. The speed of transport must be constant to assure correct sojourn of the films in each section. Spacing the rollers must be accurate to an extremely small tolerance to avoid slipping or jamming of the films. Figure 17.4 shows diagrammatically a typical automatic processor; its apparent simplicity belies the difficulties encountered during its early development.

2. **Processing Chemicals.** Although the basic principles of photographic chemistry are similar to those in manual processing, conventional solutions were found to cause insurmountable difficulties in the first experimental models. Roller spacing was so critical that the swelling of the emulsion in the developer, and its subsequent shrinkage in the fixer, caused the films either to jam between the rollers or to wrap around them. Further complications arose when an attempt was made to speed up processing through the use of stronger solutions.

Figure 17.5 shows a variation of about eleven units of gel swell of the emulsion during manual processing. This renders conventional solutions completely unsuited to automatic processing and shows that the problem is largely *physical.*

As a result of intensive experimentation, the manufacturers discovered the changes required to adapt the processing solutions to automation. These may be summarized as follows for the *original 7-min processor*:

a. *Increased Concentration of Solutions* to shorten the processing time in the various sections to 1 to 2 min. Drying time was reduced to 2 min.

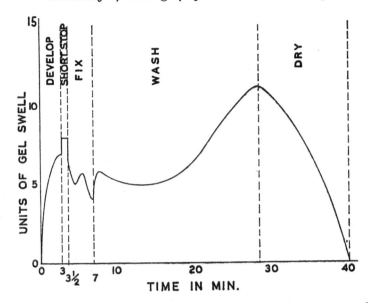

Figure 17.5. Relative swelling of a film emulsion, in arbitrary units of gel swell, during various stages of *manual processing* at 68 F. Note the large variation in the degree of swelling of the emulsion. (Courtesy of R. E. Humphries, GAF Corporation.)

b. *Increased Processing Temperatures* to help speed up processing; for example, 83 F in developer, 79 F in fixer and wash, and 95 to 115 F in dryer. To combat chemical fog at these high temperatures, antifogging agents (mainly aldehydes) were added to the developer.

c. *Hardening of Emulsion* to prevent softening and sticking to the rollers. Therefore, special hardening chemicals were added to the developer.

d. *Control of Emulsion Thickness* to keep it as constant as possible throughout the course of processing, to permit the use of constant spacing between the rollers. Compounds such as sulfates were added to the developer to minimize swelling of the emulsion.

e. *Precise Replenishment* of the developer and fixer to maintain the proper alkalinity of the developer, acidity of the fixer, and the chemical strength of both solutions. *Replenishment was adjusted to a constant rate for each film processed.*

3. **Temperature Control.** As just noted, there was a considerable increase in temperature at every step in the process. This resulted in marked shortening of processing times, but now extremely accurate temperature control had to be maintained in each section of the processor. Temperature was even more critical than in manual processing because the technologist could no longer adjust developing or fixing time for differences in temperature.

4. **Film Characteristics.** The manufacture of films had to be conducted even more precisely than before to maintain constant thickness of the base and emulsion. Curling tendency had to be eliminated to prevent wrapping around rollers or misdirection to the wrong rollers, causing jamming of the films. The emulsion characteristics had to be made compatible with the new types of processing solutions. Finally, stickiness of the emulsion had to be minimized to prevent adherence to the rollers and wraparound.

The various modifications just described have made automatic processing so dependable that it has become practicable even in the private office and small hospital. Figure 17.6 shows how constant thickness of the emulsion is maintained. The overall varia-

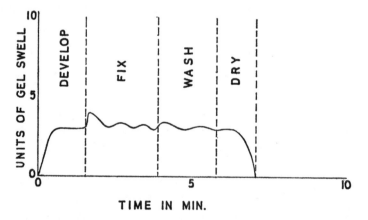

TIME IN MIN.

Figure 17.6. Relative swelling of a film emulsion, in arbitrary units of gel swell, during various phases of *automatic processing* at 81 F. Note the small variation in the degree of emulsion swelling; this is due to special additives in the solutions. (Courtesy of R. E. Humphries, GAF.)

tion in gel swell from start to finish in automatic processing is about four units, but after the initial relatively slight swelling of the emulsion in the developer, the thickness remains virtually constant well into the drying phase.

Table 17.1 contrasts the important steps in manual and automatic processing. It bears out the assertion at the beginning of this section that automation has removed one of the main bottlenecks in the average busy Radiology Department.

TABLE 17.1

COMPARISON OF MANUAL AND AUTOMATIC PROCESSING

	Manual	Automatic* 7-minute	Automatic* 4-minute	Automatic* 1½-minute
Developing Temperature	68 F	83 F	90 F	103.5 F
Fixing Temperature	68	79	90	90 – 95
Washing Temperature	68	76	75 – 85	98
Drying Temperature	90 – 110	95 – 115	115	135 – 150
Developing Time	3 – 5 min.	2 min	70 sec	25 sec
Fixing Time	2 – 10	1½	50	21
Washing Time	15 – 30	1½	50	9
Drying Time	15 – 20	2	70	20
Surface Change	Manual Agitation	Circulating Pumps		
Replenishment	Manual	Automatic		

*Data vary with type of equipment.

The question naturally arises as to the frequency of breakdown of automatic processing equipment. Experience has shown that the greatest cause of breadown is *failure to keep the rollers scrupulously clean,* as recommended by the manufacturer. Another cause is improper rate of automatic replenishment which should be carefully checked at regular intervals.

Sensitometry is coming into wider use to check the correctness of developer replenishment. It is long overdue as a method of quality control. Sensitometric testing of the developer should be done periodically, in addition to the usual test for replenishment, to assure uniformity of development during the useful life of the solution.

By a relatively simple modification, the 7-min processor can be speeded to 2 to 4 min, thereby approximately doubling its capacity from about 100 films to 200 films per hour.

Several brands of high speed processors are available with 90-sec total processing time. Occupying an extremely small space, these units can accommodate about 300 films of assorted sizes per hour. However, because of the special chemistry and the high temperature required with such high speed processing, latitude is narrower and contrast greater than with the 4-min and 7-min models. It seems at this time that 2-min processing yields significantly better results than 1½-min processing. Improvement in chemistry and film emulsion is under constant study by the various film manufacturers, especially for use with lower development temperature, to provide *optimum speed, contrast, and latitude.* In any case, the correct films and chemicals should be used for the particular processing system.

As a rule, two smaller units are preferable to one larger unit of the same total capacity because (1) in the event of equipment failure, the second unit is still available, thereby obviating the need of an auxiliary manual system; and (2) there is less waiting time in loading two processors during peak hours.

One can now select from a number of different brands of automatic processors. Although they are generally comparable in performance and tend to be competitive in price, the decision as to which brand to buy rests primarily on the availability of competent repair service.

QUESTIONS

1. What is meant by latent image? How is it produced?
2. Of what do the black areas on a roentgenogram consist? What factors determine the various densities in a radiograph?
3. What is the purpose and theory of development? Fixation?
4. List the ingredients of the developing solution, and describe the function of each.

5. List the ingredients of the fixing solution and describe the functions of each.
6. What is meant by time-temperature development? Why is it preferable to sight development?
7. Discuss developer replenishment. Why is it necessary?
8. How is fixer replenished?
9. What optional steps can be taken to speed manual processing of films?
10. Name five causes of film fogging and state how they can be avoided.
11. Describe four causes of streaking of radiographs.
12. What is included in the term "film processing"?
13. Why is it advisable to trim the corners of films after drying?
14. State two reasons why over-fixing should be avoided.
15. Discuss automatic processing on the basis of equipment, modification of solutions, and advantages.
16. What effect does 90-sec processing have on film latitude and contrast? How is it being improved?
17. What is the main cause of breakdown of automatic processors?

CHAPTER *18* RADIOGRAPHIC QUALITY

MEDICAL RADIOGRAPHY seeks to provide the maximum possible *information* about a particular anatomic area by means of an x-ray image. Its success depends on the production of radiographs of superb quality. Suffice it to say that a radiograph of poor quality may cause serious error in diagnosis through the inadequate recording of information.

What do we mean by *radiographic quality?* Basically, it refers to the fidelity with which an anatomic structure is represented in a radiograph. For our purpose, a satisfactory definition is as follows: *radiographic quality denotes the visibility and sharpness of the images of structural details.* This applies not only to the larger units making up the anatomic region but also to the images of the smallest structural components. For example, a good quality radiograph of bone shows not only its general outline conforming to the true shape of the bone, albeit in only two dimensions; but of even more importance, it depicts clearly visible bony trabeculas with distinct, unblurred borders.

Radiographic images of the smallest structures are often referred to as *details,* or simply as *the detail.* The radiologic technologist should strive to produce, as consistently as possible, radiographs with the greatest possible amount of detail—in other words, with *maximum information.* In this sense, a direct exposure radiograph is a record of the information carried by an x-ray beam that has passed through an anatomic region. With intensifying screens, the information is modified by conversion of the major portion of the x-ray beam to light.

Four principal factors influence radiographic quality: *definition (sharpness), density, contrast,* and *distortion.* These will be described in the sections that follow.

DEFINITION (SHARPNESS)

The term *definition* applies to the sharpness of structure lines or minute details in a radiograph. Sharpness is measured by *acutance*—the abruptness of the boundary between a detail and its surroundings. Good definition permits the detection of minimal changes in structure. There are three main classes of factors affecting radiographic definition: geometry, motion, and screens.

Factors Governing Sharpness

Among the conditions influencing radiographic sharpness, three are usually regarded as being *geometric* in nature. This means that they have to do with the arrangement in space of the x-ray beam, the part, and the recording medium (for example, screen-film combination). Included are (1) size of effective focus, (2) focus-film distance, and (3) object-film distance. Additional important factors are motion unsharpness and screen unsharpness.

1. **Size of Effective Focus.** In accordance with elementary geometry, the size of the tube focus has a profound influence on radiographic definition. Figure 18.1 shows the radiographic image

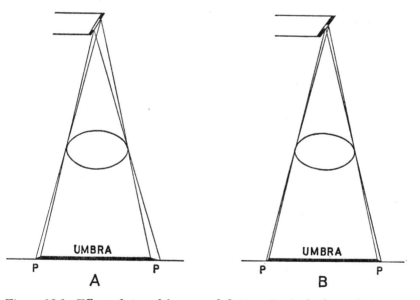

Figure 18.1. Effect of size of focus on definition. In *A*, the large focus produces a broader penumbra and a less distinct image than in *B*, where a small focus is used. (Umbra = image proper; penumbra = zone of blurring at edge of image.)

to consist of the image proper, or the *umbra,* and a lighter hazy area at the edge, *P,* the *penumbra.* The penumbra is caused by rays originating at various points on the focus and passing tangentially to the edge of a structural detail in the object (that is, skimming the edge) as shown in the figure. A wide penumbra is manifested by a *blurred* margin of the radiographic detail; in other words, a wide penumbra impairs definition, and is actually a measure of radiographic *un*sharpness or *blur.* Since the penumbral width decreases as the effective focal area decreases, one may conclude that *the smaller the effective focus, the sharper the definition.* Theoretically, a point source of x rays should give ideal definition since there would be no penumbra, but this is impossible at the present time because of the limited heating capacity of x-ray tubes. Furthermore, even if a point source of x rays were available, it would still give imperfect definition because of the inherent unsharpness of intensifying screens (the unsharpness of films is negligible in radiography).

2. **Focus-film Distance.** Definition also depends on the distance from the focus to the film. As this distance is increased, the

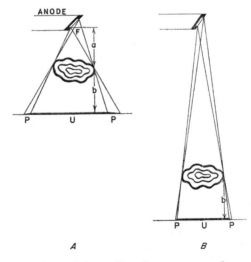

Figure 18.2. The effect of focus-film distance on radiographic definition. The *object-film* distance is the same in both *A* and *B.* However, in *B* the *focus-film* distance is greater; note that this produces a smaller penumbra and therefore a sharper image than in *A.*

effect is similar to a decrease in focal spot size. From the standpoint of the film, the farther away the focus is, the smaller it appears. As the focus-film distance increases the penumbra decreases, as shown in Figure 18.2. Therefore, the *sharpness of the radiographic image is improved by increasing the focus-film distance.*

3. **Object-film Distance.** The third geometric factor is the distance from the object to the film. If the focus-film distance remains unchanged, and only the object-film distance is altered, then by the application of the same geometric principles we find that a *decrease in the object-film distance decreases the penumbra, thereby enhancing definition* (see Figure 18.3).

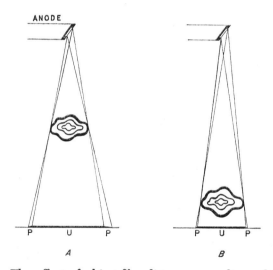

Figure 18.3. The effect of object-film distance on radiographic definition. The *focus-film* distance is the same in A and B. However, in B the *object-film* distance is smaller; note that this produces a smaller penumbra and therefore a sharper image than in A.

The above geometric factors in image sharpness may be conveniently summarized in a formula for its opposite, geometric *unsharpness* or *blur,* defined as the *width of the penumbra.* In Figure 18.2A, F is the effective focal spot width, a is the focus-object distance, b is the object-film distance, and P is the penumbra or unsharpness. Then, by similar triangles:

$$P/F = b/a$$
$$P = Fb/a$$

$$\text{unsharpness} = \frac{\text{width of focus} \times \text{object-film distance}}{\text{focus-object distance}} \quad (1)$$

In summary, then, radiographic definition (sharpness of detail) is enhanced by any factor that decreases the width of the penumbra, namely (1) small tube focus, (2) long focus-film distance, and (3) short part-film distance. Accordingly, you should place the part to be radiographed as close to the film as possible and use a reasonably long focus-film distance. However, excessive focus-film distances may require such long exposure times that motion unsharpness becomes objectionable.

4. **Motion Unsharpness.** *Motion* of the part being radiographed may be regarded as the greatest enemy of definition because it produces a blurred image and is difficult to avoid completely. Motion can be minimized in three ways: (1) by careful immobilization of the part with sand bags or compression band, (2) by suspension of respiration when examining parts other than the limbs, and (3) by using exposures that are as short as possible, generally with intensifying screens. Although screens impair sharpness, this detrimental effect is usually more than compensated for by the reduction of motion unsharpness.

5. **Screen Unsharpness.** The sharpness of detail is better in radiography with *cardboard holders* than with intensifying screens, provided *immobilization is adequate.* However, when fast exposures are needed to minimize the effects of uncontrolled motion, screens must be used; under these circumstances, the gain in sharpness with screens and short exposures usually more than compensates for the loss of sharpness that would result from the use of cardboard holders with long exposures. In general, screen unsharpness is greater than that caused by a 1 mm focal spot.

Why is radiographic definition impaired by intensifying screens even when the patient is adequately immobilized? We may summarize the reasons as follows:

a. **Crystal Size.** Recall from Chapter 15 that intensifying screens produce a radiographic image because of *fluorescence* *of crystals* in the active layer. Since each crystal, despite its very small size, produces a spot of light of corresponding size on the film, image lines are broader than when they are formed directly by x-ray photons.

b. **Active Layer Thickness.** The diffusion (spreading) of light in the active layer due to its measurable thickness contributes to loss of definition.

c. **Film-screen Contact.** Cassettes are provided with spring-metal clamps to secure close contact between the film and screens. The least amount of separation between them allows the light from any point in the screen to spread on its way to the film, thereby impairing radiographic sharpness. This was described above in greater detail (see pages 265-266).

Because of these factors, cardboard holders should be used to obtain the very sharp definition required in the radiography of small parts (hand, feet, wrists) in a cooperative patient. Should this not be feasible, we can still obtain quite good detail by using slow (ultra detail) screens. However, in the radiography of parts measuring 8 to 10 cm thick, medium speed or high speed screens should be used routinely, and above 10 cm, a grid, also.

d. **Mottle.** In examining, with a magnifying glass, a processed film that had been exposed *with intensifying screens,* you will find that it has a mottled or spotty appearance. Only a small part of this is contributed by the granular structure of the screens and by the clumps of silver in the radiographic image. More important is the fact that an x-ray beam consists of individual energy packets or *photons* which have a statistical or random distribution in space. Thus, in a radiograph that has apparently been exposed uniformly with intensifying screens, different small areas actually receive somewhat different numbers of photons because of their non-uniform distribution in the x-ray beam. With *slow imaging systems* such as slow film-screen combinations or direct-exposure film a large number of photons are needed for a particular degree of film darkening (density); under these conditions, the percentage variation in the number of photons from area to area is small and the image is very uniform. On the other hand, with *fast imaging systems* such as high-speed film-screen combinations (or image intensification), a small number of photons are needed for the same degree of darkening (density), but now the percentage variation in the number of photons from area to area is large (statistical fluctuation) and the image appears mottled. This phenomenon is called *quantum mottle* (remember that a photon is a quantum of energy). Quantum mottle impairs definition, especially when contrast is low. It is one form of *noise* (see pages 313, 315, 330).

Aside from the imaging systems themselves, there is another factor in the production of quantum mottle. This is the *kV*. As kV is increased fewer photons are required for a given degree of film darkening with a particular film-screen combination; hence, mottle increases with increasing kV.

In summary, then, quantum mottle increases with fast image recording systems and with high energy photons. In any given instance, the decision has to be made as to whether high speed or minimal mottle is more important.

At this point it might be appropriate to mention the term *resolution,* which is often confused with sharpness, although it does depend on sharpness and contrast. Resolution refers to the smallest possible distance between two objects such that they will produce two separate images on a radiograph or other imaging system such as fluoroscopy or television. Its mathematical aspects are rather complex and beyond our scope.

DENSITY

The amount of darkening of an x-ray film, or of a certain area on the film, is called *radiographic density.* It should be recalled that any region of the film emulsion exposed to x rays, or to light from intensifying screens, becomes susceptible to the action of the developer. The silver salts that have been so affected are changed by the developing agent into tiny particles of *metallic silver* which appear black because of their finely divided state. The greater the total amount of radiation that reaches the film, the greater will be the final degree of blackening. Areas receiving only a small amount of radiation undergo little or no subsequent action by the developer, such underexposed areas appearing gray or translucent in the finished radiograph. Thus, in the final analysis, *the density or degree of blackening depends on the amount of radiation reaching a particular area of the film and the resulting mass of metallic silver per unit area.* Standardization of development should produce optimal film density if the film has been exposed correctly.

For those interested in the more technical aspects of this subject, it may be stated that density is measured by an instrument known as a *densitometer,* which indicates the relationship between the intensity of light falling upon one side of a given area of a radiograph (as from an illuminator) and the intensity of the light passing through (transmitted). This relationship is shown by the formal definition of density:

$$\text{density} = \log \frac{\text{incident light intensity}}{\text{transmitted light intensity}} \qquad (2)$$

The equation can best be explained by numerical examples. If the incident light intensity is 10 times the transmitted intensity, then the density is $\log 10 = 1$; if the incident light intensity is 100 times the transmitted intensity, then the density is $\log 100 = 2$; etc. Nevertheless, one need not be conversant with logarithms to use a densitometer, since it is calibrated to read density directly. With this method, clear film base has a density of 0.06 to 0.2, depending largely on the amount and type of blue dye present. A diagnostic radiograph usually has densities varying from about 0.4 in the lightest areas to 3.0 in the darkest. Obviously, excellent radiographic quality requires optimal density for any particular anatomic structure; the correct exposure is supposed to be incorporated in the technic chart.

Density is an extremely important factor in radiographic quality because it carries *information.* Without density there is no image and therefore no detail visibility. However, density must be optimal because if it is excessive it may actually conceal information, just as too much light in ordinary photography may destroy detail.

There are five factors that profoundly affect the radiation exposure and consequent density of a radiograph: (1) kilovoltage, (2) milliamperage, (3) time, (4) distance, and (5) thickness and nature of part being radiographed.

1. **Kilovoltage.** An increase in kV applied to the x-ray tube increases both the exposure rate and the percentage of higher energy (short wave length) photons. These more penetrating photons are not so readily absorbed by the structures being radiographed, and therefore a larger fraction of the primary radiation eventually reaches the film. Thus, increasing the kV increases the x-ray exposure rate at the film and therefore increases the radiographic density. In general, an increase of 15 per cent in kV approximately doubles the exposure. For example, the exposure will be doubled at 40 kV by an increase of 6 kV, and at 70 kV by an increase of 10 kV. Note that although the radiographic density increases as the exposure increases, this relationship is not strictly proportional.

Under ordinary conditions, with grids having a ratio of 8 or less, it is unwise to exceed 85 kV because this produces an excess of *scattered radiation* which has a fogging effect on the radiograph, with impairment of contrast. On the other hand, the efficiency of x-ray equipment improves at higher kV because there is a considerably smaller heating load on the anode (see page 212) and the certainty of penetration is greater. It has been found that *high voltage radiography* at 100 to 130 kV and a high ratio grid (12 or 16) offer advantages over conventional radiography, including greater latitude and a smaller skin dose.

2. **Milliamperage.** For practical purposes, x-ray exposure rate is proportional to the mA. Hence, if the mA is doubled, the radiographic exposure rate is doubled; if the mA is tripled, the exposure rate is tripled. Tube current (mA) is governed by the number of electrons passing from cathode to anode *per second.* As the flow of electrons per second is increased, more x rays per second are produced at the target. Note that the *mA determines only the exposure rate; it has nothing to do with the penetrating power of the beam.*

3. **Time.** An increase in exposure *time* causes a proportional increase in total x-ray *exposure* of the film. Thus, doubling the time doubles the total exposure, and tripling the time triples the total exposure. A longer exposure time allows the radiation of a given exposure rate to act longer, thereby eventually affecting more silver bromide crystals in the emulsion and increasing the

density of the radiograph. In actual practice the mA and exposure time are usually multiplied and called *milliampere-seconds* = mAs. For example, if the technic is set up for 100 mA and $\frac{1}{10}$ sec, then multiplying 100 × $\frac{1}{10}$ = 10 mAs. If a faster exposure is needed, as in radiographing a small child, and the density is to remain unchanged, the mA can be increased to 400 and the time reduced to $\frac{1}{40}$ sec; 400 × $\frac{1}{40}$ = 10 mAs. In other words, the old mA times the old time in seconds equals the new mA times the new time in seconds, as expressed in the following equation:

$$MAS = mAs \qquad (3)$$

This relationship, known as the *reciprocity law*, holds reasonably well throughout the diagnostic x-ray region, provided the x-ray generating equipment has been *properly designed* and *calibrated*. You should cultivate the habit of thinking in terms of mAs because it greatly facilitates the modification of established technics, thereby increasing the flexibility of the technic chart.

4. **Distance.** The effect of distance on exposure rate is not so simple as some of the other factors, but it is easily understood by keeping in mind the following elementary geometric rules:

a. The x-ray beam originates at a "point" on the target (this is an assumption for practical purposes).

b. The photons travel in straight lines and except for the "heel effect" (to be described later), diverge almost uniformly in all directions from the focus in the form of a cone.

Since the x-ray photons diverge (spread) *the width of the beam increases as the distance from the target increases.* This means that the same amount of radiation is distributed over a larger area the farther this area is from the tube focus. Obviously, the same radiation spread over a larger area must spread itself "thinner." If, at a certain distance from the tube focus, the beam were to cover completely a film of a certain size, then at a greater distance it would cover a larger film. However, the radiographic density of the latter would be less in a given exposure time because each square centimeter of this film would have received less radiation than each square centimeter of the first film. We may therefore conclude that *radiographic exposure rate decreases as the focus-film distance increases.*

We can determine by simple geometry *how much* the exposure rate decreases as the focus-film distance increases. In Figure 18.4

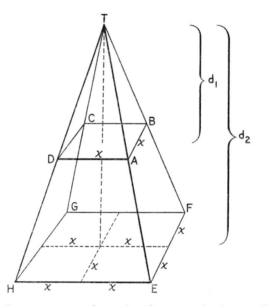

Figure 18.4. Inverse square law of radiation. The lower plane surface (*EFGH*) is selected at twice the distance from the point source of radiation (tube focus) than is the upper plane (*ABCD*). Each side of the lower plane ($x + x$) is twice as long as each side of the upper plane (x). It is evident from the diagram that the lower surface area is four times as large as the upper surface area, which means that at twice the distance from the target the x-ray beam covers four times the area and therefore the brightness of illumination must only be one-fourth as great.

the slanting lines represent the edges of a beam emerging from the focus at the target, *T*, and restricted by a square collimator. Two planes, *ABCD* and *EFGH*, are chosen at right angles to the direction of the central ray of the beam (represented by the dotted line). Both planes are assumed to be squares. Plane *EFGH* is located twice as far from the target as plane *ABCD*. Therefore each side of the lower plane, such as *HE*, is twice as long as a side of the upper plane, such as *DA*, because triangles

TEH and *TAD* are similar and their corresponding sides are proportional. To simplify the discussion, let *X* equal a side of the upper plane. Then 2*X* must equal a side of the lower plane. The area of the upper plane will then be $X \times X = X^2$, and the area of the lower plane will be $2X \times 2X = 4X^2$. Thus, the lower surface, *EFGH* has four times the area of the upper surface, *ABCD*.

It is evident that *when the distance is doubled, the same radiation is spread over an area four times as great*. Therefore, the brightness or illumination must be ¼ as great. This is an application of the *inverse square law of radiation*, which is equally valid for x rays and light and can easily be proved to hold for all distances. The law is stated as follows: *the intensity or exposure rate of radiation at a given distance from a point source is inversely proportional to the square of the distance*. For example, if the exposure rate of an x-ray beam at 20 in. from the focus is 100 R/min, what will it be at 40 in.? Let us set up the inverse square law in equation form:

$$I/i = d^2/D^2 \qquad (4)$$

where

I is the exposure rate at 40 in. = ?
i is the exposure rate at 20 in. = 100 R/min
d = 20 in.
D = 40 in.

Note that the proportion is not direct, but inverse (inverted). Substituting the above values in equation (4):

$$I/100 = (20)^2/40^2$$

$$I = \frac{\cancel{20} \times \cancel{20} \times 100}{\underset{2}{\cancel{40}} \times \underset{2}{\cancel{40}}}$$

$$I = 25 \text{ R/min}$$

Thus, at twice the distance the exposure rate is ¼ the initial value (25 as compared with 100). This means that in radiography, if the distance is doubled, and the kV and mAs are not changed, the exposure is reduced to ¼. Therefore, to keep the exposure constant the mAs has to be multiplied by 4.

Let us take another example. If the exposure rate of radiation at 60 in. is 10 R/min, what will the exposure rate be at 20 in.? Again using equation (4), let i represent the unknown exposure rate at point d located 20 in. from the focus. Then, using the data given in the above problem,

$$10/i = (20)^2/(60)^2$$

$$1/i = \frac{\cancel{20} \times \cancel{20}}{10 \times \underset{3}{\cancel{60}} \times \underset{3}{\cancel{60}}}$$

$$1/i = 1/90$$

$$i = 90 \ \text{R/min}$$

In other words, the distance has been reduced to $20 \div 60 = \frac{1}{3}$ and the exposure rate has increased by $90 \div 10 = 9$ times.

There is a *much simpler method* of applying this law. Rearranging equation (4),

$$ID^2 = id^2 \tag{5}$$

This means that for a given set of factors (kV, mA, filter, etc) the exposure rate times the square of *its* distance is a *constant.* Thus, in equation (5), id^2 is a constant. In the problem on page 309, it is stated that $i = 100$ R/min at distance, d, 20 in. Therefore,

$$id^2 = 100 \times 20 \times 20$$

The problem is now solved by substituting this constant and the *new* distance, 40 in., in equation (5).

$$I \times 40 \times 40 = 100 \times 20 \times 20$$

$$I = \frac{100 \times \cancel{20} \times \cancel{20}}{\underset{2}{\cancel{40}} \times \underset{2}{\cancel{40}}}$$

$$I = 25 \ \text{R/min}$$

In actual practice, a therapy machine should be calibrated for each treatment distance because of deviation from the inverse square law due to the large focus. However, the law is sufficiently accurate for approximation in diagnostic radiology and in protection problems.

5. **Thickness and Nature of the Part.** This will be discussed in the next section. In general, the thicker and denser the part being radiographed, the less the radiographic density.

CONTRAST

A radiograph is made up of light and dark areas; that is, it shows variations in density. The range of density variation among the light and dark areas is called *radiographic contrast.* While density represents the amount of silver deposited in a given area, contrast represents the relative distribution of silver in various areas of a radiograph. To be perceptible to the average human eye, the difference in density of adjacent areas must be at least 2 per cent.

There are three different kinds of contrast. *Radiographic contrast* refers to the contrast as it appears in a radiograph. This, in turn, depends on *subject contrast* and *film contrast.* Each will be discussed separately and also interrelated. However, the unqualified term contrast usually means radiographic contrast.

Radiographic Contrast

We usually designate radiographic contrast as *long scale* (low) or *short scale* (high). This concept may be explained as follows: suppose a radiograph were cut into small squares representing each of the different densities present, from the lightest to the darkest, and these squares were then arranged in order of increasing density. If there should turn out to be relatively many such squares with very little difference in the density of successive squares, then the original radiograph must have had *long scale or low contrast.* On the other hand, if there should be relatively few squares with a large difference in density between successive ones, then the original radiograph must have been of *short scale or high contrast.* In other words, a long scale or low contrast radiograph has a long range of tonal graduation from white, through many shades of gray, to black; whereas a short scale or high contrast radiograph has a short range between white and black (see Figure 18.5).

The function of contrast is to make detail visible. Contrast is optimal when there is sufficient difference in density among the various details to make them distinctly visible in all areas of the radiograph. This is usually achieved by *medium scale contrast.* Excessively short scale contrast tends to impair detail; an example is the so-called "chalky" radiograph which may lead to error in interpretation especially in the examination of bones. However, short scale contrast is useful under special circumstances as in angiography.

Subject Contrast

As indicated earlier, *subject contrast* is one of the factors in radiographic contrast, the other being film contrast. Before defining subject contrast, we must first introduce appropriate basic concepts.

An x-ray beam undergoes attenuation or weakening while passing through a patient. But the attenuation is not uniform throughout the cross section of the exit beam (that is, the beam leaving the patient). If we could somehow "see" the exit beam in cross section from the position of the film, we would observe that various areas have different numbers of photons per square cm. This is due to unequal degrees of absorption of x rays by various tissues. It is the spatial distribution of photons in the cross section of the exit x-ray beam that constitutes the *radiologic image* which will eventually be recorded on film (or some other system). The film record of a radiologic image is the *radiographic image.* This distinction must be kept firmly in mind.

Subject contrast is defined as the contrast existing in the radiologic image. In mathematical terms, subject contrast is the ratio of the number of photons in two or more equal areas of the radiologic image.

Note that the radiologic image contains information derived from the patient. But in addition to the wanted information, the radiologic image becomes contaminated with *noise,* this being *unwanted* radiation that impairs the quality of the information and, ultimately, degrades the quality of the resulting radiograph. Such noise factors include scattered radiation and fogging.

We shall now explore the factors that affect subject contrast,

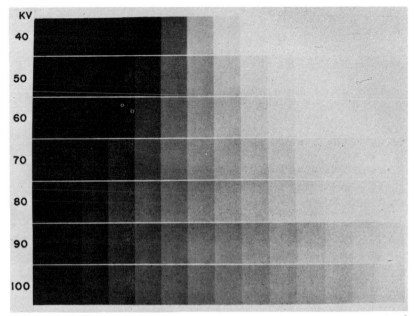

Figure 18.5. Effect of kilovoltage on contrast, using an aluminum stepped wedge. The short range (high) contrast with 40 kV, as compared with long range (low) contrast with 100 kV, is obvious.

namely, radiation quality, radiographic object, scattered radiation, and fogging.

1. **Radiation Quality** controls the penetrating ability (transmission) of an x-ray beam. We found in Chapter 12 that an increase in the kV across the x-ray tube increases the penetrating ability of the x-ray beam. A beam of high penetrating ability penetrates tissues of various densities more equally than does a beam of low penetrating ability. For example, in radiography of a leg, the bony structures absorb a much greater fraction of the primary x rays than do the soft tissues. As the kV and penetrating ability of the primary photon increase, the difference in the absorption of the radiation, as between bone and soft tissues, becomes smaller. Consequently, the transmission of x rays through bone is more nearly like that through soft tissue and there is less difference in radiographic density between them; in other words, there is a reduction in subject contrast. Figure 18.5 shows the effect of

various kilovoltages on contrast. You can see that an ***increase in kV produces longer scale or lower contrast*** in the radiograph of an aluminum stepped wedge (a bar made up of a series of increasing thicknesses of aluminum).

An additional factor at higher kV is the increase in the fraction of scattered radiation, but this is largely removed by the use of high ratio grids.

2. **Radiographic Object**—thickness and nature of object being radiographed. The human body consists of tissues which differ in transparency or *radiolucency* to x rays. The more photons a given type of tissue absorbs, the less its radiolucency and the smaller the amount of *exit* or *remnant radiation* reaching the film. Ranging from the greatest to the least density are (1) tooth enamel, (2) bone, (3) tissues of "water" density such as muscle, glands, solid nonfatty organs, (4) fat, and (5) gas. For example, a radiograph of the abdomen has a light zone in the center representing the vertebral column which absorbs more radiation than an equal mass of soft tissue. Various soft tissues are represented as shades of gray, depending on differential absorption. The fat around the kidneys, along the psoas muscles, and in the abdominal wall, appears as dark gray or nearly black lines. This differential absorption of x-ray photons produces subject contrast in the radiologic image, ultimately contributing radiographic contrast which makes it possible to distinguish various details and discover abnormalities in them. Where organs cannot be studied in detail because their density is similar to that of adjacent organs, *contrast media* are used; for example, dense media such as barium sulfate in the gastrointestinal tract, and iodinated compounds in the gall bladder, kidneys, vascular system, and bronchial tree. *Note that the photoelectric effect produces a sharp peak in absorption of x-ray photons, roughly equivalent to a 65-kV beam, because this is the binding energy of the K-shell electrons of iodine. The best contrast would therefore be obtained at about 65 kV. However, we find in practice that adequate contrast between iodine and soft tissue occurs with x-ray beams generated at 65 to 90 kV.* Sometimes a less dense medium can be used, such as air in the ventricles of the brain. At this point you should review the physical reasons for the difference

in absorption of x-rays by various tissues, discussed on pages 177 and 181.

3. **Scattered Radiation** is a noise factor, obscuring information (that is, detail) in the radiographic image. This is manifested by impaired contrast resulting from the general graying effect on the lighter areas of the radiograph. Scattered radiation is controlled by means of stationary or moving grids and by limiting the size of the x-ray beam with a collimator, cone, or aperture diaphragm (see Chapter 19).

4. **Fogging** from any cause is also a noise factor, producing a general graying of the radiograph with reduction in contrast.

Film Contrast

As we have already pointed out, film contrast plus subject contrast constitute radiographic contrast. However, we cannot separate film contrast from the manner in which the film is used; that is, the method of processing, and whether exposure is directly to x rays or with intensifying screens. It is more appropriate to speak of the *film recording system,* a designation which includes film and processing, with or without screens.

Films themselves vary in their inherent contrast, depending on their emulsion characteristics. Thus, it is possible to obtain films that are specifically designed for long, medium, or short scale contrast.

The *development process* also affects film contrast. For example, a developer containing sodium hydroxide as an accelerator produces greater contrast than one containing sodium carbonate. Within the normal range of time-temperature development, contrast is not appreciably influenced by the temperature. However, at excessively high temperature or unduly prolonged developing time chemical fog may occur, reducing contrast. Gentle agitation during development improves contrast. More important, a decrease in radiographic exposure with appropriate increase in developing time, within reasonable limits, enhances contrast. For optimum contrast, strict adherence to the correct time-temperature curve is essential. With automatic processing the danger of excessive contrast is greater than with manual processing, and it is of the utmost importance that we use the correct films, chemicals, and processing time and temperature.

Finally, you should know that *intensifying screens* modify the radiologic image by converting about 98 per cent of it to light. In other words, the radiologic image is changed almost completely to a light image which is then recorded by the film. During the conversion process, contrast is enhanced because screen type film has more inherent contrast for the light emitted by screens than for x rays directly.

There is no simple definition of film contrast. It can be understood only by studying the so-called *characteristic* or *sensitometric* curve of a particular film recording system, shown in Figure 18.6.

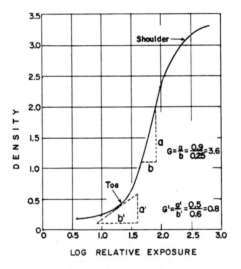

LOG RELATIVE EXPOSURE

Figure 18.6. Characteristic (H & D) curves of screen type x-ray film. *G*, the *slope* of the "straight line" portion of the curve, represents film contrast in this region of the curve. X-ray exposure should be such that this portion is utilized. *G'* is the slope of the "toe" portion; note that contrast here is smaller, a larger exposure being required for a given increase in density.

This type of curve was first used in photography by Hurter and Driffield and is often referred to as an H & D curve. Film manufacturers use such curves to monitor film quality and consistency. They show the response of film recording systems to specific exposure conditions, being obtained experimentally by plotting radiographic densities corresponding to a series of different exposures (actually, density versus logarithm of relative exposure). Referring to Figure 18.6, we note that as the exposure increases above zero, the density first increases gradually in the *toe* portion, then steeply along nearly a *straight line,* and finally more gradually along a *shoulder* portion. Film contrast is defined as the average gradient or slope of the useful *mainly straight line* portion of the characteristic curve.

The importance of the characteristic curve lies in its demonstration that at low densities (toe) and at high densities (shoulder) contrast is low. Maximum contrast occurs in the intermediate density region (steep, nearly straight line portion). Study of an average radiograph reveals that the lightest and darkest areas have the poorest contrast and it is in precisely these areas that details are least visible. Radiographic technics should be established so that wherever possible overall exposure should fall within the steep portion of the characteristic curve.

Most radiographic *technic charts* are based on the different densities of various parts of the body, and on different thicknesses of the same part of the body in different individuals. The thicker or denser the part, the greater is the required exposure in terms of kV, mA, and time. In some technic charts, the mAs value is kept constant for a given region of the body and the kV is varied according to the thickness. Others prefer to use an *optimum kilovoltage* for a given part and vary the mAs. If high kV equipment is available, we can use 100 to 140 kV with grids having a ratio of 12 or 16. High voltage radiography produces radiographs with good definition, wide latitude, and moderate contrast, provided suitable filters and grids are employed.

Soft Tissue Radiography—Mammography

In soft tissue radiography, of which mammography is a specialty, we are dealing with essentially two types of structures— those of *water density* (for example, muscle, glands, fibrous tissue, blood vessels) and *fat.* These exhibit very poor subject contrast. However, *by proper selection of technical factors we can successfully enhance radiographic contrast between fat and any tissue of water density.* But there is no way to distinguish radiographically the various tissues of water density unless they happen to be surrounded by fat, short of using contrast agents. This important principle applies directly in mammography in that lesions such as tumors (water density) can be separated radiographically from the surrounding fat which is often present in sufficient amount to act as a natural contrast agent, especially in older or obese women. Furthermore, since about one third of breast cancers contain microcalcifications (0.05 mm or less) that require viewing with a hand lens, *extremely fine detail* is essential.

Conventional radiography, especially with ordinary films, does not provide the high contrast and superb definition needed in mammography. These can be achieved only by modification of ordinary technics as follows: (1) *low kV,* (2) *minimum filtration,* (3) *small focus,* and (4) *industrial type fine grain film.*

As we found in an earlier section, *the lower the kV, the greater the radiographic contrast,* a relationship that is particularly important in soft tissue radiography. Thus, Egan's mammographic technic calls for a nominal 20 to 30 kV and 1200 to 1800 mAs. Actually, he recommends calibrating each x-ray machine to the lowest kV setting that produces x rays that just penetrate the 15 mm section of an aluminum stepped wedge (other factors, 300 mA, 36-in. focus-film distance, 6 sec, extension cylinder 12 in. long × 4 in. internal diameter, mammographic film). Exact details may be found in his book, listed in the bibliography. However, it may not be possible to do this with all machines. Furthermore, these technics may impose such a heavy load on the conventional x-ray tube that its useful life may be shortened, unless only an occasional mammogram is made. Special tungsten target tubes are available to permit operation at low kV and high mAs.

All added filtration must be removed in order to use the lowest energy radiation for contrast enhancement. Inherent filtration (that is, tube port and oil) should not exceed 1 mm Al. General radiographic equipment is usually unsuitable for mammography not only because it may not operate below 40 kV, but also because of the excessive filtration by the plastic cover of the collimator and the inherent filtration of the tube.

The smallest available tube focus should be used, certainly

never more than 1 mm. Unduly heavy loading of the tube prevents the use of an ordinary x-ray tube with a fractional focus. Extremely slow, fine grain film without screens contributes not only to contrast enhancement, but also superb definition. Modern mammographic film is a modified industrial type film that can be processed in an automatic processor.

The excellent quality of the mammograms obtained with Egan's technic could not be surpassed for a number of years. However, this became possible for the first time in 1966 with the invention of a new type of mammographic tube in France by J. R. Bens and J.-C. Delarue. Knowing that optimum contrast is achieved in soft tissue radiography with x rays ranging in wavelength from 0.6 to 0.9 Å, or average energy of about 17.5 kV, they designed an x-ray tube with a *molybdenum (Mo) target.* They selected Mo because at about 30 kV it emits characteristic K_α and K_β radiation with wavelengths 0.71 and 0.63 Å, respectively (average 0.66 Å), corresponding to energies of 17.4 and 19.5 kV (average 19 kV). Note that the energy of this radiation matches fairly closely the optimum for soft tissue radiography (in fact, it is slightly higher than the optimum 17.5 kV, and this is desirable). Furthermore, they introduced a 0.03 mm *Mo primary filter* into the beam to remove much of the radiation with energy above and below the desired range, thereby producing a *nearly monoenergetic beam.* Using the *same* metal—Mo—for *both* the target and the filter utilizes the principle of the *spectral window;* that is, a filter readily transmits the characteristic radiation produced initially by the same element, while absorbing much of the general radiation (continuous spectrum). Note that a tungsten target emits only a continuous spectrum (that is, no characteristic radiation) at 30 kV; although it is fairly monoenergetic, it is of such low intensity that a large mAs must be used.

This special type of radiographic tube, having both the target and filter made of molybdenum, is used in the Senograph* and the Mammorex.** In the Senograph, the tube has a stationary anode at one end, grounded and water cooled. The anode has a 0.7 mm focus (for better definition than with a 1.0 mm focus)

* CGR Medical Corporation
** Picker Medical Products Division

and is recessed within a hood, as shown in Figure 18.7, to trap scattered and secondary electrons which might otherwise generate

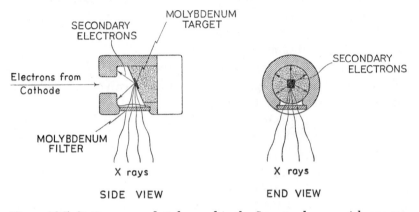

Figure 18.7. Stationary anode tube used in the Senograph, a special mammographic unit. Note the hooded molybdenum target which helps minimize off-focus radiation by trapping scattered and secondary electrons. (Courtesy of CGR Medical Corporation.)

off-focus x rays that could impair radiographic contrast. Thus, both radiographic definition and contrast are significantly improved, thereby facilitating the detection of small lesions and microcalcifications.

The Mammorex uses the Mammotron tube (Dunlee) provided with a rotating anode having 1 and 2 mm focal spots; as noted above, it has a molybdenum target and filter. Like the Senograph, this tube has a beryllium window to include very low energy radiation in the beam. This is not desirable because the soft radiation is absorbed in the breast without being image-producing; besides, it is largely removed by the filter.

Another mammographic tube that is rated up to 70 kV and can be used with conventional equipment is the Dynamax M60 (Machlett). It has a rotating anode tube with a molybdenum target and 1 and 2 mm focal spots. However, it does not have a beryllium window, since the molybdenum filter is optional (with the other mammographic tubes the molybdenum filter is built into the housing, since skin burns may result if it is accidentally omitted in the presence of a beryllium window).

Before leaving this subject you should turn to the discussion of xeroradiography on pages 391-393. With this system excellent mammograms are obtainable with conventional radiographic equipment.

DISTORTION

A radiographic image is not a true representation of the anatomic part, but differs from it in varying degrees of size and shape. Such *misrepresentation of the true size and shape of an object is called distortion.* The amount of distortion depends on several factors, to be discussed in this section. While distortion generally has a degrading effect on radiographic quality, we cannot completely eliminate it because the image lies in a single plane and is two dimensional, whereas the object, being solid, is three dimensional. In fact, we often use distortion deliberately in order to bring out a structure that would otherwise be hidden; for example, in oblique radiography of the gall bladder to separate it from the vertebral column.

1. **Size Distortion—Magnification.** The formation of a radiographic image follows simple geometric principles because x rays originate at the focus of the tube and diverge to form a cone-shaped beam. These geometric rules prevail in the realm of x-ray image formation just as they do with ordinary light. We can easily show that when an object is held between a source of light and a white surface, the size of the shadow enlarges as the object is moved nearer the light, and shrinks as it is moved closer to the white surface. This is due to the *divergence of the light rays* in the beam. In Figure 18.8, the divergent beam of light *magnifies* the shadow more in *A* than in *B;* that is, the shorter the distance between the object and the source of light the greater the magnification. *The law of image magnification states that the width of the image is to the width of the object as the distance of the image from the light source is to the distance of the object from the light source.*

$$\frac{\text{image width}}{\text{object width}} = \frac{\text{image distance}}{\text{object distance}} \qquad (6)$$

Precisely the same law applies to the formation of x-ray images.

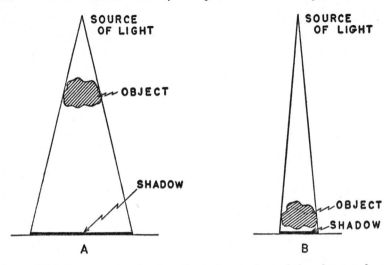

Figure 18.8. Image magnification. In *A* the shadow of the object is larger than in *B*, because in *A*, the object is nearer the source of light. The same principle applies to the formation of x-ray images.

For example, in cardiac radiography the standard procedure is to use a six-foot focus-film distance to avoid size distortion or magnification of the heart, so that measurement of the transverse cardiac diameter on the radiograph gives virtually the true diameter. On the other hand, contrary to what is usually taught, the cardiothoracic ratio (that is, the ratio of the transverse cardiac diameter to the transverse thoracic diameter) undergoes no significant change as the focus-film distance is varied between 40 and 72 in. This is due to the fact that as the focus-film distance is changed, these two diameters change at about equal rates. Figure 18.9 shows these relationships.

Thus, magnification, as the manifestation of size distortion, is inherent in radiography because of the geometry of image formation. As a rule *magnification or size distortion can be decreased by either reducing the object-film distance or increasing the focus-film distance.*

Radiographic magnification is easily calculated by applying the geometric relationship of similar triangles (see page 19). For example, an object measures 15 in. in diameter and lies 10 in. above the level of the film. The focus-film distance is 40 in. What is the magnification of the radiographic image? We must

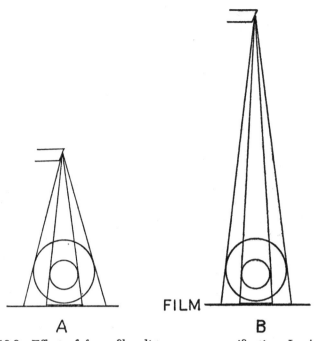

FILM

A   B

Figure 18.9. Effect of focus-film distance on magnification. In *A*, with a shorter focus-film distance, magnification is greater than in *B*. However, in this example, the *ratio* of the diameter of the *image* of the inner circle to that of the outer circle is essentially the same in both *A* and *B*.

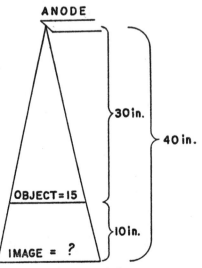

Figure 18.10. Magnification of the image in radiography. Image width/object width $= 40/30$. $x/15 = 40/30$. $\therefore x = 20$ in.

first determine the width of image by constructing a diagram as shown in Figure 18.10 and applying proportion (6):

$$\frac{\text{image width}}{\text{object width}} = \frac{\text{focus-film distance}}{\text{focus-object distance}}$$

$$\frac{\text{image width}}{15} = \frac{40}{30}$$

$$\text{image width} = \frac{40 \times 15}{30} = 20 \text{ in.}$$

The *linear magnification of the image,* or the enlargement of its width relative to that of the object, may now be expressed in one of two ways:

a. **Magnification Factor.** This is defined as the ratio of image width to object width:

$$\text{magnification factor} = \frac{\text{image width}}{\text{object width}} \qquad (7)$$

In the above example,

$$\text{magnification factor} = \frac{20}{15} = 1\tfrac{1}{3}$$

b. **Percentage Magnification.** This is defined as the percentage enlargement of the image as compared with the object, and is *not a true ratio.*

$$\text{percentage magnification} = \frac{\text{image width} - \text{object width}}{\text{object width}} \times 100 \qquad (8)$$

In the above example,

$$\text{percentage magnification} = \frac{20 - 15}{15} \times 100$$

$$= \tfrac{1}{3} \times 100$$

$$= 33\tfrac{1}{3}\%$$

The same geometric law of image formation is used to find the *diameter of the object* being radiographed when the image diameter, the focus-film distance, and the object-film distance are known. A diagram is prepared as in Figure 18.10, the known values inserted, and the object diameter determined from the proportion. This is the basis of most methods of finding the pelvic measurements of the mother with relation to the size of the fetal skull, a procedure known as *cephalopelvimetry.*

In *magnification radiography* we deliberately place the film at some distance from the object to obtain an enlarged radiographic image. With an ordinary x-ray tube (1 mm focus) the increase in image size would be more than offset by the increase in unsharpness. However, we can retain a reasonable degree of sharpness by decreasing the size of the tube focus, as might be anticipated from the discussion of geometric unsharpness (see above). Such *fractional focus tubes* are available with 0.3 mm and even smaller focal spots. The relationship of the focus-film distance to the focus-object distance for a desired magnification can be obtained from equation (6). Thus, for a magnification factor of 2 (that is, 100 per cent) we place the object midway between the tube focus and the film. With a 0.3 mm focus this arrangement produces radiographs of adequate sharpness. For larger degrees of magnification a 0.1 mm focus is needed for acceptable sharpness. At present, magnification radiography is finding its greatest usefulness in angiography, although it is by no means limited to such procedures.

2. **Shape Distortion** is caused by improper alignment of the object with relation to the tube focus and film. In Figure 18.11, an oval piece of lead placed in the x-ray beam nonparallel to the film, casts a circular shadow. A rectangular piece of lead can be placed in a beam so that it casts a square shadow. If these objects are placed perpendicular to the central ray and parallel to the film, the images will, of course, show no appreciable shape distortion and appear oval and rectangular, respectively. Under certain conditions, distortion is purposely employed to bring out parts of the body that are obscured by overlying structures; for example, in radiography of the gall bladder the patient is often rotated to displace the gall bladder shadow away from the vertebral column.

We can summarize the data on distortion as follows: *there are*

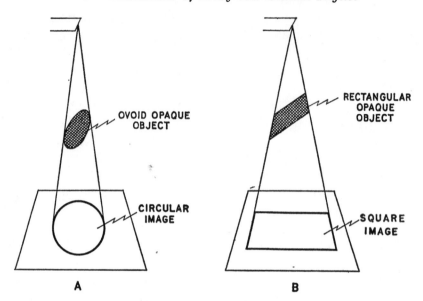

Figure 18.11. Shape distortion. In *A* the projected image of an oval opaque object is circular because the object is not parallel to the film. In *B* the image of a rectangular opaque object is almost square.

two types of distortion—size distortion and shape distortion. *Size distortion* is magnification caused by progressive divergence (spread) of the x rays in the beam: the shorter the object-film distance and the longer the focus-object distance, the less the degree of size distortion.

Shape distortion is due to improper alignment of the tube, object, and film. It is, of course, possible to have shape and size distortion occurring together if the causative factors of both conditions are present.

Since distortion and definition are influenced by similar factors —focus-film distance and object-film distance—*the greater the distortion of the radiographic image the poorer the definition* (see Figure 18.3). You can readily demonstrate this principle by observing the shadow cast by a pencil on a white surface. As the pencil is moved toward the source of light, the shadow not only becomes larger (distortion) but also becomes more blurred (poorer definition). As the pencil is moved nearer the white surface, the shadow becomes smaller and sharper.

QUESTIONS AND PROBLEMS

1. Define the four factors in radiographic quality, and explain how they affect it.
2. How can distortion be minimized?
3. An object being radiographed is located 4 in. above the film. The focus-film distance is 40 in., and the diameter of the object is 9 in. What is the size of the image? What is the percentage magnification? What is the magnification factor?
4. Define umbra; penumbra. Which impairs definition?
5. Name and discuss briefly the five major factors that influence definition (sharpness).
6. What is the relationship between penumbra and definition (sharpness)? What is unsharpness?
7. What is the greatest enemy of radiographic definition, and how is it minimized?
8. What five factors determine radiographic density? What is density?
9. State the inverse square law in your own words.
10. At a point 50 cm from the anode, the exposure rate of a beam is 20 R per min. What will the exposure rate be at 25 cm?
11. A certain technic requires an exposure of 70 kV and 100 mAs at a distance of 60 in. If we wish to decrease the distance to 20 in., what new mAs value is required to maintain the same radiographic density?
12. A certain technic calls for 30 mA and $\frac{1}{10}$ sec at 65 kV. If we have an uncooperative patient and wish to reduce motion by using an exposure time of $\frac{1}{60}$ sec, what will the new mA value be?
13. What determines the penetrating power of an x-ray beam? How can penetrating power be changed?
14. How do intensifying screens affect density? Contrast? Definition?
15. List the main materials in the human body in decreasing order of their radiographic density.
16. What is meant by the radiologic image? How is it formed?
17. What is meant by the radiographic image?

18. Discuss radiographic contrast; subject contrast; film contrast. Describe and explain the characteristic curve of a film recording system.
19. What is the function of radiographic contrast?
20. Explain information and noise in radiography.
21. What is quantum mottle? State its importance in radiography. Under what conditions is it accentuated?

CHAPTER 19 DEVICES FOR IMPROVING RADIOGRAPHIC QUALITY

IN THE PRECEDING chapter, we described the important factors affecting radiographic quality. We shall now show how the quality of a radiograph can be improved by the use of various special devices.

As already noted, the rotating anode tube, with its small focus, provides excellent radiographic definition. Furthermore, this tube has made possible the use of high mA with very short exposures, thereby minimizing motion, the greatest enemy of definition. Thus, while the inherent construction of modern equipment should provide radiographs of superb quality, certain auxiliary devices are needed to reduce *scattered radiation,* the remaining obstacle to be overcome in securing the best possible radiographic quality.

Scattered Radiation

You should recall from Chapter 12 that the primary x-ray beam leaving the tube focus is heterogeneous in that it contains photons of various energies. The primary beam is made up of *brems rays* resulting from the conversion of the energy of the electrons as they are stopped by the target; and *characteristic radiation* emitted by the target metal due to excitation of its atoms. As the primary beam passes through the patient, some of the radiation is absorbed, while the rest is scattered in many directions. In the diagnostic range of 30 to 140 kV the scattered radiation consists of characteristic photons resulting from photoelectric collision, and scattered photons produced by the Compton Effect. The characteristic photons are of extremely low energy and are absorbed locally in the tissues; but the *scattered photons, having an energy nearly equal to that of the primary radiation, are able to pass through the body and eventually approach the film from many directions.*

The multidirectional, scattered radiation is a noise factor which seriously impairs radiographic quality by its *fogging effect*, diffusing x rays over the surface of the film and thereby reducing contrast (see Figure 19.1). You can see that the larger the per-

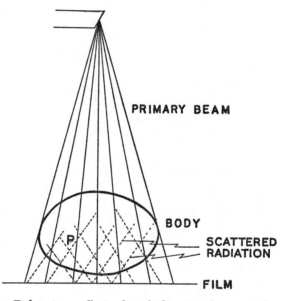

Figure 19.1. Deleterious effect of multidirectional scattered radiation on radiographic quality. This causes loss of radiographic contrast of a detail such as *P*.

centage of scattered radiation relative to the primary radiation, the greater will be the loss of contrast of a detail such as *P*.

Of the total radiation reaching the film, the *ratio of scattered radiation relative to primary radiation increases with the following factors:*

a. An *increase in the area of the radiation field and the thickness of the part* traversed by the beam. As will be shown below, restriction of beam size by means of a collimator has a limited beneficial effect in decreasing the amount of scattered radiation.

b. *An increase in the kilovoltage applied to the tube.* As the kV is increased, there is an increase in the importance of the Compton interaction *relative to* the photoelectric interaction (see page 179); recall that the Compton interaction is responsible for the production of scattered radiation. Furthermore, at higher kV the scattered photons are more energetic, that is, more penetrating and therefore more likely to reach the film. In general, the energy of the scattered photons is only slightly less than that of the primary photons, as shown in Table 19.1. Finally, as the

TABLE 19.1

RELATIONSHIP BETWEEN PRIMARY PHOTON ENERGY AND
ENERGY OF PHOTONS SCATTERED AT 45°

Primary Photon	*45° Scattered Photon*
kV	*kV*
40	39
60	58
80	76
100	94
120	112

applied kV is increased, more and more of the radiation is scattered in a forward direction, that is, at an angle of 45° or less with relation to the direction of the primary beam. Thus, increasing kV increases not only the energy of the scattered photons but also the number directed toward the film, so that a larger fraction of the scattered radiation can reach the film and impair contrast.

c. An *increase in the density of the tissues* traversed by the primary beam; for example, as between air and soft tissue. In non-grid exposures of the chest and abdomen, the following are approximate relative values of the scattered and primary radiation reaching the film:

chest	50% scattered	50% primary
abdomen	90% scattered	10% primary

REDUCTION OF SCATTERED RADIATION
BY MEANS OF A GRID

Since scattered radiation cannot be entirely eliminated because of the very nature of x rays and their interaction with matter, we can only reduce it to a minimum. This is especially important in the radiography of large anatomic areas such as the abdominal viscera, because of the greater amount of scattering by such a large volume of tissue. Most effective in reducing scattered radiation are the *stationary grid* and the *moving grid*. Although the latter is by far the more frequently used of the two, the stationary grid will be discussed first because of its historical priority and relative simplicity.

Stationary Grid

In 1913, Gustave Bucky designed and built the first stationary grid, a flat, thin, rectangular device the same size as the film, placed **between the patient and the cassette** for the purpose of reducing the amount of scattered radiation reaching the film. The grid is composed of alternating parallel strips of lead and plastic. In modern grids, there are 60 to 110 such lead strips per inch, arranged vertically as shown in Figure 19.2. The alternating strips of lead and translucent material (aluminum preferred) are extremely thin.

How does a radiographic grid function? Figure 19.3 shows a

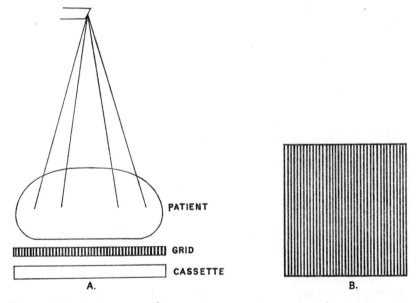

Figure 19.2. Stationary grid, seen in cross section in *A*, and top view in *B*.

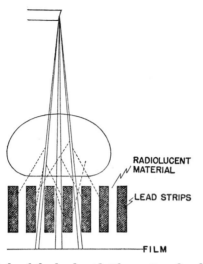

Figure 19.3. Principle of the lead grid. The scattered radiation, represented by the dotted lines, proceeds in various directions and is absorbed in large part by the lead strips. The rays that pass straight down through the x-ray transparent (radiolucent) material between the lead strips reach the film.

magnified diagram of a section of a grid. Only those x-ray photons passing directly through the narrow spaces will reach the film. Since the scattered photons are directed at various angles, most of them will strike the lead strips and be absorbed, never reaching the film. Such a grid absorbs as much as 90 per cent of the scattered radiation and produces a startling improvement in radiographic contrast and detail perceptibility. The stationary grid has one main disadvantage—the lead strips cast shadows on the radiograph as thin white lines; however, at the usual radiographic viewing distance, and especially with high quality grids, the lines are almost invisible. Besides, their presence is more than compensated for by the improved quality of the radiographic image. Because of the interposition of the grid in the x-ray beam, a significant fraction of both primary and scattered radiation is absorbed, and therefore the exposure must be increased according to the type of grid being used.

Efficiency of Radiographic Grids. Certain criteria have been established to designate the quality or efficiency of radiographic grids; that is, *their ability to remove scattered radiation*, or *"cleanup."* These apply to both stationary and moving grids (to be described later), and include *physical* and *functional* aspects.

1. *Physical Factors in Grid Efficiency.* The two physical factors mainly responsible for grid efficiency are *grid ratio* and *number of lead strips per inch.*

a. *Grid Ratio* is defined as *the ratio of the height of the lead strips to the distance between the strips,* as expressed in the following equation:

$$r = \frac{h}{D} \tag{1}$$

where r is the grid ratio, h is the height of the lead strips, and D is the width of the spaces between the strips. These factors are shown diagrammatically in Figure 19.4. As an example, if the lead strips are 2 mm high and the space separating them is 0.4 mm, the grid ratio is $2/0.4 = 5$.

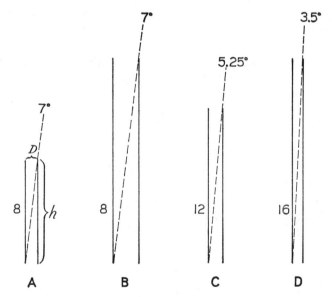

Figure 19.4. Effect of grid ratio on maximum angle of obliquity of a scattered ray that can get through the grid. In A and B, it is obvious that the ratio may be the same in grids of different thickness, and 7° is the maximum angle that a ray can make with the vertical and still pass between the lead strips. In D, with the grid ratio twice that in A, the maximum obliquity of a transmitted ray, 3.5°, is half as large as that in A.

Notice in Figure 19.4 that the ratio in *A* is the same as in *B*, despite the difference in height of the two grids. In Figure 19.4*D* the grid has the same height as in *B*, but because of the closer spacing of the lead strips in *D* the ratio is higher. Further comparison of the grids in Figure 19.4 reveals that the most divergent scattered photon in *A* and *B* makes the same angle with the vertical, while it makes half that angle in *D*. Thus, the higher the grid ratio, the "straighter" the rays have to be to get through the grid. All other factors being equal, the higher the grid ratio, the better will be the "cleanup" of scattered radiation and the resulting radiographic contrast and clarity of detail.

b. *Number of Lead Strips per Inch* has an important bearing on grid efficiency. In general, the more strips per inch the thinner the strips, and the greater will be the likelihood of scattered photons passing through the strips, especially at high kV. With grids of equal ratio, the one having fewer strips per inch possesses superior quality but at the same time the visibility of the grid lines becomes more objectionable (except in moving grids). Conversely, *as the number of strips per inch increases, the grid ratio must also be increased to maintain the same efficiency.*

In recent years, the work of Swedish investigators has suggested that the weight of lead per square inch of grid may be the most important factor in grid quality. However, this has not been generally accepted. Studies in the United States have shown that this criterion is only a rough approximation, and that the *number of lead strips per inch and grid ratio together provide a better indication of the cleanup of scattered radiation by a grid.* For those interested in this subject, the International Commission on Radiologic Units and Measurements (ICRU) *Handbook 89* provides abundant, well-documented material.

2. *Functional Factors in Grid Efficiency.* Knowing that a grid has a particular ratio and number of lead strips per inch, how can we judge its ability to minimize scattered radiation? In other words, is there a quantitative measure of grid efficiency based on its actual function? Fortunately, there are several criteria of grid function, the best ones being *selectivity and contrast improving ability,* as defined by the ICRU in *Handbook 89.*

a. *Selectivity.* Since the grid lies in the path of the x-ray beam, its lead strips absorb a fraction of both primary and scattered radiation. The absorption of primary radiation by a grid is known as **grid cutoff**. Obviously, a grid should transmit as large a fraction of the primary radiation as possible, to keep the exposure of the patient small; and at the same time, absorb the maximum amount of scattered radiation, to provide the best possible improvement in radiographic contrast. As specified by the ICRU, if Greek letter Σ represents selectivity, T_p the fraction of *primary* radiation transmitted through the grid, and T_s the fraction of *scattered* radiation transmitted through the grid, then

$$\Sigma = \frac{T_p}{T_s} \qquad (2)$$

Thus, the larger the ratio of the unabsorbed (transmitted) fraction of the primary radiation to the unabsorbed fraction of the scattered radiation reaching the film, the greater will be the selectivity of the grid. At the same time, it must be remembered that the larger the fraction of the *total radiation* (primary + scattered) that is transmitted and reaches the film, the less the exposure required for a given radiographic density.

b. *Contrast Improvement Factor.* The ultimate criterion of the efficiency of a grid is its actual performance in improving radiographic contrast, as explained earlier in this chapter. The contrast improvement factor, K, is defined as the ratio of the x-ray contrast with grid divided by the contrast without grid:

$$K = \frac{x\text{-}ray \ contrast \ with \ grid}{x\text{-}ray \ contrast \ without \ grid} \qquad (3)$$

Unfortunately, this factor is so dependent on kV, field size, and thickness of part that it cannot be specified for a given grid under all conditions. (The ICRU has given special recommendations for measuring K in a 20-cm thick water phantom at 100 kV.)

We must emphasize that selectivity and contrast improvement factor have certain limitations because they are measured in a water phantom which is inherently different from the human body. In making these measurements, certain conditions of standardization have to be laid down, and these may or may not hold during practical radiography. However, selectivity and contrast improvement are still two useful criteria for the intercomparison of grids.

Practical Aspects of Stationary Grids. These are available in various sizes. If a grid larger than the casette is used, a wooden frame must be placed around the cassette to avoid bending the grid. Since stationary grids are extremely light, they facilitate portable radiography, and radiography with a horizontal beam.

A special *crisscross grid* is available, consisting of two sets of grid lines arranged perpendicular to each other. However, its practical superiority over conventional high ratio grids has not been proved. A more versatile arrangement is the superimposition of two lower ratio grids, one of which is turned 90° relative to the other; the grid ratio of such a combination of crossed grids equals the sum of the individual grid ratios. For example, if grids with ratios of 5 and 8 are crossed, the resulting cross-grid combination has a ratio of 13, although centering of the x-ray beam need not be any more perfect than with an 8 ratio grid. On the other hand, the grids can be used separately when a lower grid ratio suffices.

Almost invariably, grids are used with intensifying screen cassettes because of the relatively heavy exposures that would otherwise be required. Because the grid is very thin and does not appreciably increase the object-film distance, it causes little or no distortion of the radiographic image. The **grid cassette** whose

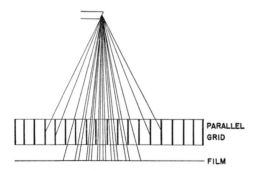

Figure 19.5. The effect of excessively short target-film distance on efficiency of a parallel grid. The x-ray beam is sharply divergent, and a large fraction of the peripheral rays will be absorbed by the lead strips, with reduction in density near the outer margins of the radiograph. This is one form of *grid cutoff.*

front has a built-in grid is much more convenient to use than an ordinary cassettee with a separate grid.

Some stationary grids are *parallel* or *not focused;* that is, the lead strips are all parallel to each other. Others are *focused* (see page 340 and Figure 19.7). With the grid at excessively short focus-film distances, the outer, more slanting rays of the primary x-ray beam strike the sides of the lead strips and cause underexposure of the periphery of the radiograph, that is, *peripheral grid cutoff* (see Figure 19.5). If the beam is not exactly perpendicular to the grid, with a tilt across the direction of the lead strips, a larger amount of radiation will be absorbed on one side of the beam so that one edge of the radiograph will be lighter —another example of **grid cutoff** (see Figure 19.6). However, with

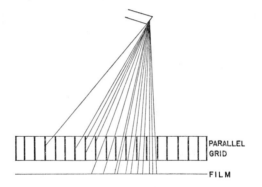

Figure 19.6. The effect of having the x-ray beam inclined toward the grid. One portion of the radiograph will be underexposed, as indicated at *A*, an example of *grid cutoff*.

the beam tilted parallel to the length of the strips, there will be no variation in density across the radiograph (except for the heel effect).

Moving Grid

To eliminate the grid lines produced in a radiograph by a stationary grid, Dr. Hollis Potter, in 1920, conceived the idea of moving the grid between the patient and the film during x-ray exposure, in a direction perpendicular to the lead strips. The moving grid has come to be called, loosely, a Potter-Bucky diaphragm, but the ICRU considers this to be too confusing; they recommend instead the term *moving grid,* reserving the name Potter-Bucky for the mechanism that activates the grid. However, the term Bucky for the grid and mechanism together has become so standardized in diagnostic radiology that it would seem futile to try to discard it.

The direction in which the grid moves is important. As just mentioned, it must move in a direction perpendicular to the long dimension of the lead strips. Thus, in Figure 19.7 the grid would

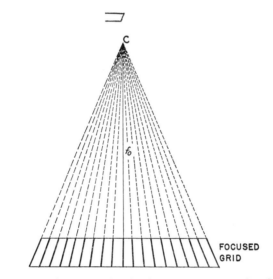

Figure 19.7. Focused type grid. The lead strips are inclined so that their extensions, indicated by the broken lines, meet along a specified imaginary line, *C,* above the grid, called the *convergence line* (seen end-on). The vertical distance from the center of the grid surface to the convergence line is called the *focusing distance, f_o.*

move from left to right, or from right to left. Obviously, if it were to move "into" or "out of" this page, the images of the lead strips would not be blurred and the effect would resemble that of a stationary grid.

The grid consists of alternating strips of lead and radiolucent material, such as plastic or aluminum, but these can be thicker than in the stationary grid, because grid motion during the exposure blurs out the lead shadows ("grid lines") on the radiograph.

Construction of Moving Grids. Most moving grids (and some stationary grids) are *focused*; that is, the lead strips are tilted more and more toward the edges of the grid, as shown in Figure 19.7. If one extends the planes of these tilted strips upward, the imaginary planes intersect along a line called the *convergence line*, which is parallel to the surface of the grid. Naturally, when seen end-on as in the figure, the convergence line appears to be a point. The vertical distance between the convergence line, C, and the center of the grid is called the *focusing distance, f_o*, of the grid. This term should not be confused with the focus-film distance.

Precautions in the Use of a Focused Grid. Care must be taken if the beneficial effects of a moving grid on radiographic quality are to be fully realized. Some of the precautions apply equally to parallel and focused grids, whether they are stationary or moving. The important ones are as follows:

1. *Focus-film Distance.* A grid with a ratio of 8 or less is efficient if the focus-film distance is no greater than the focusing distance plus 25 per cent, and no less than the focusing distance minus 25 per cent. Thus, if the focusing distance is 40 in., focus-film distances of 30 to 50 in. are satisfactory with an 8-ratio grid. Below or above these limits, peripheral grid cutoff will be appreciable, as shown in Figure 19.8. *With grid ratios of 12 or 16, the focusing distance must coincide with the focus-film distance, and centering must be precise, to avoid grid cutoff.*

2. *Angulation of the Tube.* With both stationary and moving grids, the tube must not be angled across the lead strips, otherwise cutoff is likely to occur near one edge of the radiograph. The tube may, however, be angled in a direction parallel to the strips without causing grid cutoff. In actual practice, the Potter-Bucky moving grid is placed under the table top with the lead strips oriented parallel to the long axis of the table, as shown in Figure 19.9. Consequently, the tube may be angled in the direction of the long axis of the table, but should not be angled across the table.

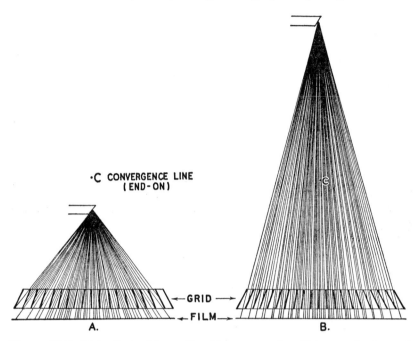

·C CONVERGENCE LINE
(END-ON)

←GRID→
←FILM→

A.   B.

Figure 19.8. The effect of focus-film distance on the efficiency of a focused grid. In *A*, the focus-film distance is less than the grid focusing distance minus 25 per cent, and more and more rays are absorbed by the lead strips near the edge of the film. In *B*, the focus-film distance is more than the grid focusing distance plus 25 per cent, and again there is relatively greater absorption of the beam at the edges of the film. In either case, the periphery of the radiograph will show reduced density—*grid cutoff*.

3. *Centering of the Tube.* The tube must be centered to the central axis of the grid for maximum efficiency. If the tube is off-

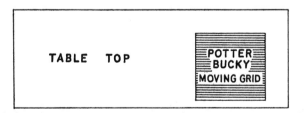

TABLE   TOP

POTTER
BUCKY
MOVING GRID

Figure 19.9. Top view of a radiographic table equipped with a Potter-Bucky diaphragm. Note that the lead strips are parallel to the long axis of the table.

center, across the direction of the lead strips, the *effect on radio-graphic density depends on the focus-film distance relative to the grid focusing distance.* If the tube focus is off-center and at a distance above the grid equal to the focusing distance, there will be uniform reduction in the density of the radiograph, as in Figure 19.10A. This can be readily proved by elementary geometry. If the tube focus is at an appreciable distance *above* the convergence line, and is at the same time off-center, the area of the film directly beneath the target will be lighter than the remainder of the film. If the tube focus is *below* the convergence line and off-center, then the area of the film directly below the target will be darker than the rest of the film. These relationships are illustrated in Figure 19.10.

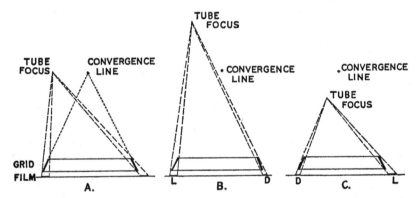

Figure 19.10. The effect of off-center position of the tube focus, on the efficiency of a focused grid. In *A*, the tube focus is off-center but level with the grid convergence line. The projections of the lead strips on the film are equal in size, but are broader than they would be if the tube were not off-center. Therefore, the radiographic density is uniformly decreased over the entire radiograph.

In *B*, the tube focus is off-center but is *above* the level of the grid convergence line. The projection of the lead strip lying more directly under the tube is broader than that of the corresponding lead strip on the opposite side (*L* is broader than *D*). Therefore the radiograph will show less density on the side toward which the tube is shifted.

In *C*, the tube focus is off-center but at a level below the grid convergence line. The result is opposite that in *B*. *L* = lighter. *D* = darker.

4. *Tube Side vs. Film Side.* The focused grid has a "tube side" and a "film side." In all the above figures, the grid is shown correctly with the tube side toward the tube target. If the grid is inserted reversed, as in Figure 19.11, the tilted strips near the

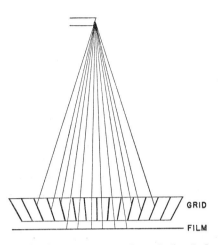

Figure 19.11. The focused grid is inverted, with the "tube side" away from the tube focus. Therefore, the outer regions of the radiograph will show decreased density or grid cutoff.

edges of the grid will absorb progressively more radiation, causing severe peripheral cutoff.

Practical Application of Moving Grids. Because of the absorption of both primary and scattered radiation by a grid, the exposure must be increased beyond non-grid technic. Moving grids with a ratio of 8 are most often used in conventional radiography at 85 kV or less. Such grids require an increase of about 15 kV above nongrid technic. However, these corrections are approximate because of the many factors governing the percentage of radiation absorbed by any particular grid; this can be determined by trial exposures. Since grids must be used in the radiography of parts measuring 11 cm or more in thickness, intensifying screens are required to reduce the patient's exposure and minimize the effect of motion.

At this point it would be advantageous to state the relative merits of grids with ratios of 12 and 16 *in general medium-kV radiography.* A grid with a *ratio of 12 is preferred* for the following reasons:

1. Its efficiency of cleanup of scattered radiation is not significantly less than that of a grid with ratio 16 in routine general radiography below 100 kV.

2. Centering is less critical, so there is a slightly larger permissible margin of error.

3. Less exposure of the patient is required.

4. Lead strips can be thicker, thereby absorbing more efficiently the higher energy scattered radiation, although this can be compensated for with a 16 ratio grid by having more strips per inch.

Above 100 kV a 16 ratio grid is needed for radiographs of optimum quality. Table 19.2 compares the characteristics of various types of grids from the practical standpoint. In general,

TABLE 19.2

CHARACTERISTICS OF VARIOUS TYPES OF GRIDS, STATIONARY AND MOVING

Ratio	Type	Recommended kV	Positioning Latitude	Cleanup
16	linear	above 100 kV	very poor	superlative
6	crisscross	above 100 kV	good	
12	linear	up to 100 kV	poor	excellent
5	crisscross	up to 100 kV	excellent	
8	linear	up to 90 kV	fair	good
6 5	linear	up to 80 kV	excellent	moderate

positioning latitude (that is, permissible error in centering, angulation, and distance) is worse with parallel than with focused grids of like ratio especially at distances under 40 in.

Efficiency of Moving Grids. The factors described for stationary grids also apply to moving grids—grid ratio, number of lead strips per inch, selectivity, and contrast improvement factor. These have been discussed on pages 333-337.

Specification of Grids. In 1963 the ICRU, in *Handbook 89*, recommended that manufacturers include certain data, in addition to grid ratio, to specify the characteristics of focused grids, both stationary and moving:

1. *Number of lead strips per inch;* for example, 80-strip, 100-strip.

2. *Focusing distance*—the distance between the convergence line and the grid surface.

3. *Focus-grid distance limits*—the limits between which the distance between the tube focus and grid surface can be varied without excessive loss of primary radiation (cutoff).

4. *Contrast improvement factor* under standard conditions. The first three items are to be included in grid labeling, the fourth is also included in grid specification.

Patient Dosage with Grids. Various combinations of kV and grid ratio affect the radiation exposure of the patient. Table 19.3

TABLE 19.3

EFFECT OF GRID RATIO AND CORRESPONDING kV AND mAs ON ENTRANCE SURFACE EXPOSURE OF PATIENT IN LATERAL PROJECTION OF LUMBAR SPINE. ALL RADIOGRAPHS OF COMPARABLE DIAGNOSTIC QUALITY*

Grid Ratio	kV	mAs	Entrance Exposure	Entrance Exposure as % of Maximum
			R	%
16	70	1,000	18.9	100
8	70	500	9.5	50
16	100	160	5.9	31
8	100	120	4.5	25

* University of Rochester School of Medicine, Department of Radiology.

compares these factors, obtained experimentally at the University of Rochester School of Medicine, in the radiography of the lumbar spine in the lateral projection. All the radiographs were diagnostically equivalent. It is evident that in radiography at 70 kV, the radiation exposure of the patient is twice as big with a grid ratio of 16 as with a ratio of 8. Therefore, *grids with a ratio of 16 should not be used in conventional radiography*—below about 90 kV. On the other hand, *above 100 kV* a 16 ratio grid produces far better cleanup of scattered radiation than lower ratio grids without a significant increase in patient exposure. For general radiography *up to 100 kV* a 12 ratio grid seems to be a satisfactory compromise both from the standpoint of grid efficiency and patient exposure.

Potter-Bucky Mechanisms. As mentioned before, the Potter-Bucky grid moves during the x-ray exposure. In the almost obsolete *single-stroke* type, one first "cocks the Bucky"; that is, pulls it to one side of the table by a lever, putting a spring under tension. When the grid is released (either by a string or by an electromagnetic tripping device), the spring pulls it across the table, its motion being smoothly cushioned by a piston acting against oil in a cylinder.

There are several causes of *grid lines* in a radiograph made with a moving grid. Since moving grids have relatively thick lead strips, the appearance of these grid lines in the radiograph as alternating light and dark stripes is especially objectionable. The common causes of grid lines with moving grids of various types include:

1. Exposure starting before grid has reached full speed, or continuing after grid travel has slowed down or stopped. The proper relationship between grid speed and exposure time is important in avoiding grid lines. An electric contactor starts the exposure only when grid travel has reached the correct speed, and stops the exposure before the grid stops moving. Grid lines appear whenever exposure occurs with the grid stationary.

2. Uneven or irregular movement of the grid.

3. Tube focus not centered to the center of the grid, especially with grids of high ratio.

4. Synchronism. We found in Chapter 12 that the x-ray beam is not generated continuously, but in intermittent showers corresponding to the peaks in the voltage applied to the tube. If the travel of the grid is such that a different lead strip always happens to be above a given point on the film at the same instant that a kV peak is reached, the images of different lead strips will be superimposed on the same point on the film, even though the grid is moving (see Figure 19.12). The net effect is the same as though the grid were stationary. Synchronism occurring in single stroke Bucky units can often be eliminated by setting the Bucky timer slightly beyond the point that produces synchronism.

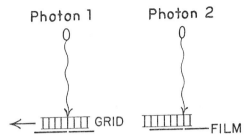

Figure 19.12. Grid synchronism. If the grid moves to the left at certain speeds, a different photon is intercepted by a different lead strip over the *same* point on the film. Hence, a shadow of the lead strip appears on the film just as though the grid were stationary.

The *reciprocating* Potter-Bucky mechanism has no separate timer, since it oscillates continuously without manual cocking. It is moved rapidly forward (0.3 sec) by a solenoid, while tension is placed on a spring, and returns slowly (1.7 sec) by action of this spring against oil in a chamber as in the single-stroke Bucky. Minimum exposure is 1/20 sec.

The *recipromatic* Potter-Bucky is activated entirely by an electric motor, the speed of both strokes being identical; exposure times may be as short as 1/60 sec.

The moving grid has its greatest application in the radiography of thick parts, and since relatively long exposures are required, cassettes with intensifying screens should be used routinely.

REDUCTION OF SCATTERED RADIATION BY RESTRICTION OF THE PRIMARY BEAM

We can decrease the fraction of scattered radiation by modifying the primary x-ray beam. This depends on two well-known facts. In the first place, *an increase in the kV to a high value with conventional equipment increases the fraction of scattered radiation.* Therefore, the kV for a given part should be just sufficient to penetrate it. Excessively high kV should be avoided, unless heavier filtration and a grid with a high ratio, such as 12 or 16, is used as explained above. Under ordinary conditions, with a grid ratio of 6 or 8, the peak potential should not exceed about 85 kV.

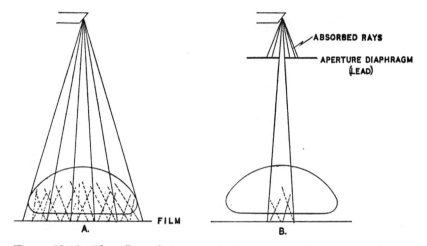

Figure 19.13. The effect of aperture diaphragm C on the volume of tissue irradiated and the resulting scattered radiation. In A there is no restriction of the x-ray beam; a large volume of tissue is irradiated, with resulting abundance of scattered radiation. In B the aperture diaphragm narrows the beam, a small volume of tissue is irradiated, and there is less scattered radiation.

In the second place, *the fraction of the radiation scattered increases as the volume of irradiated tissue increases.* Therefore, the larger the area and the thicker the part being radiographed, the greater will be the amount of scattered radiation; and hence, the greater will be the impairment of radiographic quality. The practical lesson to be learned from this is the need of restricting or *collimating* the x-ray beam to encompass the smallest area that will include the selected anatomic region. We do this with *collimators, cones,* or *aperture diaphragms* placed in the path of the x-ray beam, *as close to the tube port as the housing permits.* Beam restriction is most effective when the field is narrowed to a diameter of less than 6 in., especially in non-grid radiography. However, with larger areas, such as the abdomen, the *grid* is the most efficient means of removing scattered radiation; beam restriction adds to radiographic quality when large areas are to be covered only if the collimator collar reaches and surrounds the tube port.

1. *Aperture Diaphragms.* Figure 19.13 shows how an aperture

diaphragm decreases scattered radiation, thereby enhancing contrast. However, about 50 to 90 per cent of the density of a radiograph may be due to scattered radiation, so that restriction of the primary beam requires an increase in exposure, to compensate for the loss of density. The increase in exposure must be found by trial or in published tables. In setting up a technic chart, we must state the dimensions of the beam (or film size) for each technic. Whenever the beam is further restricted or enlarged, the exposure must be increased or decreased, respectively.

The aperture diaphragm would be very efficient if the x rays originated at a true point source. However, since the tube focus has a measurable area, the x rays actually originate at innumerable points in the focus so that some x-ray photons "undercut" the diaphragm opening (see Figure 19.14A). Thus, the diaphragm does not adequately collimate the beam and, consequently, its use should be strictly limited to situations in which a collimator cannot be adapted to existing equipment, such as older mobile units.

2. *Cones.* The radiographic cone is a conical metal tube placed in the x-ray beam to restrict its area. However, the cone is subject to the same criticism as the aperture diaphragm in that it permits undercutting of the penumbral radiation, as shown in Figure 19.14. A modification of the cone, the *extension cylinder*, has apertures of equal size at the top and bottom; and because of the increased distance of the lower aperture from the tube focus, penumbral undercutting is reduced (see Figure 19.14B).

Another serious shortcoming of the cone is the excessive circular area of the irradiated field required to cover a rectangular film of any given dimensions (see Figure 19.15). This is objectionable because it exposes the patient to unnecessary radiation. A partial solution would be to have cones with rectangular openings just sufficient to cover each film size at various focus-film distances, but the collimator, as shown below, does this much more efficiently, provided it is used correctly.

3. *Collimators.* These are really variable aperture devices which, if properly designed, combine the best features of the cone and the aperture diaphragm, at the same time minimizing penumbral width. The best type is a box-like apparatus provided

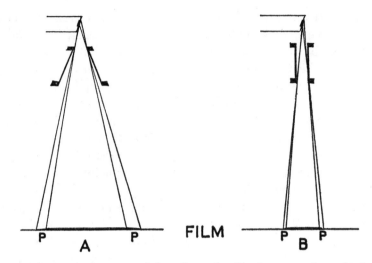

Figure 19.14. Comparison of the effect of a flared cone and a cylinder on the size of the penumbra. In *A,* the upper aperture limits the beam, similar to an aperture diaphragm. In *B,* the lower aperture limits the beam. Note the larger penumbra in *A* than in *B.* This shows why a cylinder is a more efficient collimator than a flared one.

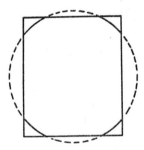

Figure 19.15. The x-ray beam cuts the corners of the film.

with at least two sets of adjustable diaphragms placed one above the other (corresponding to the upper and lower apertures of a cone) that can be opened or closed to delimit accurately a beam of any desired *rectangular size.* Furthermore, the double diaphragm *collimates* the beam (gives it a sharp margin) by severely narrowing the penumbra, as shown in Figure 19.16B. The best modern collimators have a collar extending to and surrounding the tube port for the purpose of further improving detail

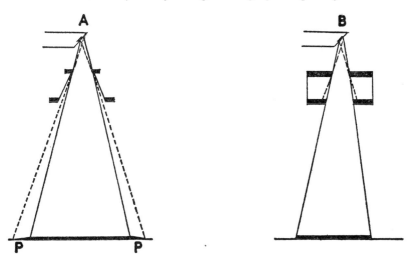

Figure 19.16. Effect of a well-designed collimator on penumbra. In *A*, an ordinary cone reduces the size of the beam, but there is "undercutting" of the upper aperture by penumbral radiation *P*. In *B*, a double diaphragm collimator eliminates penumbra.

sharpness. On the other hand, circular beams obtained with cones and single aperture diaphragms irradiate beyond the zones of interest (see Figure 19.15), unnecessarily exposing the patient; and, besides, they do not provide adequate collimation (see Figure 19.14A).

Most collimators are equipped with some type of illuminated beam-centering device. They also have calibrated scales to permit adjustment of both apertures for films of various sizes at different focus-film distances. In the most advanced types of modern equipment the collimator adjusts itself *automatically* to the size of any cassette clamped in the Bucky tray.

Film Coverage by Aperture Diaphragms and Cones. As just noted, collimators are calibrated, thus obviating the need of calculations. However, with aperture diaphragms and cones, we can compute film coverage by apertures of various sizes at desired focus-film distances (or the data can be obtained from technic manuals). To do this, we need the following information:

1. Distance from focus to aperture.

2. Distance from focus to film.
3. Diameter of aperture.

The calculation of film coverage by a beam limited by an aperture diaphragm is not difficult. In Figure 19.17, the opening is 1 in. in diameter and is 4 in. from the focus. The focus-

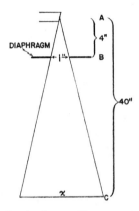

Figure 19.17. Method of calculating film coverage with an aperture diaphragm. In this example, $x/1 = 40/4$. Therefore, $4x = 40$ and $x = 10$.

film distance is 40 in. What is the diameter of the beam at the surface of the film, disregarding penumbral radiation? Note that the beam diverges from the focus (that is, the beam gets broader and broader) and continues to diverge as it passes through the aperture. In longitudinal section, this forms a larger triangle with its apex at the focus and its base on the film; and a smaller triangle with its apex at the focus and its base at the aperture. These are similar triangles, so by the application of elementary geometry we obtain the proportion:

$$\frac{AB}{AC} = \frac{d}{x}$$

where $AB =$ focus-aperture distance

$AC =$ focus-film distance
$d =$ diameter of aperture
$x =$ diameter of film surface covered by beam

Substituting the numerical values in the proportion,

$$\frac{4}{40} = \frac{1}{x}$$

Crossmultiplying,

$$4x = 40$$

$$x = 10 \text{ in.}$$

Therefore, the x-ray beam at the level of the film is a circle with a diameter of 10 in. This is obviously too small to cover a 10 x 12 film, but it will cover the long and short sides of an 8 x 10 film. However, since the diagonal of the film measures 12.8 in., its corners will be clipped (see Figure 19.15). The diagonal of the film can be found by measurement or by calculation.

The same principles apply to the *cone*, but we must choose the correct measurements for calculating film coverage. A cone is usually so designed that the upper opening narrows the beam, as shown in Figure 19.14A. Therefore the calculations must be made as though an aperture diaphragm were at that level, the lower opening being completely disregarded.

With a *cylinder* the lower aperture always limits the size of the beam (see Figure 19.18), and therefore the calculations are made as though an aperture diaphragm were present at the level

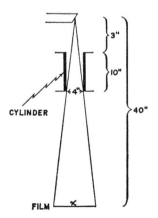

Figure 19.18. The effect of a cylinder in limiting the size of an x-ray beam.

of the lower opening. The following proportion may be used to find the film coverage diameter (that is, the diameter of the beam at the film):

$$\frac{\text{film coverage diameter}}{\text{cylinder diameter}} = \frac{\text{focus-film distance}}{\text{focus-to-lower aperture distance}}$$

where the "focus-to-lower aperture distance" is obtained simply by adding the distance between the focus and the top of the cylinder, to the length of the cylinder. Substituting the corresponding values from Figure 19.18 in the above equation,

$$\frac{\textit{film coverage diameter}}{4} = \frac{40}{3+10}$$

$$\textit{film coverage diameter} = \frac{4 \times 40}{13} = \frac{160}{13} = 12.3 \text{ in.}$$

Thus, an 8 x 10 film will have about ¼ in. cut off all four corners under these conditions, provided the beam has been centered exactly to the film. In general, the closer the cylinder is to the patient, the better is the collimation of the beam.

OTHER METHODS OF ENHANCING RADIOGRAPHIC QUALITY

The Heel Effect

As it leaves the focus, the primary beam does not contain a uniform distribution of radiation. The exposure rate at any point in the beam depends not only on the distance from the focus (inverse square law) but also on the *angle the rays make with the face of the target.* Figure 19.19, shows a 20° anode with the angular distribution of the exposure rates in an x-ray beam, based on the assumption that the exposure rate on the central axis of the beam (at a given distance) is 100 per cent. Thus, if the beam falls on a 14 x 17 film placed lengthwise with relation to the long axis of the tube, at a focus-film distance of 40 in., the exposure rate at the end of the film below the cathode will be about 105 per cent, and at the end of the film below the anode about 80 per cent of the exposure rate at the center of the film. This dis-

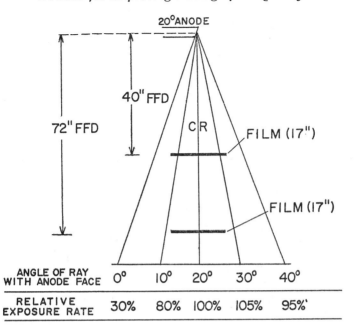

ANGLE OF RAY WITH ANODE FACE	0°	10°	20°	30°	40°
RELATIVE EXPOSURE RATE	30%	80%	100%	105%	95%˙

Figure 19.19. Angular distribution of radiation in an x-ray beam. The exposure rate is maximal in the central portion of the beam, decreasing toward the edges. The fall-off in exposure rate is greater toward the anode side of the beam; this is the *heel effect*. Note that the heel effect becomes more pronounced as the size of the film increases and the focus-film distance decreases.

crepancy becomes less pronounced as the focus-film distance is increased. *The variation in exposure rate with the angle of emission of the radiation from the focus is called the heel effect.* The smallest exposure rate is at the anode end of the beam. This is useful in certain procedures; for example, in radiographing a region of the body which has a wide range of thicknesses, we can take advantage of the heel effect by placing the thickest part toward the cathode end of the beam (where the exposure rate is greatest) and the thinnest part toward the anode (where the exposure rate is least). Accordingly, a more uniform radiographic density is recorded despite varying thicknesses of the object. For example, in radiography of the thoracic spine in the anteroposterior projection the patient should be placed with his head toward the anode and his feet toward the cathode because the upper part of the thorax is thinner than the lower part.

COMPENSATING FILTERS

Various filtering devices can be used in radiographing parts of the body that differ greatly in thickness or density, in order to record them with satisfactory radiographic density on a single film. Suitable filters include aluminum and various barium-plastic compounds. An illustration of this principle is the use of an *aluminum wedge* in radiography of the foot in the anteroposterior projection. Placing the thickest part of the wedge under the toes and the thinnest part under the heel produces radiographs of more uniform density. The disadvantage of such a system is the increased object-film distance resulting in distortion and some impairment of definition. We can avoid this by the insertion of a *compensating filter* in the filter slot of the x-ray tube housing, an arrangement that is particularly useful in the radiography of long segments of the spine, and of the pregnant uterus.

Another, and perhaps simpler, method of compensating for differences in tissue opacity requires the insertion of a sheet of *black paper between one screen and the film* in that part of the cassette which is to lie behind the more radiolucent tissues. A heavy exposure can then be given to penetrate the denser structures without overexposing the less dense ones, provided the difference in opacity is not excessive. For example, in soft tissue studies for placental localization in the lateral projection, black paper can be used to block one-half of one screen in that half of the cassette which is to lie beneath the anterior portion of the abdomen and pelvis. As a result, one can use an exposure that adequately portrays the more opaque posterior abdominal and pelvic structures, at the same time providing excellent soft tissue detail of the anterior structures. A similar procedure can often be used in the radiography of the chest when one side is denser than the other because of disease.

SUMMARY OF RADIOGRAPHIC EXPOSURE

In Figure 19.20 a typical arrangement of x-ray equipment and patient in radiography is shown to summarize the important items in radiographic exposure.

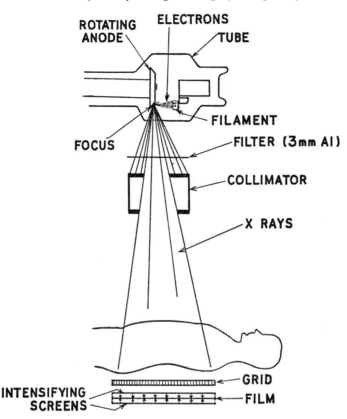

Figure 19.20. Summary of radiographic exposure.

QUESTIONS AND PROBLEMS

1. Discuss the three factors affecting the percentage of scattered radiation relative to primary radiation in the beam striking the film.
2. Describe the principle and construction of a radiographic grid.
3. Explain the physical factors in grid efficiency, including grid ratio and number of lead strips per inch.
4. Discuss the functional characteristics of grid efficiency, including selectivity and contrast improvement factor.
5. What is meant by the convergence line of a grid? The focusing distance of a grid? Describe briefly the three main types of Potter-Bucky mechanisms.

6. Discuss four precautions in the use of a grid.
7. What is the preferred grid ratio in conventional radiography? In high voltage radiography?
8. What causes decreased density of the edges of a radiograph when a grid is used? What is it called?
9. Name four causes of grid lines with a moving grid and state how each can be corrected.
10. Under what conditions may the x-ray beam be angled with respect to a radiographic stationary or moving grid?
11. Describe four causes of grid cutoff.
12. What is meant by grid ratio? What is the relationship between grid ratio and grid efficiency?
13. How does an aperture diaphragm diminish scattered radiation?
14. An aperture diaphragm measuring 2 in. in diameter is located 5 in. below the tube focus, What will be the diameter of the beam at 40 in.?
15. Describe an adjustable rectangular-field collimator. Why is it the most desirable type of beam-limiting device?
16. A cylinder measuring 12 in. length is used to narrow an x-ray beam. The lower opening is 15 in. from the focus and is 4 in. in diameter. What is the largest standard film that will be covered (except the corners) at a focus-film distance of 40 in.?
17. What is the most effective method of eliminating scattered radiation in radiography?
18. What is meant by the heel effect? How is it influenced by the focus-film distance?
19. Draw a simple diagram summarizing radiography.

CHAPTER 20 SPECIAL EQUIPMENT

A NUMBER OF USEFUL modifications of radiographic equipment are available for special purposes. Among them are image intensification, stereoscopic radiography, tomography (body-section radiography), and photofluorography. These will be described in the following sections.

BRIGHT FLUOROSCOPY (IMAGE INTENSIFICATION)

Fluoroscopy is an invaluable adjunct in diagnostic radiology, permitting the visualization of organs in motion; positioning of the patient for spotfilming; instillation of opaque media into hollow organs; insertion of catheters into arteries; and other important procedures. Unfortunately, ordinary fluoroscopy, utilizing a zinc cadmium sulfide screen, is an extremely inefficient process, both physically and visually. This is due not only to the very low brightness level of the fluoroscopic image, about $1/30,000$ that of a radiograph on a fluorescent illuminator, but also to the nature of human vision which we shall explore briefly in introducing the subject.

The human eye is a living camera. Light reflected from an object enters the eye and passes through the lens which focuses it on the *retina*, a membrane lining the inside surface of the back of the eye (see Figure 20.1). Within the retina there are two types of light-sensitive cells, the *cones* and the *rods*. Concentrated in a small spot at the center of the retina, the cones are sensitive to light at ordinary brightness levels; since they are not concerned with vision in dim fluoroscopic light, they will be ignored for the present. The rods are microscopic rod-shaped cell bodies distributed throughout the retina, except at its center. They are responsible for light perception at the *low brightness*

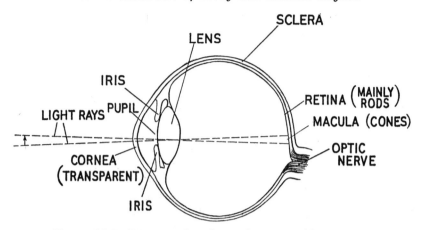

Figure 20.1. Diagram of midsagittal section of human eye.

levels prevailing in conventional fluoroscopy, but they exhibit very poor visual acuity (that is, ability to discriminate small images). In fact, we know that the contrast (difference in brightness) of two adjacent images must be approximately 50 per cent before they can be perceived by rod vision, but cones can discriminate images with a contrast of about 2 per cent in high level radiographic illumination. Since fluoroscopic contrast averages about ¼ to ½ that in radiography, it is very difficult to see detail in the dim fluoroscopic image. Another factor in rod vision is the prolonged dark adaptation, 20 to 30 min, required for the rods to attain adequate sensitivity, which then has to be preserved by conducting fluoroscopy in an almost completely darkened room. Fortunately, the rods are particularly sensitive to the green light emitted by the zinc cadmium sulfide screen.

In summary, then, ***conventional fluoroscopy is an exceedingly inefficient process*** because of the low brightness level, the low contrast, the dependence on rod vision with its poor visual acuity, and the need of prolonged dark adaptation and a darkened room.

In 1942 Dr. W. E. Chamberlain, in a classic paper, predicted that if the brightness of the fluoroscopic image could be increased about 1,000 times, cone vison would be brought into play and visual acuity immeasurably improved. Besides, dark adaptation and a completely darkened room would become unnecessary.

After extensive research by various manufacturers, intensification of the fluoroscopic image has become a reality through the development of the *x-ray image intensifier*. Its main component is the image tube, an electronic device which converts instantaneously, in several steps, an x-ray image pattern into a corresponding visible light pattern of significantly higher energy per square cm of viewing screen. Figure 20.2 shows the basic con-

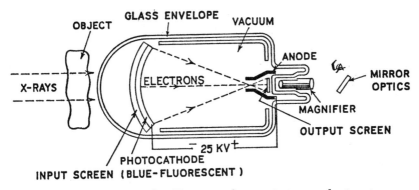

Figure 20.2. Image intensifier. X-rays produce an image on the *input screen,* whose light liberates electrons from the *photocathode.* Their concentration is proportional to the light intensity at any given point. The electrons are accelerated at 25 kV and focused on the *output screen,* where the final intensified image is formed.

struction of a typical image intensifier. After passing through the patient, the x-ray beam enters the *highly evacuated glass envelope* of the image tube and produces a conventional fluorescent image on the *input screen* which resembles an ordinary fluoroscopic screen except for its finer grain. The light from this image then falls on the closely applied photo-cathode, liberating electrons in the form of an *electronic image* that is a duplicate of the original but is made up of various densities of electrons instead of photons. The electrons forming the image are then *accelerated* toward the anode by an applied high voltage—about 25 kV—and focused, by means of a series of electrostatic "lenses," on the *output screen.* Here the electrons' energy is converted to a corresponding visible light pattern or image, whose brightness is about 4,000 to 7,000 times that of an ordinary fluoroscopic screen operating under similar conditions, a comparison that is designated as *brightness gain.* The image finally passes through an optical system either for direct viewing, for closed circuit television, or for cine (motion picture) recording.

Brightness gain depends upon two main properties of the image intensifier. In the first place, in an image tube with an input screen of 9 in. diameter, and an output screen with a 1 in. diameter, the output screen has about $\frac{1}{80}$ the area of the input screen. Since the light photons have essentially been "crowded" into an area $\frac{1}{80}$ as large, the brightness will increase eighty times simply through reduction in size or *minification* of the image. In the second place, the acceleration of the electrons increases the brightness by a factor of about 50 times. Hence, using these figures as an example, we arrive at an increase in brightness of the output screen as compared with the input screen of about $80 \times 50 = 4,000$ times. Further increase in brightness can be achieved by application of higher accelerating voltage, but this is not unlimited. When the final image is passed through the optical lens and mirror system (see Figure 20.3) it regains about 80 to 90 per cent of its original size without appreciable loss of brightness.

The ICRU has proposed a standard for specifying the intensifying ability of an x-ray image intensifier, based on strictly defined conditions. This is the *conversion factor,* defined as the *ratio of the brightness* (strictly speaking, *luminance*) *of the output screen to the exposure rate of radiation at the input screen.* The conversion factor is a more precise concept than brightness gain and should eventually supersede it.

Certain defects inherent in the image intensifier can be improved by the use of auxiliary equipment. For example, we can enhance image contrast by means of television monitoring.

There are several advantages in the image intensifier, making it an indispensable part of the equipment in the modern Radiology Department.

1. Increases brightness level and image contrast. Cone vision is brought into play by the higher brightness level, resulting in improved visual acuity.

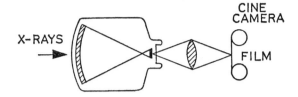

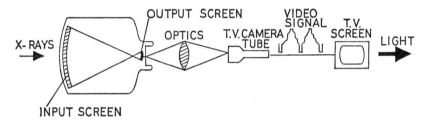

Figure 20.3. Two common auxiliary methods of recording the image in an image intensifier. A movie camera is used in the upper diagram, and closed circuit television in the lower diagram. (Adapted from *ICRU Handbook 89*.)

2. Eliminates need for dark adaptation and an almost totally darkened fluoroscopic room.

3. Decreases radiation exposure rate to patient, although this is limited at present because a certain irreducible minimal brightness is needed at the input screen to provide adequate detail visibility in the final intensified image.

4. Can be used to localize certain structures such as the optic foramina for spot filming.

5. Can be picked up by a television camera for more convenient viewing and for enhancement of contrast (see Figure 20.3). Note that the image intensifier by itself cannot increase the inherent contrast of the image presented to it on the input screen.

6. Can be picked up by a cinecamera for cineradiography (see Figure 20.3) or by a video tape system.

The basic x-ray image intensifier has been so improved in its maneuverability, optical system, and adaptability to routine fluoroscopic procedures, that it is rapidly becoming an essential part of the equipment, not only in the hospital Radiology Department, but also in the private radiology office. The improvement in the auxiliary systems, such as television, cine, and video tape, has immeasurably widened its field of usefulness beyond that of just a few years ago.

STEREOSCOPIC RADIOGRAPHY

Stereoscopy is the process of "seeing solid"; that is, seeing in three dimensions. An ordinary photograph, having only two dimensions, lacks depth because the camera has "one eye" and projects the image on the film in only two dimensions. On the other hand, when a pair of human eyes sees the same subject, the brain perceives it as a three-dimensional solid body. How is this brought about? Each eye sees the object from a slightly different angle depending on the distance of the object and the distance between the pupils of the eyes (see Figure 20.4). Hence, slightly

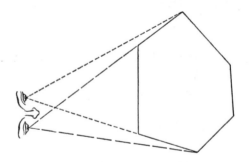

Figure 20.4. Stereoscopic vision. Each eye sees a slightly different view of the solid object. These two images are blended in the brain to give the impression of one image having a solid appearance; that is, height, width, and depth.

different images are formed on the retinas of the respective eyes and are carried separately over the optic nerves to the brain. Here the two images are fused and perceived as one having three dimensions—height, width, and depth.

Stereoscopic vision also permits us to judge distance. You can easily verify this by trying to touch an object with one eye closed, and then trying to do the same thing with both eyes open. Still, it is possible for a one-eyed person to learn to judge distance fairly well, although stereoscopy is impossible.

We are not born with stereoscopic vision but develop this faculty during infancy. Dr. O. V. Batson has shown that we differ in our ability to fuse stereoscopic images, and that in many instances this ability can be improved by corrective glasses, exercise, and practice.

Principle. It is possible, and indeed desirable in some cases, to produce a stereoscopic image radiographically. To do this we must attempt to duplicate ordinary stereoscopic vision. First, we must make *two separate exposures of the same object from different points of view;* the x-ray tube takes the place of the eyes, exposing two films from slightly different positions (see Figure 20.5). The part being examined is immobilized while the tube is shifted to obtain the two stereoradiographs. Second, we must *view properly* the pair of stereoradiographs if we are to achieve stereoscopic or depth perception.

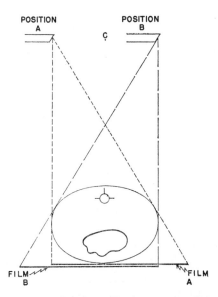

Figure 20.5. Stereoradiography. Two films are exposed separately, one with the tube target at *A*, and the other with the target at *B*. Note that the object is not moved. The target positions are at equal distances from the center *C*.

Stereoradiography—Tube Shift. The question arises, "How far must the tube be shifted between exposures to obtain a satisfactory pair of stereoradiographs?" Under ideal conditions, the stereoscopic image should appear to lie about 25 in. in front of the eyes for optimum viewing. Furthermore, the interpupillary distance (that is, the distance between the pupils of the eyes) normally averages about 2.5 in. These two values are in the ratio of $\frac{1}{10}$, and the same ratio holds for the relationship between tube shift and focus-film distance:

$$\frac{\text{tube shift}}{\text{focus-film distance}} = \frac{1}{10}$$

Thus, according to this equation, stereoradiography at a focus-film distance of 40 in. requires a tube shift of 4 in.; for a focus-film distance of 72 in., the tube shift is about 7 in. We may conclude that the *tube shift is theoretically one-tenth the focus-film distance when the viewing distance is 25 in.* While, in most instances, this ratio lays down the conditions for excellent stereoradiographs, some observers find it difficult to fuse the images unless the ratio is reduced to $\frac{1}{13}$, or even $\frac{1}{16}$. However, the resulting apparent depth of the stereoscopic image will be less than that with the larger ratio of $\frac{1}{10}$.

The *direction* of the tube shift relative to the anatomic area is important. In general, the tube is shifted perpendicular to the dominant lines; thus, chest stereoradiography requires a tube shift across the ribs. When the tube is shifted perpendicular to the grid strips, there should be no difference in the density of the two radiographs if the tube has been correctly centered to the grid. But when the tube is shifted parallel to the grid strips, there may be a difference in the densities of the radiographs due to the heel effect; and there will also be a difference in sharpness of the two radiographs because the size of the effective focus will be different in the two positions. This is shown in Figure 20.6. However, with a *high ratio Bucky grid such as 16,* because of the criticality of centering, the shift *must be parallel to the long axis of the table* to obtain stereoscopic pairs of nearly equal density.

We must be careful to open the collimator so as to cover the films in both positions of the tube. If the beam is collimated

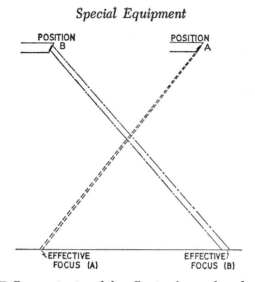

POSITION B   POSITION A

EFFECTIVE FOCUS (A)   EFFECTIVE FOCUS (B)

Figure 20.6. Difference in size of the effective focus when the tube is shifted parallel to the long axis of the tube. Note that in position *A* the effective focus is smaller and therefore definition is better than in position *B*. The density of the radiograph will be reduced in position *A* because of the heel effect. But with a grid having a ratio of 16, these deficiencies have to be ignored because they are less serious than those existing when the shift is across the grid strips.

excessively, the tube will have to be tilted to cover both films. The steps in stereoradiography may now be summarized:

1. Center accurately the part to be radiographed.
2. Determine the total shift of the tube for the given focus-film distance. Suppose this shift is to be 4 in. (that is, $\frac{1}{10}$ of 40 in.).
3. Determine the correct direction of shift.
4. Now shift the tube one-half the total required distance, away from the center. In this case, $\frac{1}{2} \times 4 = 2$ in. Make the first exposure.
5. With the patient immobilized in the same position, change the cassette. (For chest stereoradiography, patient should suspend respiration.)
6. Now shift the tube the total distance, in this case 4 in., in a direction exactly opposite the first shift, and make the second exposure. (This is the same as shifting the tube one-

half the total distance to the opposite side from the center point.) The exact method of shifting is shown in Figure 20.7.

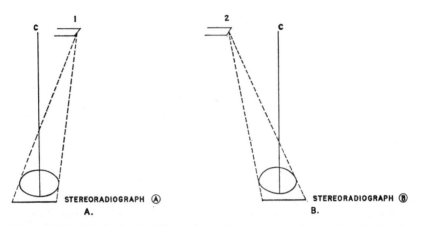

STEREORADIOGRAPH Ⓐ

A.

STEREORADIOGRAPH Ⓑ

B.

Figure 20.7. Method of shifting the tube in stereoradiography. In *A*, the tube has been shifted one-half the total distance to one side of the center, *C*, for the first exposure. In *B*, it has been shifted the full distance in the opposite direction for the second exposure.

7. Process the films.
8. The stereoradiographs are now ready for viewing.

Viewing Stereoradiographs. There are various methods of viewing a pair of stereoradiographs so that they are perceived three-dimensionally. These will now be described.

The films can be viewed stereoscopically without auxiliary equipment. They are placed side by side in illuminators, in the same relative position as they were "seen" by the x-ray tube, and with a horizontal shift. The observer then attempts to fuse the images by "crossing" his eyes. However, not many are able to do this successfully.

Usually, optical devices are necessary to facilitate stereoscopic vision. Their basic purpose is to displace the image of one of the stereoradiographs onto its mate. The simplest device is a 30° or 40° *glass prism*, which is available at optical supply houses. Figure 20.8 shows how a prism bends the light rays coming from one radiograph so that they seem to come from the other one, thereby fusing the images to achieve depth perception. To be

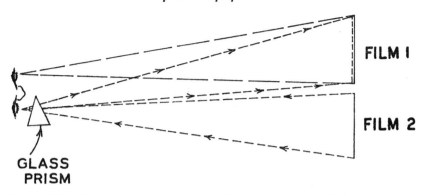

Figure 20.8. Basic principle of stereoradiographic viewing. The glass prism bends the light rays from stereoradiographic *2* in such a way as to superimpose its image on stereoradiograph *1*. This facilitates the fusion of the images to provide depth perception.

viewed properly, the stereoradiographic pair is held up with the tube side facing the observer; in other words, the eyes see the radiograph just as the tube did. At the same time, they are oriented so as to represent a *horizontal shift of the tube.* If the tube shift were vertical, as in radiography of the chest, the films would have to be rotated 90°, the chest then appearing to lie on its side. On one radiograph, the image will be seen to be *farther from the left hand edge;* this radiograph was made with the tube in the left position, and is therefore placed in the *left illuminator.* The other film was made with the tube in the right position and is placed in the right illuminator. The observer now holds the prism between the thumb and index finger of the right hand in front of the right eye, with the *thinnest portion near the nose,* as shown in Figure 20.8. (If preferred, the prism can be held with the left hand in front of the left eye, again with the thinnest edge near the nose.) By slightly adjusting the position of the prism and stepping forward and backward, the observer soon succeeds in fusing or blending the images. This method requires very little practice, is inexpensive, and provides a convenient *portable monocular stereoscope.* If preferred, two 20° prisms may be used, one being held before each eye, again with the thinnest part near the nose; this is the simplest type of *portable binocular stereoscope.*

The more elaborate stereoscopes, such as the floor model *Stan-*

ford and *Wheatstone* units, are modifications of the types just described, making use of prisms or mirrors to displace the images. Because of their bulk and high cost, they are gradually being replaced by various types of small *hand-held binocular stereoscopes*. Some of these operate on the principle of the Stanford unit, while others use the Wheatstone method. With the Stanford type, the stereoradiographs are placed in the illuminators in the same way as for the 40° prism. The same procedure holds for the Wheatstone unit, except for the *added step* of flipping each radiograph over, around a vertical axis; the image in the left hand illuminator is now nearer the left edge of the film, whereas that in the right hand illuminator is nearer the right edge.

One must be careful to place the stereoradiographs in the illuminators correctly, as described above, for true stereoscopy. Only then will an anteroposterior radiographic projection appear to be anteroposterior, and conversely. For example, with a chest stereoradiographed in a posteroanterior position, the anterior ribs should appear to curve away from the observer, if the films are viewed correctly. This verification, which is important in the localization of lesions or foreign bodies, can be simplified by taping an identifying marker—"stereo"—to the table top or cassette holder, to indicate the part that was farthest from the tube during radiography.

The frequency with which stereoradiography is used varies widely in different departments, depending on the preference of the individual radiologist. It has its greatest value in localizing lesions more accurately and in giving a better perception of the shape, structure, and relationship of a lesion to adjacent anatomic areas.

TOMOGRAPHY
(BODY-SECTION RADIOGRAPHY)

In ordinary radiography a three-dimensional object—the body—is represented on a two-dimensional film. In effect, the object has been compressed virtually to zero thickness in the direction of the x-ray beam so that the images of a number of structures are superimposed on each other in the radiograph. Not infrequently, this makes radiographic interpretation difficult, if not impossible.

We have at our disposal three methods of separating such superimposed images through modification of ordinary radiography:

1. *Increased Distortion.* We can separate confusing images by simply rotating the patient or angulating the x-ray tube. Thus, we use an oblique projection of the gall bladder to separate its shadow from that of the spine, or from that of gas-filled bowel.

2. *Decreased Focus-film Distance.* When structures are relatively far apart along the axis of the x-ray beam, we can use a very short focus-film distance. This increases the geometric blurring (see page 300) of the images of structures farthest from the film, while keeping the images of structures nearest the film reasonably sharp. An example is radiography of the mandible in the posteroanterior view with a very short focus-film distance, blurring the image of the cervical vertebrae.

3. *Motion of a Part.* When two structures are directly aligned in the x-ray beam and the undesired one happens to be movable, we can set it in motion and blur its image while the desired image remains sharply defined. For example, the dens (odontoid process) of the second cervical vertebra can be well shown in the anteroposterior projection by having the patient move his jaw during the exposure.

While the methods just described are frequently successful, they are not always so. Furthermore, they fail completely when the desired structure is buried within a denser one, as is the case with a cavity inside a densely consolidated or scarred part of the lung. Here we make use of a technic which is a radical departure from ordinary radiography, a technic called *body-section radiography* or *tomography.*

Tomography provides a reasonably sharp image of a particular structure, while images of structures lying in front of or behind it are blurred beyond recognition. Ideally, the desired image stands out sharply in a hazy sea of blurred shadows. *Tomography simply consists of motion of the tube and film in opposite directions in a prescribed manner during the exposure.*

It is of interest that body-section radiography, one of the outstanding innovations in radiology, was invented independently in France by Bocage (1921), and Portes and Chasse (1922); in the Netherlands by Ziedses des Plantes (1921); and in the United States by Kieffer (1929), a radiologic technologist.

Although various systems of body-section radiography have been developed—planigraphy, stratigraphy, laminagraphy, tomography—the ICRU, in Report 10f, has proposed the term *tomography* for all these systems. The device is called a *tomograph,* and the radiographic set of films *tomograms.* We shall use this terminology.

Principles of Tomography. This can be most readily understood by first considering how a *moving* source of light casts a shadow of an opaque object. The model used here is a pen flashlight, a pencil, and a sheet of paper. If these items are arranged in a darkened room as in Figure 20.9A, and the light is moved to the right as in Figure 20.9B, the shadow of the pencil will be seen to move to the left across the paper. Suppose we replace the paper with a sheet of film and repeat the experiment. Since the shadow of the pencil moves across the film, its image on the film

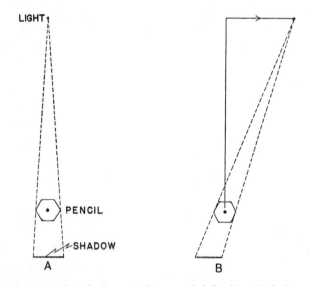

Figure 20.9. In *A* the shadow of the pencil falls directly below. In *B* the light has moved to the right, displacing the shadow to the left. If the shadow were to fall on a moving film which moved to the left as the same speed as the *shadow,* the latter would have a sharp (unblurred) image.

will be blurred. But *if we move the film* to the left *at the same speed as the shadow*, the position of the shadow on the film remains constant and its image will be sharp.

Suppose, now, we line up three pencils in the path of the light from a stationary source as in Figure 20.10. They will cast a single shadow representing the superimposed shadows of the three pencils. But if the light source is moved to the right, as shown in Figure 20.10, while the film is moved to the left at the same speed as the shadow of one of the pencils, such as *1*, its shadow image will remain sharp on the film. At the same time, the shadow image of pencil 2, moving more rapidly than the film, and the shadow image of pencil 3, moving more slowly than the film, will both be blurred. This, basically, is the tomographic principle.

Substituting an x-ray beam and two structures in the body for the pencil-light model, we find an analogous situation. As shown

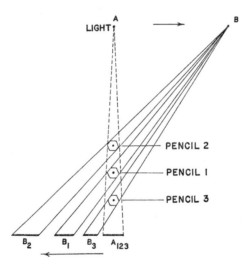

Figure 20.10. The effect of motion of a source of light on three objects lined up as shown. If pencil *1* is at the pivot level and if the film is moved to the left at the same speed as shadow B_1, this shadow will remain sharp. The shadow of pencil 2, B_2, moves faster than the film and is blurred. The shadow of pencil 3, B_3, moves slower than the film and is also blurred out. Note that the shadow of pencil 2 is larger than that of pencil *1*, whereas the shadow of pencil 3 is smaller than that of pencil *1*.

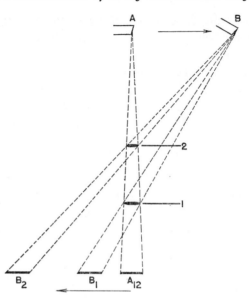

Figure 20.11. Principle of tomography. At position A the images of structures *1* and *2* are superimposed on the film at A_{12}. When the tube is shifted
to B and the film moves in the opposite direction so that the axis of rotation
(pivot) of the x-ray beam is at level *1*, the image of *1*, B_1, will have a
constant position on the film and will be *sharp*. In other words, A_1 and B_1
will strike the same spot on the film. But image B_2 will shift relative to B_1
and will blur out. (In practice, the tube is moved from position B through
A to a corresponding point on the opposite side of A during the exposure.)

in Figure 20.11, with the tube at A, the images of structures *1*
and *2* are superimposed on the film. If the tube is now shifted to
the right (from A to B) *during the exposure* while the film is
shifted in the opposite direction at the same speed as the image
of the desired structure, *1*, this image will maintain a constant
position on the film and be sharply defined. On the other hand,
the image of structure *2* will move faster than the film and be
blurred.

By applying the tomographic principle, we can radiograph a
"slice" of desired thickness, called a *tomographic section,* at various depths in the body. The image of the slice is reasonably
sharp, whereas the images of structures lying in front of and
behind it are blurred beyond recognition. As we shall see, this

is especially important in the radiography of certain anatomic areas such as the first cervical vertebra, the larynx, the sternum, and the temporomandibular joints. It is also indispensable in uncovering pulmonary cavities in dense lung, and delineating constricted bronchi and mediastinal invasion in lung cancer.

Let us now describe the basic principles of any *genuine* system of tomography. During tomographic exposure, these conditions must be satisfied:

1. *Reciprocal motion of the tube and film* (that is, in opposite directions at any instant).

2. *Parallelism of the film plane and the plane containing the desired structure* (that is, the *objective plane*).

3. *Proportional displacement of the tube and film.* The ratio of the focus-film distance to the focus-objective plane distance must be constant throughout the excursion of the tube.

To satisfy these conditions, the tube and the Bucky tray must be connected by a *rigid rod* which is free to rotate (within limits) about a pivot known as the *fulcrum.* There must also be movable *joints* or *linkages* between the rod and the tube, and the rod and the Bucky tray. A simple drawing of a *rectilinear tomograph* is shown in Figure 20.12. Only images of points in the plane of the

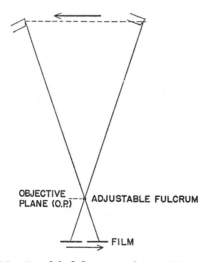

Figure 20.12. Simplified diagram of a rectilinear tomograph.

fulcrum parallel to the film—that is, the *objective plane*—remain in constant position on the film and are sharply defined. Images of points in planes above or below the objective plane are blurred in proportion to their distance from the objective plane. Such blurred images vary in size and shape and, ideally, are less visible than the relatively sharp image of the objective plane detail. Figure 20.13 shows how the radiographic image of a point, *P*, outside the objective plane shifts on the film relative to the position of the image of a point, *E*, inside the objective plane.

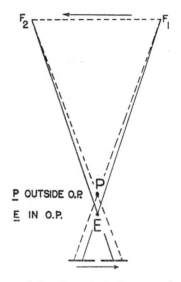

Figure 20.13. Relative shift of images of points lying outside (*P*) and within (*E*) the objective plane during tomographic excursion. Note that the image of point *E* retains the same location on the film, while the image of point *P* moves across the film and blurs.

In Figure 20.14 you see a simple rectilinear tomograph with the standard terminology and symbols used by the International Commission on Radiologic Units and Measurements (ICRU).

Thickness of Section. A tomogram is really a radiograph of a *slab* of tissue rather than of a plane of zero thickness. Since blurring gradually increases at greater distances from the objective plane (that is, level of fulcrum), there is no distinct boundary between the tomographic section (slab) and its surroundings.

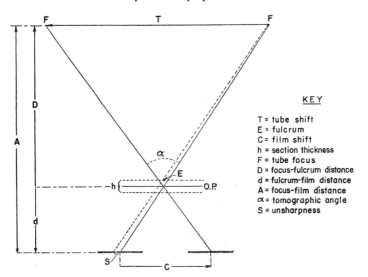

Figure 20.14. Standard terminology of the *ICRU* applicable to tomography.

Therefore we arbitrarily choose the boundary planes so that images of points in them have a certain amount of blurring and no more. Thus, *we define the tomographic section as the slab between these two planes of maximum permissible unsharpness.* Obviously, the tomographic images of all points in the tomographic section will be sharper than the images of points beyond the boundary planes. If the maximum permissible unsharpness has been properly chosen, say 0.5 mm, the images of all points within the slab will have satisfactory definition, (Note that unsharpness is the actual width of the border of a detail image; thus, if unsharpness is 0.5 mm, it means that the image border measures 0.5 mm.)

Two factors affect the thickness of section, *h*, for a given maximum permissible unsharpness, *S*:

1. *Tomographic Angle*—the angle described at the fulcrum by the excursion of the tube. The *larger the tomographic angle, the thinner the section,* and conversely (compare Figure 20.15A and 20.15B).

2. *Focus-film Distance*—with a constant *exposure angle,* a change in focus-film distance does *not* affect section thickness

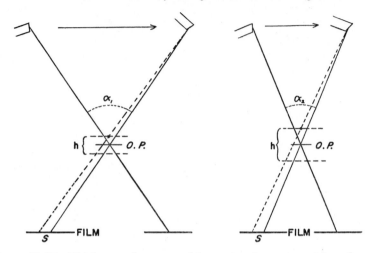

Figure 20.15. Thickness of tomographic section increases with a decrease in tomographic angle.

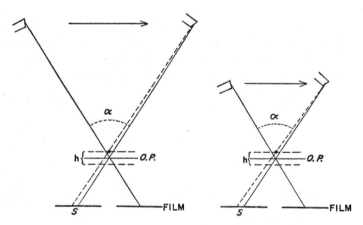

Figure 20.16. Tomographic section thickness is *not* affected by a change in focus-film distance if the *tomographic angle remains constant*.

(compare Figures 20.16A and 20.16B). However, when the tube shift or *amplitude* is measured in inches (rather than degrees as with exposure angle), the section thickness *does* depend on the focus-film distance. If the amplitude is held constant, section thickness decreases as focus-film distance decreases, and conversely (compare Figures 20.17A and 20.17B). For this reason, *we prefer to use tomographic angle* rather than amplitude in designating tube shift in tomography. In any case, focus-film distance influences image sharpness in tomography just as in ordinary radiography (see page 300).

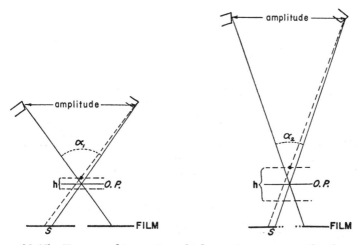

Figure 20.17. Tomographic section thickness increases as the focus-film distance is increased only if the *amplitude or linear shift of the tube remains constant* (compare with Figure 20.16).

A formula for section thickness derived by Kieffer may be simplified as follows by the use of simple trigonometry:

$$h = \frac{S}{tan\ \tfrac{1}{2}\alpha}$$

where h = section thickness
S = maximum permissible unsharpness
α = tomographic angle in degrees

If we select a reasonable maximum unsharpness such as 0.5 mm, then Table 20.1 shows the approximate section thickness for various tomographic angles. When plotted graphically, these data produce the curve shown in Figure 20.18. Note that *as the exposure angle increases there is a disproportionate reduction in section thickness.* It is important to know that tomographic image contrast decreases with decreasing section thickness.

Tomographic Image. A tomogram is essentially the radio-

TABLE 20.1

RELATION OF SECTION THICKNESS TO TOMOGRAPHIC ANGLE

Angle	Section Thickness
	mm
50°	1.1
40°	1.4
30°	2.0
20°	3.0
10°	6.0
5°	11.0

graphic representation of a particular tomographic section, *h.* It consists of two parts. First, it contains a *relatively sharp image of the tomographic section* whose definition, contrast, and density largely determine the qaulity of the tomogram. But the tomogram also contains the blurred images of points outside the tomographic section, called *redundant shadows,* which contribute significantly to the quality of the tomogram. Since the redundant shadows are generated by a moving x-ray tube focus, tomographic quality ultimately depends in large part on the *path of the tube focus during the exposure.* The more uniform the blurring of the redundant shadows, the better is the quality of the tomogram.

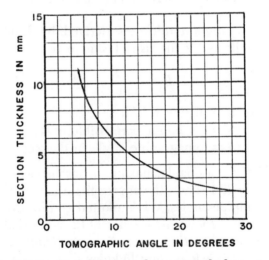

Figure 20.18. Relationship of tomographic section thickness to tomographic angle. Note that the *section thickness increases as the tomographic angle decreases.*

Thus, *tomographic quality* depends ultimately on (1) the sharpness and contrast of the image of the tomographic section and (2) the uniformity of blurring of the redundant shadows.

Types of Tomographic Motion. The four most important types of motion are shown in Figures 20.19. They may be summarized briefly as follows:

1. *Rectilinear*—tube and film move in straight lines parallel to each other, but in opposite directions (see Figure 20.19A). In other words, the movement is *plane parallel.*

2. *Curvilinear*—tube and film in an arc in the same plane (see Figure 20.19B).

3. *Vertical*—tube and film move in opposite directions along a line perpendicular to (or at some angle to) the objective plane (see Figure 20.19C).

4. *Pluridirectional*—tube and film move in parallel planes in circular, elliptical, hypocycloid, spiral, sinusoidal, or random paths (see Figures 20.19D). This movement is also *plane-parallel.*

Effect of Type of Motion on Blurring of Redundant Shadows.

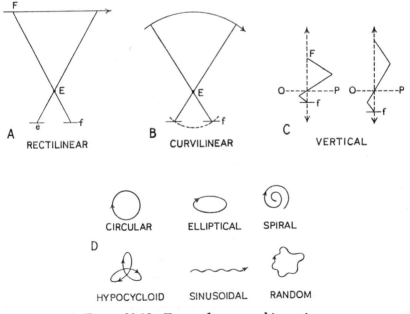

Figure 20.19. Types of tomographic motion.

As we have already stated, the blurring pattern of the redundant shadows is an important factor in tomographic quality. Since the blurring pattern, in turn, depends on the type of tomographic motion used, we shall now discuss this aspect of tomography.

Rectilinear tomography is the simplest and most commonly used system. Here, maximum blurring and most effective tomography occur when the dominant axis of the tomographed structure lies *perpendicular* to the tube shift (see Figure 20.20).

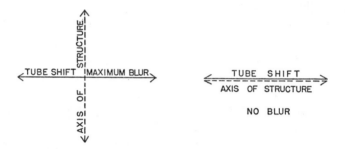

Figure 20.20. Relationship of degree of blurring to direction of tube shift. Note that maximum blurring occurs when the tube is shifted perpendicular to the long axis of the object and that no blurring occurs when the tube is shifted parallel to the axis.

Conversely, minimum blurring and least effective tomography occur when the dominant axis is parallel to the tube shift. For example, in anteroposterior tomography of the vertebral column, the lateral vertebral margins, being parallel to the tube shift, will appear sharp in *all* sections without blurring of structures outside the section; therefore, these margins are not truly tomographed. But the vertebral plates, oriented perpendicular to the tube shift, will be sharp only in the selected layer, other layers being blurred; hence, these margins will be truly tomographed. If the vertebral column were arranged with its long axis across the table, the lateral margins would be tomographed and the plates would not. A reasonable compromise would be to tomograph such elongated objects as the vertebral column, sternum, and long bones with their long axis placed *diagonally* on the table.

From the preceding discussion we can see that rectilinear tomography, despite its simplicity and easy availability, has certain deficiencies. In practice, we find:

1. *Incomplete blurring of points outside the tomographic section.* This becomes worse as the orientation of elongated structures approaches parallelism with the direction of tube shift. It accounts for the streaking that is so objectionable with rectilinear tomography.

2. *Elongation of the section image in the direction of tube shift* due to distortion from angular displacement of the image.

3. *Incomplete formation of the image of the tomographic section.* Radiographic image formation generally follows the *Law of Tangents;* that is, image-forming rays are those which pass tangential to (skim) the edges of object elements in the body. The greater the number of such tangential rays, the more closely will the image conform to the object. An ideal tomograph would therefore be one in which the blurring rays arise from an infinite number of sources, providing at the same time the greatest number of image-forming rays for points in the tomographic section. This situation does not exist in rectilinear tomography, so the tomographic image is incompletely formed. (See Littleton in Bibliography.)

Pluridirectional tomography uses complex motion of the tube and film. As already described, the incomplete image formed in rectilinear tomography is accompanied by linear streaking in the direction of tube shift due to parallelism of some of the object elements in the tomographic section. This can be overcome by moving the tube and film in a complex pattern while maintaining their paths parallel to each other. During such excursion of the tube, some component of motion is always perpendicular to the border of any structure in the path of the x rays, regardless of its orientation. But such equipment is extremely expensive and is necessarily limited to certain highly specialized centers. Therefore, most Radiology Departments must depend on rectilinear tomography, which is quite satisfactory, in general, if we bear in mind its limitations.

The most common types of pluridirectional motion are shown in Figure 20.19D and include:

1. *Circular*—redundant shadows may be ring-like, giving rise to double margins and fictitious sharp cutoff at image margins due to incomplete blurring of rounded objects. This is especially true if the diameter of the x-ray path approaches that of the object.

2. *Elliptical*—results are intermediate between circular and rectilinear paths.

3. *Spiral*—excellent blurring with a minimum of objectionable redundant shadows. However, it is limited mechanically to small exposure angles.

4. *Hypocycloid*—provides the most uniform blur and sharpest tomographic sections, with a good range of tomographic angles. There is a gradual falloff of density at image margins. This type of motion provides sections of about 1.3 mm with a tomographic angle of 48°, but exposure time may be as long as 6 sec unless three-phase equipment is used. The Law of Tangents is fulfilled in that x rays originate from virtually an infinite number of points during tube travel.

Limitations of Tomographic Equipment. One of the main problems in the design of tomographic equipment is that of lost motion or "play" at the linkages (joints). Each linkage causes a certain amount of lost motion because it is impossible to have an absolutely tight fit which still permits the required motion at the linkage. As the number of linkages increases so does the total lost motion, since the individual lost motions are additive. Thus, the least amount of lost motion may occur in the rectilinear system, and the greatest amount in the pluridirectional systems.

We are concerned with lost motion because it gives rise to loss of sharpness of the section image. Since there are many more linkages in the pluridirectional systems, much care must be used in their construction to minimize play at the joints. This has been accomplished to a surprising degree in modern pluridirectional tomographs.

As shown in Figure 20.21, the effect on image sharpness depends on the location of the lost motion. When it occurs:

1. *At the fulcrum*—maximum impairment of definition occurs

because the effect is magnified in the film plane (see Figure 20.21A).

2. *At the Bucky tray linkage*—unsharpness equals the lost motion because the latter is essentially at the film plane (see Figure 20.21B).

3. *At the x-ray tube linkage*—minimum impairment of definition occurs because it is minified at the film plane (see Figure 20.21C).

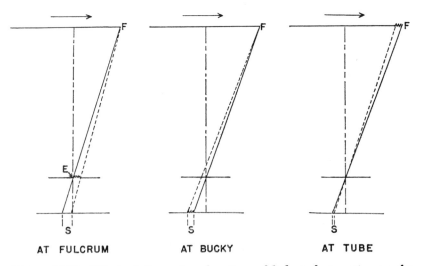

AT FULCRUM   AT BUCKY   AT TUBE

Figure 20.21. Effect of the various locations of linkage lost motion on the unsharpness of the tomographic section.

Limitations of Materials. There may be significant deterioration of tomographic image sharpness due to the thickness of the film emulsion with tomographic sections measuring less than 1 or 2 mm thick (that is, with large tomographic angles). Under these conditions, single emulsion films should be used for optimum tomographic quality. However, in ordinary tomography where section thickness is in the range of 5 to 10 mm, emulsion thickness does not significantly influence the sharpness of the tomographic image.

Applications of Tomography. As we observed at the beginning of this section, tomography is indicated whenever a structure to be radiographed would be obscured by the images of structures lying in front of or behind it, or whenever the desired structure lies within a denser structure. Included are lesions of the chest, sternum, dens, temporomandibular joints, upper thoracic spine (lateral), thoracic spine pedicles, cholecystography and cholangiography, nephrotomography, orbits, middle and inner ears, and others. Figure 20.22 shows a typical setup in rectilinear

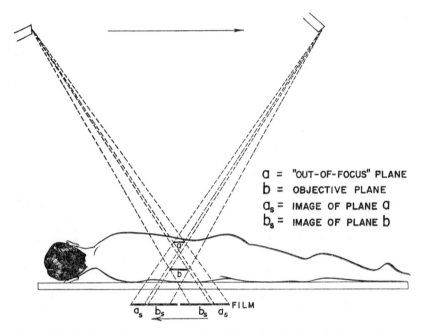

a = "OUT-OF-FOCUS" PLANE
b = OBJECTIVE PLANE
a_s = IMAGE OF PLANE a
b_s = IMAGE OF PLANE b

FILM
a_s b_s b_s a_s

Figure 20.22. Rectilinear tomography. The tube is shifted during the exposure in the direction of the upper arrow, while the film moves in the opposite direction. The fulcrum level is selected at the level of the desired objective plane, b. The image of b, b_s moves with the film, always striking the same spot. Its image is therefore sharp. But the image of a, a_s, constantly moving over the film during the exposure, is blurred out.

tomography. A metal rod (behind the table) connects the tube and Bucky tray through a movable linkage at each end. The rod is provided with a pivot or *fulcrum* which is adjustable at various heights above the table and determines the *objective plane.* Because the rigid rod pivots at the fulcrum, motion of the tube in one direction causes the Bucky tray to move in the opposite direction with proportional displacement of the tube and film during the entire range of motion. For protection of personnel, this motion should be activated by a motor through a switch located in the control booth.

To obtain a tomographic series, the patient is placed in the appropriate position and the desired structure localized by appropriate radiographs. Next, the fulcrum is adjusted for various heights above the table, corresponding to the series of objective planes needed to include the entire structure. At each level of the fulcrum a tomogram is exposed—commonly called a cut. Usually, the separation of successive fulcrum heights (and objective planes) is 0.5 or 1.0 cm, depending on the size and nature of the structure being investigated. Ordinarily, the tomographic angle is 30°, but there are certain exceptions which will be discussed later. In this manner, a series of tomograms is obtained in which at least one should show the structure with optimum definition (sharpness). For example, a series of tomograms of the dens (odontoid process) should include one that depicts the dens with optimum sharpness, while the remaining cuts will display it with varying degrees of unsharpness.

The *exposure* in tomography is greater than that in ordinary radiography of the same region because of the obliquity of the beam during most of the tube excursion, with consequent increase in the average focus-film distance and thickness of part. Although the increase in tomographic exposure varies with the area of interest and the type of tissue, it is usually 1½ to 2 times the corresponding conventional exposure.

Multitomography (multisection radiography). First proposed by Ziedses des Plantes in 1931, multitomography produces a set of tomograms of different sections simultaneously, with one sweep of the x-ray tube. A special multisection or "book" cassette is used, consisting of a series of pairs of graded intensifying screens, a film being placed between each pair of screens. Thus, a multisection cassette accommodates multiple films. Each pair of screens is separated from its neighboring pair by 0.5 or 1.0 cm balsa wood or polyester foam spacers, depending on the desired separation. Because of the decrease in intensity of the x-ray beam as it passes through the cassette (mainly by absorption in the films and screens), the speed of the screens is graded so that the slowest are on top and the fastest at the bottom.

Figure 20.23 shows the principle of multitomography Because of divergence of the rays, the spacers between the screen pairs

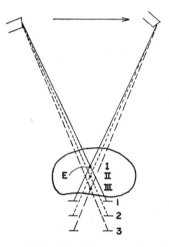

Figure 20.23. Multitomography. Simultaneous "cuts" of objective planes I, II, and III.

are slightly greater in thickness than the separation of the objective planes. The thickness of the spacers (more strictly, the separation of the films) can be calculated by the formula:

$$S_F = \frac{AS_L}{D}$$

where S_F = separation of films
        A = focus-film distance
        S_L = separation of objective planes
        D = focus-fulcrum distance

Commercially available multisection cassettes provide three or five pairs of screens with 0.5 or 1.0 cm spacers.

Multitomography has the advantage of saving time, since several tomograms can be obtained with a single excursion of the x-ray tube. However, it has the distinct disadvantage of producing tomograms with poorer contrast and definition, especially toward the bottom of the casette, as compared with a set of individual tomograms.

It is erroneously believed by many that multitomography saves exposure of the patient. Actually, radiation exposure of the patient is approximately the same for a multitomographic series (with screens) and for a set of individual tomograms containing the same number of films. This is explained by the fact that there is a limit to the speed of the fastest available screens which are at the bottom of the book cassette; hence, to balance the radiographic exposures, the uppermost screens must be very slow, with resulting overall increase in exposure. The *apparent* reduction in exposure of the patient during multitomography is due to the higher kV, but this could just as well be used in ordinary tomography with similar saving in patient exposure.

Plesiotomography (simultaneous plesiosectional tomography. Introduced by McGann, a radiologic technologist, plesiotomography is an important refinement of multitomography. The prefix *plesio* means *near* or *close,* indicating that the tomographic cuts are close to each other. A plesiotomographic cassette contains four pairs of speed-graded screens *without* added spacers; the screens themselves provide a spacing such that separation of the tomographic sections is approximately *1 mm.* Intensifying factors of the screen pairs, from top to bottom, are about 22.5, 32, 48, and 68. These cassettes are available in 8 x 10 and 10 x 12 in. sizes. The application of plesiotomography in the diagnosis of eighth nerve tumors in the internal auditory canal is described in McGann's paper. However, the procedure can be simplified by dividing the cassette in half (instead of quarters) to show both canals at the same time.

Plesiotomography is used principally in studies of the internal and middle ear. Most of the significant structures are beautifully demonstrated, including the external auditory canal, malleus, incus, semicircular canals, internal auditory canal, and mastoid antrum. Eighth nerve tumors and cholesteatomas can be shown by this method when conventional methods fail. A plesiotomographic cassette is relatively inexpensive and, although not a complete substitute for pluridirectional tomography, it should be available in the Radiology Department.

Zonography. First proposed by Ziedes des Plantes in 1931, zonography has recently been rediscovered. It is simply tomography with a *small tomographic angle*—5° to 10°—producing a thick section. For example, with a tomographic angle of 10° section thickness is 6 mm, and with 5°, 11 mm. Such a thick section, called a *zonogram,* superficially resembles an ordinary radiograph. It has better contrast and definition than a conventional tomogram made with a tomographic angle of, say, 30°. Zonography is helpful when (1) the obscuring structures are far from the objective plane and (2) thin sections are not needed. Examples of structures usually suitable for zonography are the sternum (thoracic spine far removed), temporomandibular joint (opposite joint far removed), nephrotomography (thin sections usually not needed), and cholangiography (thin sections not needed). Zonography is ordinarily ill-advised in studies of the lungs because small cavities may be missed. The paper by Ettinger discusses the practical use of zonography.

Autotomography. Mention should be made of this variety of tomography which does not require special equipment. Instead of moving the tube and film during the exposure, we can sometimes move the anatomic part and produce excellent tomographic effects. This is especially applicable when the desired structure is centrally placed. In 1949 Ziedses des Plantes introduced head movement in pneumoencephalography, which has since come to be known as *autotomography.* In recent years, Schechter has revived autotomography for improved delineation, during air encephalography, of the third ventricle, aqueduct, and fourth ventricle. With ordinary radiography these structures are obscured by the dense bone at the base of the skull. Since they are located virtually at the axis of rotation of the head, they will blur very little when the head is rotated during radiographic exposure in the lateral view; whereas structures lying farther laterally will blur in proportion to their distance from the center. Figure 20.24 shows diagrammatically how autotomography is accomplished.

Patient Exposure in Tomography. In general, the surface dose for a given tomographic exposure is of the same order of magnitude as that for ordinary radiography, but is distributed over a larger area of the body. As already mentioned, patient exposure with multitomography is about the same as for a sequential series of individual tomograms with the same number of

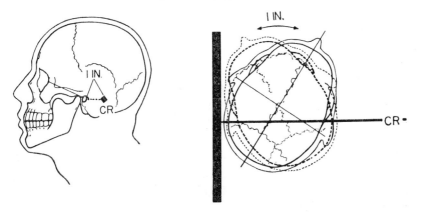

Figure 20.24. Autotomography. The head is rotated during the exposure, and only images of points in the axis of rotation of the head appear sharp on the film. This procedure is especially useful in air studies of the aqueduct and fourth ventricle.

films. Precautions during tomography are the same as for conventional radiography—adequate filtration (2 to 3 mm Al), high kV, collimation, shielding of gonads, and optimum processing of films. Under these conditions, and because its use is relatively limited, tomography adds an insignificant amount of radiation to the general population.

XERORADIOGRAPHY (XEROGRAPHY)

In 1937, C. F. Carlson invented *xerography* (pronounced "zero-"), a radically different kind of photocopying process. Not long afterward the xerographic method was applied as a substitute for film in radiography where it became known as *xeroradiography*.

The xerographic *principle* depends on the fact that semiconductors such as selenium, while ordinarily behaving as insulators, become conductors when struck by light or x rays. A typical xerographic plate consists of *selenium* coated on a sheet of metal, usually aluminum. In use, the aluminum is electrically grounded.

To make a *xeroradiograph* or, in short, a *xerogram*, a positive electrostatic charge is first applied evenly to the selenium surface. Next, the charged plate is placed under the anatomic part and an appropriate x-ray exposure given, just as in film radiography, although the exposure factors are different. In any area of the plate struck by x-ray photons, the selenium leaks off positive charges through the grounded aluminum backing, the amount of charge leakage depending on the x-ray exposure. Thus, the selenium loses much of its charge in areas beneath soft tissues which transmit large numbers of photons, whereas it loses relatively little charge in areas underlying bone which transmits few photons.

With this system you can see that we obtain an *electrostatic latent image* consisting of areas varying in the density of positive charges remaining on the plate after exposure. These densities are then converted to a visible or manifest image by spraying the plate with a toner—a fine cloud of negatively charged blue plastic powder which sticks to the various areas on the plate in amounts depending on the size of the residual positive charge. A permanent image is obtained by pressing a sheet of clear plastic or of plastic-coated paper against the plate, thereby transferring the blue powder to the plastic which is then peeled off the plate.

The plate can be reused after being cleaned and heat-treated for 45 sec, a process called *relaxation.* Heating is necessary to prevent a residual image that may form a "ghost" when the plate is reused.

At present xeroradiography is applicable only to smaller parts and soft tissues, especially mammography. It has the following significant *advantages* over film radiography:

1. **Detail Visibility.** Xerograms are unique in showing the so-called *edge effect,* manifested by accentuation of the borders between images of different density. This produces an *apparent* increase in local contrast. The edge effect is due to the "theft" of powder by the denser area from the less dense area along the border (electrical "fringe" phenomenon). As a result, details stand out strikingly, resembling bas relief. The edge effect becomes more pronounced as overall contrast is increased; therefore, *low or medium kV* technics are preferred.

2. **Low Overall Contrast.** Despite the apparently high contrast caused by the edge effect, the overall contrast of xerograms is *low* in comparison with ordinary radiographs. The low contrast and the edge effect conspire to produce xeromammograms in

which details stand out boldly and all structures, from fat to bony ribs, are clearly seen with a single exposure. However, even though xeromammograms are easier to interpret than film mammograms, comparative studies show them to yield about the same degree of diagnostic accuracy.

3. **Wide Latitude.** Xeroradiography is characterized by wide exposure latitude, so that variation of about 10 kV at 80 kV in nonmammogram technic has only a slight effect on xerogram quality. At the same time, a wider range of densities can be depicted on a single plate, facilitating interpretation.

4. **Magnification Effect.** Not entirely explained is the phenomenon of magnification of microcalcifications and other fine structures, making them more readily detectable.

5. **Rapid Process with no Liquid Solutions.** Processing is complete in 90 sec and is done in daylight in the radiographic room. No plumbing is required, since this is entirely a dry process.

6. **Less Patient Exposure in Mammography.** Exposure is reduced to about ⅓ in xerographic mammography as contrasted with film mammography. Excellent xeromammograms are obtained at 50 to 70 kV and 150 mAs, making possible a reduction in exposure time and thereby decreasing motion unsharpness. (Recall that film mammography requires about 20 to 30 kV and 1200 mAs.)

7. **Positive or Negative Image.** By proper selection of charges, we can obtain either positive or negative xerograms. Furthermore, these can be transferred to film or paper depending on the preference for viewing by transmitted or reflected light.

There are several *disadvantages* in xerography. An important one is in mammography where artefacts caused by dust or imperfections in the plate may produce shadows resembling calcifications. Serious interpretational errors may result if these are not taken into account. Another disadvantage is that xerography cannot be used at present in x-ray examinations of large parts, such as the trunk, because of the large exposures needed in comparison with film-screen radiography. However, continual improvement in equipment may eventually widen the field of usefulness of xerography beyond that of mammography and the radiography of small parts.

PHOTOFLUOROGRAPHY (PHOTORADIOGRAPHY)

In photofluorography, the x-ray beam, after traversing the chest, falls on a special fluoroscopic screen at the end of a light-tight hood. A high speed camera at the other end photographs the fluoroscopic image, producing a miniature photoradiograph. This provides an accurate and economical method of conducting mass surveys of the chest.

A photofluorographic apparatus (see Figure 20.25) includes:

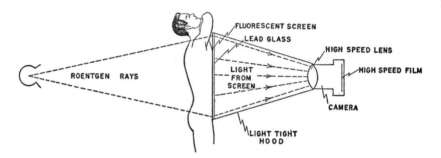

Figure 20.25. Essential components of a photofluorographic unit.

1. **X-ray Control, Transformer, and Tube** having a capacity of 200 to 500 mA at 100 kV. A rotating anode with high rating is essential for a large patient load. A phototimer expedites mass survey photofluorography.

2. **Fluorescent Screen** emitting *blue light* of sufficient intensity to be photographed. The screen measures 14 in. x 17 in. and is backed by lead glass to absorb x rays, at the same time allowing the light from the screen to pass to the camera.

3. **Light-tight Hood,** placed between the fluorescent screen and the camera, is shaped like a truncated pyramid, larger at the screen and tapering toward the camera.

4. **High-speed Camera** receiving the light from the fluorescent screen. The lens has tremendous light-gathering ability (aperture) because, at best, the light emitted by the screen is of low intensity.

5. **Photographic Film,** high speed, loaded into the camera to photograph the fluorescent image. Various film sizes are available: 35 mm, 70 mm, 4 in. x 5 in., and stereoscopic 4 in. x 10 in.

Excellent radiographs are obtained with exposures of $\frac{1}{10}$ to $\frac{1}{2}$ second. The grain of the fluorescent screen is effectively reduced by minification of the image, since the film image is much smaller than the screen image.

The advantages of photofluorography include: (1) the large patient load, (2) the excellent film quality, (3) the small film storage space required, (4) the transportability of the equipment (mounted in a truck or bus), and (5) the economy of small film size.

QUESTIONS

1. Discuss fluoroscopic image intensification. Compare it with ordinary fluoroscopy.
2. What is meant by stereoscopic vision?
3. What determines the correct tube shift in stereoradiography?
4. List the steps in making and viewing stereoradiographs.
5. Describe the principle of tomography (body-section radiography).
6. Define tomographic section; fulcrum; objective plane; redundant shadows; tomographic angle; amplitude.
7. Discuss the various types of tomographic motion. Compare their advantages and disadvantages.
8. Explain the main modifications of tomography: zonography; multitomography; plesiotomography; autotomography.
9. What is the comparative exposure with multitomography and with a series consisting of an equal number of individual tomograms?
10. How should body-section radiographic equipment be activated, for maximum safety of personnel?
11. What is meant by "tomographic section"? How is its thickness defined? Explain the factors that influence section thickness.
12. Show by diagram, completely labeled, the essential parts of a photofluorographic unit for mass chest radiography.

CHAPTER *21*   *RADIOACTIVITY AND RADIUM*

ONE OF THE OUTSTANDING ACHIEVEMENTS of the late nineteenth century was the discovery that certain elements had the unique property of giving off penetrating radiation. In 1896, Henri Becquerel reported his experiments showing that the heavy metal *uranium* emitted rays which passed through paper and darkened a photographic plate in a darkened room. Soon after this, Marie Curie and her husband Pierre began their search for other elements which might have the same peculiar property. In 1898, after two years of extraordinary hardship and personal sacrifice, they announced their discovery of *polonium* and *radium,* new elements with the property of emitting penetrating radiations. Not only did this discovery open new therapeutic opportunities in medicine, but it also expanded immeasurably our knowledge about the structure of matter. The term *radioactivity* was applied by Marie Curie to the special behavior of these elements.

RADIOACTIVITY

Unstable Atoms

What happens when an atom of a radioactive element such as uranium or radium emits radiation? There are two important phenomena associated with this sort of atomic behavior: (1) *the nucleus of the atom disintegrates spontaneously,* undergoing a change in its structure, and (2) *penetrating radiations are given off.*

Why do some elements undergo such spontaneous breakdown? The precise answer is unknown, but there is a rapidly growing body of information about nuclear structure.

The accepted designation of nuclear structure is *nuclide, defined as a nuclear species with a particular number of protons and neutrons, and having a particular energy state.* We know that all atomic nuclei contain protons and neutrons in rapid motion. In naturally occurring nuclides with a high atomic number—above 81—nuclear energy is large enough to cause random escape of nuclear particles with or without accompanying electromagnetic radiation. Such nuclides are called *radioactive nuclides* or *radionuclides.* They are unstable and tend to "explode," throwing out certain types of radiation and changing to other nuclides. It is uncertain which particular nucleus of a sample of nuclide will break down at any particular instant, but a *constant fraction of those present will decay in unit time* (for example, per sec). This fraction, called the *decay constant* (other terms, *disintegration* or *transformation constant) is the same for all samples of a given radionuclide, but is different for different radionuclides.* Furthermore, the decay constant, for all practical purposes, *cannot be changed.*

Radioactivity, then, may be defined as the ability of certain nuclides—radionuclides—to (1) undergo spontaneous uncontrolled decay, and (2) emit penetrating radiation. There are two general types of radioactivity, natural and artificial; but as we shall see, the same laws apply to both. *Natural radioactivity* is the kind associated with naturally occurring radionuclides, such as radium. On the other hand, *artificial radioactivity* is the kind associated with artificial radionuclides such as radioiodine; these are nuclides that have been made radioactive by the bombardment of ordinary, stable nuclides, with subatomic particles (neutrons, deuterons, etc) in special high energy physical devices such as the nuclear reactor and cyclotron.

The *artificial radionuclides,* to be discussed more fully in the next chapter, constitute an impressive array of material that is finding ever greater application in biologic research, clinical medicine, and industry.

Radioactive Series

The naturally occurring radionuclides can be grouped in three series or families, based on their parent elements: (1) *thorium,*

(2) *uranium,* and (3) *actinium.* Each of these gives rise, by spontaneous decay, to a series of nuclides characteristic of that series. The chain of breakdown of the successive nuclides always occurs in exactly the same way and at a fixed rate. Some daughter nuclides, such as thorium A, disintegrate at an extremely rapid rate and are therefore difficult to isolate; others, such as radium, in the uranium series, have a relatively long life and can therefore be obtained in pure form. At present, radium is one of the most widely used radionuclides in medicine. Thus, the uranium family has been a very important one for radiology because radium and its offspring comprise the last descendants of the uranium series.

RADIUM

Properties

Since radium is used extensively in therapy and since *its behavior exemplifies that of the other radioactive nuclides,* the discussion will deal mainly with this element. It is a heavy metal having a silvery-white appearance when pure, and behaving chemically like barium and calcium. It belongs to the uranium radioactive series. Radium has atomic number 88, and mass number 226. Since its radioactivity is not affected when it is combined with other elements, it is available only in the form of one of its salts—*radium bromide* or *radium sulfate*—which are more easily and more cheaply obtained than the pure metal.

Types of Radiation

There are three different kinds of radiation emitted by radionuclides: (1) alpha particles, (2) beta particles, and (3) gamma rays. A given radionuclide may emit all of these radiations, or may emit only one or two of them. Various members of the radium series, as will be shown below, give off one or more of these radiations.

1. **Alpha particles** are *helium nuclei* (that is, helium ions) consisting of 2 protons and 2 neutrons. Since this gives them an atomic number of 2 and a mass number of 4, they can be represented by the symbol ^4_2He. Alpha particles carry two positive

charges and therefore strongly ionize the atoms of matter through which they pass. Upon accepting two electrons, an alpha particle becomes a neutral helium atom:

$$\text{alpha particle} + 2 \text{ electrons} \longrightarrow \text{helium atom}$$
$$He^{++} + 2e^- \longrightarrow He$$

Alpha particles leave the nucleus at high speeds, 9,000 to 18,000 miles per sec, but do not penetrate matter for any appreciable depth because of their strong ionizing ability; in fact, they are absorbed by an ordinary sheet of paper.

2. **Beta Particles.** Often called beta rays, they consist of high speed *electrons,* represented by the symbol $_{-1}^{0}e$. Ejected from the nucleus at speeds approaching that of light rays, they are able to penetrate matter to a greater depth than alpha particles, their maximum range in tissue being 1.5 cm. In fact, 0.5 mm of platinum or 1 mm of lead is required for almost complete absorption of beta particles. They are used in therapy, mainly for superficial skin lesions in small areas. Beta particles are strongly ionizing, but less so than alpha particles.

3. **Gamma rays** are electromagnetic waves (photons), usually of high energy. For example, the gamma rays of radium have energy equivalent to 1.5 million volt (MV) x rays. They travel with the speed of light and are much more penetrating than alpha or beta radiation. Gamma rays are physically identical with x rays, the difference between them being that gamma rays originate in atomic *nuclei,* whereas x rays are produced outside the nuclei. Since they are electromagnetic waves, they are not deflected by a magnetic field. The application of radium in therapy is most often based on gamma radiation, the alpha and beta particles usually being removed by appropriate absorbing filters.

The Radium Series

When an atom of radium disintegrates, it does not disappear, but changes to another nuclide, a phenomenon called *radioactive transformation.* Let us "look into" an atom of radium and see what happens during its radioactive breakdown. The atom ejects an alpha particle with a mass number of 4. Since the mass number of radium is 226, the mass number of the atom remaining after the emission of an alpha particle is $(226 - 4) = 222$. The atomic number of radium is 88 and the charge on an alpha particle is $+2$. Therefore, the atomic number of the new nucleus is $(88 - 2) = 86$. Thus, there remains a new atom with mass number 222 and atomic number 86; this element is *radon,* the first breakdown product of radium.

$$1 \text{ radium atom} \longrightarrow 1 \text{ radon atom} + 1 \text{ alpha particle}$$

$$^{226}_{88}\text{Ra} \longrightarrow {}^{222}_{86}\text{Rn} + {}^{4}_{2}\text{He}$$

Note that the sum of the mass numbers (superscripts) of the products radon and alpha particle equals the mass number of radium. The same holds for the atomic numbers (subscripts). Therefore, this is a balanced equation.

The radon atom is radioactive and disintegrates to another kind of atom, which in turn breaks down to still another kind of atom, and so on. This chain of disintegration continues until a stable atom is left, in this case a metal, *lead.* Figure 21.1, showing

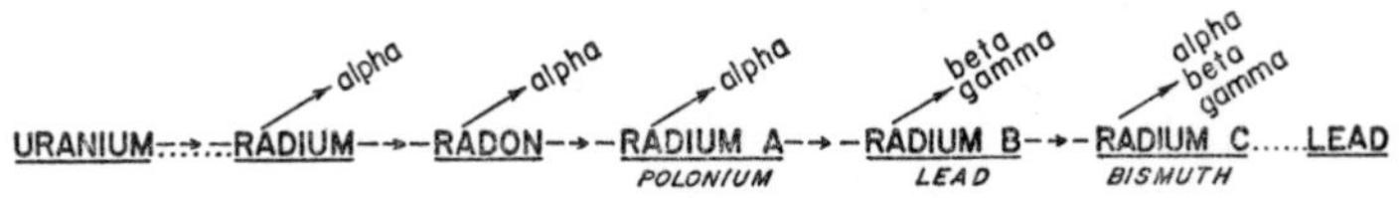

	URANIUM	RADIUM	RADON	RADIUM A	RADIUM B	RADIUM C	LEAD
				POLONIUM	LEAD	BISMUTH	
Atomic No.	92	88	86	84	82	83	82
Mass No.	238	226	222	218	214	214	206
Half Life	4.5×10^9 yr	1622 yr	3.8 da	3 min	26.8 min	19.7 min	stable

Figure 21.1. Successive disintegration products of radium, shown from left to right. These make up the radium series which is the last part of the uranium series. Data on radium D, E, and F have been omitted.

the radium series, is included mainly to indicate pertinent data, and not to tax your memory. Note that the radiation emitted by the radium element itself is limited to alpha particles; radium element also emits a gamma ray, too weak to be of use in therapy. Beta and penetrating gamma rays first appear with radium B (lead isotope) and C (bismuth isotope). Since a sealed source of radium reaches a state of *equilibrium* with its descendants after about 30 days (that is, the same number of atoms of a particular daughter is appearing as is disappearing in a given time interval) and then actually consists of a mixture of these descendants, we speak loosely of radium as emitting all three types of radiation. It is more correct to say that the *radium series* emits all three types of radiation. Whether we start with radium or radon gas there is soon a sufficient accumulation of radium B and C to give appreciable intensities of beta and gamma radiation. Therefore, radon as well as radium can be used in therapy.

Half Life

The caption *half life* appears in Figure 21.1. This is a term which *denotes the length of time required for one-half the initial amount of a radionuclide to disintegrate.* Another way of regarding this is as follows: suppose we have 1 gram of radium at the start. Referring to Figure 21.1, note that its half life is 1622 years. This means that it will have gradually decayed to one-half its present mass in 1622 years; or, in 1622 years, there will be only ½ g of radium left. In another 1622 years, it will again be reduced by one-half; that is, it will be one-fourth the original amount. Notice that the half life is constant for any given radionuclide, and that the half lives of different radionuclides show wide differences. Thus, the half life of uranium is 4.5 billion years, while that of radium A is three minutes.

In any radioactive series, after equilibrium has been reached among all the daughter nuclides, the quantity of any one member of the series is larger if its half life is longer. Thus, if we start with a radium sample, radium with the longest half life in its series will be present in largest amount, whereas radium A (polonium) with a half life of 3 minutes will be present in smallest amount.

As we have already indicated, there is no way of predicting which radium atom will disintegrate at a given instant. However, we know statistically that in 1622 years one-half the atoms in a given amount of radium will have decayed. Thus, half life is really a *statistical* concept indicating the time it takes for disintegration of one-half the atoms on the basis of pure chance.

RADON

When an atom of radium emits an alpha particle, an atom of radon remains:

1 radium atom − 1 alpha particle ⟶ 1 radon atom

Radon is an element existing under ordinary conditions as a colorless, heavy gas. Its *radioactivity* and a convenient half life (3.8 days) make it useful in therapy. In practice, the radon is pumped from a container of radium, purified by passage through an intricate system of glass tubes and chemicals, and collected in thin glass or gold capillary tubes. The tube is pinched off and sealed in short segments. After several hours, the radon reaches equilibrium with its offspring elements since they cannot escape form the tiny capillary tubes or *implants.* The preparation now emits alpha, beta, and gamma radiation. The exposure rate of radiation given off by any such preparation can then be determined by a physicist. Since radon decays rapidly, the activity remaining at a particular time after a sample has been calibrated must be obtained from a table or graph. Figure 21.2 shows the

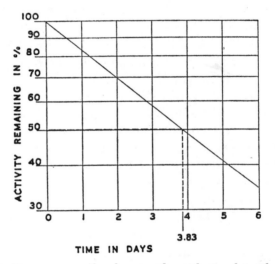

Figure 21.2. Decay curve of radon, semilogarithmic plot. The half life is 3.83 days.

decay curve of radon as well as the method of finding the half life. It should be noted that decay *occurs at a constant rate; that is, the same fraction of any quantity of radon decays per day.* This rule applies to all radionuclides, but the decay rates of different radionuclides are different and cannot be altered by any practicable means.

RADIUM DOSAGE

The effect of penetrating radiations on a given tissue depends on the amount of radiant energy absorbed by the tissue; and this, in turn, is related to the radiation exposure to which the tissue has been subjected. In order to specify radiation exposure we must have certain precisely defined units which provide all the essential information. Such a system of dosage must be sufficiently accurate to permit the comparison of various treatment methods and the evaluation of treatment results. Besides, it must provide the radiologist with a means of exchanging information with his colleagues.

The first step in calculating *gamma-ray dosage* is the determination of the *time-intensity factor.* With radium, this consists of multiplying the exposure time by the amount of radium in the applicator. Thus, if 10 mg of radium element are applied for 24 hours, the time-intensity factor is $10 \times 24 = 240$ milligram-hours (usually abbreviated mg-hr). The gamma-ray activity of a *radium* source is essentially constant during any treatment period because of its very long half-life. However, *radon* deteriorates rapidly and we must take into account the progressive loss of exposure rate of radiation during a treatment period lasting more than a few hours. Thus, with a radon treatment applicator the exposure rate drops to one-half the initial value in 3.8 days, and to one-fourth in 7.6 days. The radon gamma-ray exposure rate reaches a negligible level at the end of one month.

The time-intensity factor for *radon* therapy requires a more general unit, the *millicurie* (mCi), a unit of activity which represents 37 million disintegrations per second (3.7×10^7 dps). This unit applies to all radionuclides. Thus, 1 mCi of radon has the same activity, physically, as 1 mCi of radium, and it also happens that 1 mg of radium has an activity of practically 1 mCi. For relatively short treatment times, the gamma-ray exposure with 1 mCi of radon is equivalent to that with 1 mg of radium. However, for treatment periods longer than a few hours, the rapid decay rate of radon requires a correction factor, obtained from published decay tables. These are based on the general rule that 1 mCi of radon, when it has completely disintegrated, is equivalent to 1 mg of radium acting for 133 hours. Thus,

1 mCi radon destroyed = 133 mg-hr (or mCi-hr) radium

After calculation of the time-intensity factor, that is, the number of mCi-hr of radon or radium, we must then determine the radiation *exposure* at the site of the lesion. The radiation exposure depends not only on the time-intensity factor, but also on the *distribution* of the radium in the tissues, the *volume of tissue irradiated,* and the *filtration* of the applicator. In surface treatment, one must consider the *area* of the surface, its *shape,* the *distribution* of the radium in the applicator, the *distance* of the applicator above the surface, and the *filtration* of the applicator. The exposure rate of gamma radiation at a distance of 1 cm from a point source of 1 mg of radium filtered by 0.5 mm of platinum is 8.25 R/hr. This observation has enabled the physicist and radiologist to express radium and radon exposure in roentgens. If the radium in a surface applicator, or in needles inserted into the tissues, is distributed according to certain plans, such as those of Paterson and Parker or of Quimby, the minimum number of roentgens delivered to the lesion per 1000 mg-hr or mCi-hr exposure can be found from the tables published by these authors. The *absorbed dose* in rads can then be derived by multiplying the exposure in R by the appropriate conversion factor, f, for the *gamma rays of radium* ($f = 0.95$ for soft tissue; 0.92 for bone):

$$\text{absorbed dose (rads)} = 0.96 \text{ R } \textit{in soft tissue}$$
$$= 0.92 \text{ R } \textit{in bone}$$

An obsolete method of expressing radiation dosage, of historical interest only, was based on the so-called threshold erythema dose —the smallest radiation exposure that causes faint reddening of the skin of the forearm. Since it was obtained by trial and error and was too imprecise, it was replaced by the modern method of expressing radiation dose as described above. In general, a single exposure of approximately 1000 R with 1 to 2 million volt radiation delivers a threshold erythema dose. To get the same result with x rays having a half value layer of 1 mm Cu would require 680 R; and with half value layer 1 mm Al, 270 R.

TYPES OF APPLICATORS

There are several different kinds of applicators for the administration of radium or radon treatment. The planning and execution of such therapy is the direct responsibility of the radiologist, but since the technologist sometimes aids in preparing the applicators, we shall describe briefly some of the more frequently used forms.

At present, most radium applicators are made of platinum, although brass, silver, gold, or monel is sometimes used. These metals serve two main purposes: (1) they act as filters and (2) they permit the application of radium in the desired form.

Filtration

The radium series emits three types of radiation differing in penetrating ability. Any metal container will absorb all alpha particles because of their low penetrating power. This is desirable because the alpha particles are strongly ionizing and would produce severe local reactions without reaching very far into the lesion being treated. The beta particles penetrate better and, for the treatment of superficial lesions, the radium may be placed in a container with a monel metal filter 0.05 mm thick. If we need gamma radiation only, then the alpha and beta radiation can be removed by filtration with 0.5 mm platinum or gold, 1 mm of lead or silver, or 2 mm of brass. Most containers are available with appropriate wall thicknesses for any desired filtration. Heavier filtration than that required to absorb the beta radiation tends to harden slightly the emergent gamma rays by absorbing some of the softer gamma rays much as the filter in an x-ray beam absorbs relatively more soft than hard x rays. However, filtration of radium by more than 0.5 mm Pt is of questionable value in practice.

Containers

The various types of containers for radium and radon will now be described.

1. **Implants** ("seeds"). These are tiny, sealed capillary tubes containing radon. The dimensions of these "seeds" are usually about 0.75 mm × 3 mm, with a wall thickness of 0.3 mm of gold. The activity of the radon in mCi is usually measured by the supplier for *groups* of seeds rather than individually. This may result in sizable differences in activity of various seeds in the

same batch. Implants can be left in the tissue permanently because the gold is innocuous and the radon loses almost all its activity in about one month; dosage is therefore calculated for a treatment time of one month. Sometimes removable implants are used, being left in the tissues for a predetermined time.

2. **Needles.** These are hollow platinum-iridium tubes with a point at one end and an eye for threading at the other. Radium chloride is first sealed in platinum cells which are then encased in the hollow needle shaft, to form the *permanent type needle* (see Figure 21.3). Total filtration of the needle and cell usually equals 0.5 mm of platinum.

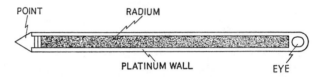

Figure 21.3. Radium needle. The wall is actually a platinum-iridium alloy, stronger than pure platinum. Sealed within the needle is the radium in the form of a salt, usually radium chloride.

3. **Tubes.** These resemble needles, but are usually larger and are rounded at both ends. Their usual wall thickness is 0.5 or 1.0 mm platinum-iridium, an alloy that is harder and stronger than platinum. One end of the tube is a screw cap with an eye for threading. Tubes can be loaded with radium needles or with sealed radon sources.

4. **Plaques.** These are flat, hollow applicators, having one face made of thin monel metal (0.05 mm), on the inner surface of which is permanently spread a layer of radium chloride. Plaques can be obtained with any desired activity, and filters can be added to remove beta radiation. Plaques are usually employed in the treatment of superficial skin lesions by beta radiation.

PROTECTION

The protection of personnel from radium radiation will be described in greater detail in Chapter 23. However, it must be emphasized that the best protection is *distance.* Radium applicators should never be picked up with the fingers. Loading of applicators, or other manipulations, must be performed by the use of long forceps. Radium should be stored in a lead container of proper thickness, and should be transported in a lead-lined carrying case provided with a long handle.

Because of the long half life of radium, the form in which it is used (usually radium chloride crystals), and the accumulation of radon gas, accidental breakage of an applicator or needle is fraught with serious radiation hazard. For this reason, artificial radionuclides that can be prepared in solid form as wires and beads and that give off gamma rays of suitable energy are being used more and more as radium substitutes. At present, cobalt 60 and cesium 137 are available as encapsulated needles and tubes, , ~ iridium 192 as tiny beads enclosed in nylon tubes, and, as a radon substitute, gold 198 in the form of implantable seeds. The dosage *distribution* in tissue from these radionuclides resembles that from radium, at least for the first few centimeters. However, they have different gamma-ray energies, and dosage has to be adjusted accordingly in terms of milligram-radium-equivalent (mg-Ra-eq).

LOSS OF RADIUM

In some cases, the technologist may be held responsible for the transportation or storage of radium. This requires constant checking, since lost radium may be very difficult to trace. The services of a radiation physicist or other qualified personnel may be required; he uses some type of *survey meter* usually provided with a *Geiger counter* (see pages 422-424). This is an electronic instrument which clicks rapidly when it is brought into the vicinity of radioactive material. It is very important that lost radium be found, not only because it is so expensive—about twenty dollars per milligram—but also because it is dangerous to persons unaware of its presence. Nurses and other personnel are always to be instructed that no dressings from patients who are receiving radium therapy should be discarded without first consulting the radiologist.

QUESTIONS

1. Define radioactivity. By whom was it discovered?
2. Why are some elements radioactive? Where do the penetrating radiations arise?

3. Discuss radioactive series.
4. How long should a newly manufactured radium needle be stored before it is used in therapy? What is radioactive equilibrium?
5. What is the difference between natural and artificial radioactivity?
6. Of what radioactive series is radium a descendant?
7. Why is radium of value in radiology?
8. Name and describe briefly the three types of penetrating radiations emitted by the radium series. Which members of the series emit gamma rays that are useful in therapy?
9. Describe all the steps in calculating radium gamma-ray dosage. Why and how does the calculation have to be modified for radon therapy?
10. What is the main purpose of the filter in a radium applicator? Which metal is most often used to filter radium, and why?
11. Define half life. Why is this important in therapy?
12. Describe the physical properties of radon.
13. What is a millicurie?
14. What are the four main types of radium or radon applicators? Describe each.
15. Why is it important to find lost radium?
16. Discuss the use of artificial radionuclides as radium substitutes. Which ones are now available, and in what forms? What advantages do such substitutes have over radium?

CHAPTER *22* *RADIONUCLIDES*
AND ARTIFICIAL
RADIOACTIVITY

In Chapter 4 we learned that matter is made up of extremely small particles called *atoms*. The structure of the atom has attracted a tremendous amount of scientific investigation, but for our purpose we must present it in simplified form—a central core or *nucleus* containing almost the entire mass of the atom and, surrounding the nucleus, the *orbits* or paths in which the negative *electrons* are in continual motion. The nucleus has two main constituents, *protons* and *neutrons*. A proton is a positively charged particle whose mass is nearly 2000 times that of an electron. A neutron has nearly the same mass as a proton but carries no charge.

All matter is made up of one or more simple entities called *elements*. The atoms of any particular element have a characteristic and distinct number of positive charges or *protons* in the nucleus, and an equal number of negative charges or *electrons* in the orbits around the nucleus. The number of nuclear protons is called the *atomic number*. Notice that the atomic number is specific for a given element; all atoms of any one element have the same atomic number. Atoms of different elements have different atomic numbers.

The mass of an atom depends on its nuclear protons and neutrons, and so the *mass number* of an atom has been defined as the total number of nuclear protons and neutrons. For example, since hydrogen has one proton in its nucleus, its mass number is 1. Helium has two protons and two neutrons in its nucleus, so it has a mass number of 4. Protons and neutrons are frequently referred to as *nucleons*.

Isotopes. As we have stated before, all the atoms in a given sample of a particular element have the same atomic number, that is, the same number of nuclear protons (their chemical behavior is also identical). However, all the atoms in this sample do not necessarily have the same mass number—total number of protons *and* neutrons. Since the number of protons is the same, the difference in mass number must be due to a difference in the number of *neutrons.* Such atoms whose nuclei contain the same number of protons, but different numbers of neutrons, are called *isotopes.*

It should be pointed out that there is not an unlimited variety of isotopes of any particular element; as they occur in nature, there may be only two, or three, or several. However, the ratio of the different isotopes is relatively constant for each element.

Since all the isotopes of an element have the same atomic number, they have the same chemical properties. The gas hydrogen, for example, has two naturally occurring isotopes: hydrogen with atomic number 1 and mass number 1; and deuterium (heavy hydrogen) with atomic number 1 and mass number 2 (see Figure 4.5). These two forms of hydrogen have identical chemical properties but differ in mass (that is, number of neutrons).

The term *nuclide* refers to a species of atom having in its nucleus a particular number of protons and neutrons. Accordingly, we can say there are three hydrogen nuclides.

Radionuclides. In recent years, physicists have been able artificially to prepare isotopes which heretofore had been nonexistent. More interesting still is the fact that many of these isotopes are *radioactive,* disintegrating into entirely different elements with release of energy. These radioactive isotopes have the same chemical properties as their nonradioactive counterparts, differing only in their mass number and in their radioactive nature. It is preferable to speak of *radioactive nuclides* or *radionuclides* rather than radioisotopes, since they are nuclides that happen to be radioactive.

The artificially radioactive nuclides are produced by the irradiation of stable nuclides by subatomic particles such as deuterons or neutrons in a cyclotron or nuclear reactor, respectively. A neutron, it will be recalled, is an uncharged subatomic particle having mass number 1. A deuteron is a heavy hydrogen nucleus having a single positive charge and a mass number of 2. The nucleus of the bombarded atom captures a neutron or a deuteron, becoming unstable and exhibiting the property of spontaneous breakdown known as *radioactivity*, a property also characteristic of the naturally radioactive elements such as radium.

An example of the *transmutation* of an ordinary, stable element to a *different* element which is radioactive will make the above process more easily understandable. Ordinary sulfur is converted, during neutron bombardment, to radioactive phosphorus:

$$\text{Ordinary} + \text{Neutron} \rightarrow \text{Radioactive} + \text{Hydrogen}$$
$$\text{Sulfur} \qquad\qquad\qquad \text{Phosphorus}$$

$$^{32}_{16}\text{S} \;+\; ^{1}_{0}\text{n} \;\rightarrow\; ^{32}_{15}\text{P} \;+\; ^{1}_{1}\text{H}$$

In each term of the equation, the lower number or *subscript* represents the atomic number, and the upper number or *superscript* represents the mass number. The sum of the superscripts on one side of the equation equals that on the other side. The same holds true for the subscripts. Knowing that the atomic number of phosphorus is always 15, we can represent radioactive phosphorus as ^{32}P. Since it has a specific number of protons (15) and neutrons (17) in its nucleus and is at the same time a particular species of atom, it is an example of a *radionuclide* or *radioactive nuclide*. When ^{32}P is formed, its nucleus is unstable because of the extra neutron (ordinary nonradioactive phosphorus is ^{31}P) and it therefore disintegrates to ordinary sulfur, emitting a beta particle (negative electron) in the process:

$$^{32}_{15}\text{P} \;\rightarrow\; ^{32}_{16}\text{S} \;+\; ^{0}_{-1}\text{e} \;\; (\textit{beta particle})$$

At the present time, there is no way of altering the type of radiation emitted by a given radionuclide, natural or artificial.

Another method of inducing artificial radioactivity is to irradiate stable atoms with deuterons (heavy hydrogen nuclei) in a cyclotron. However, this is not a commercial source of medical radionuclides for general use.

Nuclear Reactor. As already mentioned, radionuclides can be produced by subjecting stable elements to irradiation by neutrons. In fact, this is now the chief method by which artificial radioactivity is induced, both for medical and industrial purposes. One type of nuclear reactor (see Figure 22.1) depends on the

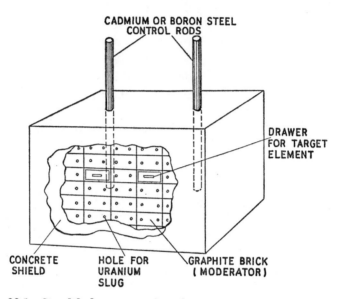

Figure 22.1. Simplified concept of nuclear reactor using uranium 235 as the fuel.

fission (splitting) of nuclei of uranium 235 (^{235}U), an isotope of ordinary uranium (^{237}U) by extremely slow, environmental neutrons (see Figure 22.2). These neutrons have a very small average kinetic energy, about the same as that of the molecules of air at the prevailing temperature and are therefore called *thermal neutrons*. When a thermal neutron is captured by a ^{235}U nucleus, the nucleus splits into fragments of various sizes and at the same time emits 1 to 3 neutrons plus 200 million electron volts (MeV) of energy:

$$^{235}U + {}^{1}_{0}n \longrightarrow \text{sum of masses of nuclear fragments} + 1 \text{ to } 3{}^{1}_{0}n + 200 \text{ MeV}$$

The splitting of the ^{235}U atom in this manner is termed *fission*. The neutrons emitted during fission are *fast neutrons;* these must

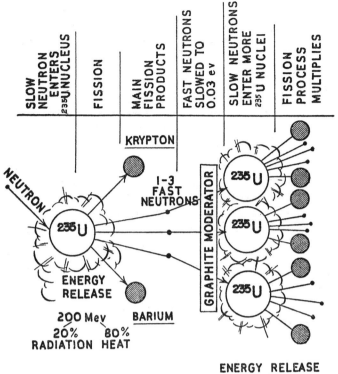

SLOW NEUTRON ENTERS 235U NUCLEUS | FISSION | MAIN FISSION PRODUCTS | FAST NEUTRONS SLOWED TO 0.03 ev | SLOW NEUTRONS ENTER MORE 235U NUCLEI | FISSION PROCESS MULTIPLIES

KRYPTON

NEUTRON

I-3 FAST NEUTRONS

235U

235U

235U

ENERGY RELEASE

200 Mev   BARIUM
20%   80%
RADIATION HEAT

GRAPHITE MODERATOR

ENERGY RELEASE TRIPLED

Figure 22.2. Fission of uranium 235 by slow (thermal) neutrons, resulting in a nuclear chain reaction. This differs from radioactive decay in the production of large nuclear fragments, as well as radiation, during fission.

be slowed to thermal energy to facilitate their capture by other ^{235}U atoms to repeat the fission process. Graphite blocks are used as moderators to slow the neutrons. The liberation of one to three neutrons each time a neutron is captured and the rapid multiplication of this process together constitute a *chain reaction;* it is accompanied by the release of a tremendous amount of energy in an exceedingly small interval of time.

If a piece of ^{235}U is large enough to permit more neutrons to be liberated than can escape from the surface, the mass will explode. Thus, for the reactor to operate properly, the mass of ^{235}U must be just large enough for the fission process to continue, but yet not so large as to go out of control or *go critical,* as it is called. The speed of fission can be varied by the use of boron or cadmium *control rods;* these have a strong affinity for neutrons, and adjustment of the depth to which the rods are inserted into the reactor regulates the size of the *neutron flux* (flow). Normally, the reactor's operation is so adjusted by the cadmium control rods as to maintain the fission process at the desired level without letting the reactor go out of control (that is, explode). The "fuel" of the reactor is not pure ^{235}U, but rather natural uranium, of which only about $\frac{1}{140}$ is ^{235}U, enriched by the addition of ^{235}U.

Since the neutron is uncharged, it is not repelled by the positively charged nucleus; in fact, as the neutron approaches the nucleus there is actually an attractive force between them. Neutrons entering stable nuclei render them unstable. Thus, when stable nuclides are placed in suitable containers and subjected to neutron flux in a reactor, they are converted to radionuclides through *neutron capture.*

The use of the nuclear reactor for the production of radionuclides will now be exemplified by typical equations.

1. *Transmutation Reaction*—a different element is produced from the one entering the reaction, the latter being called the *target element* (see Figure 22.3).

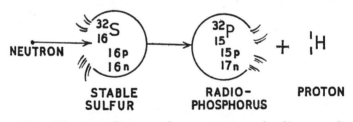

STABLE   RADIO-   PROTON
SULFUR   PHOSPHORUS

Figure 22.3. Schematic diagram of transmutation of sulfur to radiophosphorus by neutron capture.

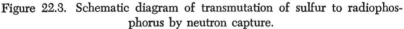

$$^{32}_{16}\text{S} \quad + \quad ^{1}_{0}\text{n} \quad \rightarrow \quad ^{32}_{15}\text{P} \quad + \quad ^{1}_{1}\text{H}$$

stable   *neutron*   *radioactive*   *proton*
sulfur   *phosphorus*

This reaction is designated as an (*n,p*) reaction because a neutron enters the nucleus, and a proton (hydrogen nucleus) leaves it. The ^{32}P can be separated chemically from the sulfur that remains, so that the resulting ^{32}P is pure, that is, *carrier free.*

2. *Radiative Neutron Capture*—the product nuclide is a radioactive isotope of the target nuclide, and the two can be separated only with great difficulty. The radionuclide is impure, being contaminated by its stable isotope; thus, it is not carrier free.

$$^{59}_{27}\text{Co} \quad + \quad ^{1}_{0}\text{n} \quad \rightarrow \quad ^{60}_{27}\text{Co} \quad + \quad \gamma$$

| *stable cobalt* | *neutron* | *radioactive cobalt* | *gamma ray* |

This is an (n,γ) reaction because a neutron is captured by the nucleus and a gamma ray is emitted.

3. *Fission Products*—various radioactive nuclides, representing the fragments of the ^{235}U nuclei that have undergone fission. These fragments are known collectively as *waste products* or *ashes* of the nuclear reactor. They are usually separated chemically and, in fact, are the main source of supply of iodine 131 (^{131}I), cesium 137 (^{137}Cs), and strontium 90 (^{90}Sr), all three of which are important in medicine.

Properties of Radionuclides

Some of the more important properties of radionuclides will now be discussed. In general, they obey the same laws as does radium (see Chapter 21). However, many of the medical radionuclides decay fairly rapidly and also differ in their decay rates. Furthermore, they have widely different applications in medicine.

Types of Radiation. The radiations emitted by radioactive elements have already been described on pages 398-399. They consist of three main types: alpha and beta particles, and gamma rays.

1. *Alpha particles* are helium ions carrying two nuclear positive charges, and having atomic number 2 and mass number 4. They are *not* emitted by artificially radioactive nuclides of medical interest.

2. **Beta particles** are fast moving electrons ejected by certain radionuclides, and arise from nuclear neutrons as follows:

$$\begin{array}{ccc} {}_{0}^{1}n & \longrightarrow & {}_{1}^{1}H + {}_{-1}^{0}e \end{array}$$

In the vast majority of instances, these electrons are negatively charged. For example, ^{32}P is a pure negative beta emitter. However, some radionuclides emit positrons—positively charged electrons. Beta particles cause significant ionization in tissues where they are for the most part absorbed. They cannot be detected directly by special external counting devices, although the brems radiation resulting from interaction with atomic nuclei can be so detected.

3. **Gamma rays** are electromagnetic in nature and are identical to x rays. Since they are uncharged, they are not directly ionizing, but do release primary electrons in the tissues by interaction with atoms. Their emission is invariably accompanied by beta particles; thus, ^{131}I gives off both gamma and beta radiation. External radionuclide counting by Geiger or scintillation counters is based on gamma radiation which usually has sufficient penetrating ability to pass completely out of the body.

Units of Activity. The curie (Ci) is the standard unit of *activity. One curie is an activity of 3.7(10)^{10} disintegrations per second (dps).* Smaller units that are decimal fractions of the curie are usually more convenient to use. One millicurie (mCi) is an activity of $3.7(10)^7$ dps. One microcurie (μCi) is an activity of $3.7(10)^4$ dps.

Specific Activity. A given sample of a radionuclide may not be carrier free; that is, it may also contain the stable isotope. The *specific activity* of such a sample is defined as its activity (due to the radionuclide present) divided by its total weight. For example, if 10 g of stable cobalt is irradiated in a nuclear reactor for a specified time, and the resulting activity of the ^{60}Co is 55 Ci, then the specific activity of this sample is $55/10 = 5.5$ Ci/g. In actual practice, the specific activity of ^{60}Co in teletherapy machines ranges from about 75 to 200 Ci/g.

Radioactive Decay. Every radionuclide decays at a constant rate characteristic of that particular nuclide. By constant decay rate we mean that regardless of the number of such atoms initially present in a sample, a certain fixed *fraction* will disintegrate in a given interval of time. For example, if the initial number of atoms is 1 million, and the decay rate for the radionuclide is 0.5 per sec, then 500,000 will have decayed in 1 sec. In the next second, 0.5 of the remaining 500,000 atoms or 250,000 will have decayed, etc. At present, there is no practicable method of changing the decay rate of radioisotopes.

Half Life. In nuclear medicine there are three kinds of half life: *physical, biologic,* and *effective.* These depend on the method of measuring a radioactive sample over a period of time. Such curves for ^{131}I are shown in Figure 22.4.

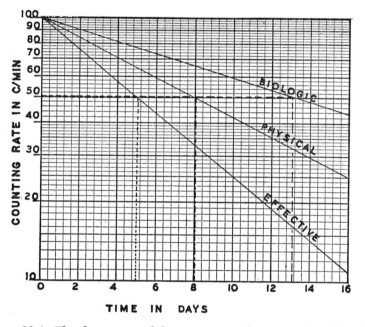

Figure 22.4. The three types of decay curves and corresponding half lives for [131]I. In this example, the effective half life is about five days and the biologic half life about 13 days. The physical half life is constant—8.05 days.

1. *Physical Half Life* (T_p) is defined as the time required for a given sample of a radionuclide to decay to one-half its initial activity. For example, if the initial activity is 10 mCi, the

physical half life is the time it takes for the activity to decrease to 5 mCi, as measured outside the body. Examples of physical half lives are: ^{131}I 8.1 days; ^{32}P 14.3 days; ^{60}Co 5.2 years.

Let us see how physical half life is actually determined. First the radioactivity of a sample of the selected radionuclide is measured by counting the number of disintegrations per min as detected by a suitable counter (see page 422) under standardized conditions. This is the initial count. The counting of the same sample is repeated at suitable intervals over a period of time longer than the half life. Then the data are plotted, with the counting rate as a function of the time that has elapsed since the initial count; the resulting curve is shown in Figure 22.5. To find

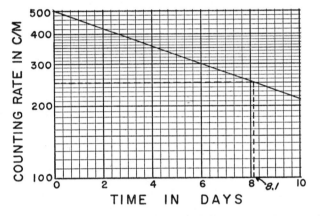

Figure 22.5. Physical decay of ^{131}I (semilogarithmic plot to obtain a straight-line curve.) With an initial counting rate of 500 counts per min, the half life is the time required for the counting rate to decrease to 250 c/m. For ^{131}I this is 8.1 days.

the half life, we locate the point on the vertical axis corresponding to one-half the initial counting rate, then trace a horizontal line to the curve and, from the point of intersection, a vertical line to the horizontal axis where the final point of intersection denotes the half life. The physical half life is *constant* for a particular radionuclide.

There is a useful formula for the fraction of a radioactive sample remaining after n half lives: $1/2^n$. For example, if a sample of ^{131}I has an activity of 100 mCi, what fraction remains after twenty-four days? Since the half life of ^{131}I is 8.1 days, the stated time represents three half lives. The fraction remaining at the end of this time is

$$1/2^n = 1/2^3 = 1/8$$

The activity remaining is then 100 mCi $\times$ ⅛ $= 12.5$ mCi.

2. *Biologic Half Life* (T_b) is the time required for a given deposit of radionuclide to decrease to one-half its initial maximum activity, due only to biologic utilization and excretion, and disregarding physical decay. It is obtained from a curve plot of a series of comparative counting rates over a part of the body, relative to that of a standard of the same radionuclide (see page 438); physical decay of the radionuclide in the body and in the standard cancel each other, leaving only biologic decay. This half life varies from person to person, and in the same person at different times.

3. *Effective Half Life* (T_e) is that due to *both* physical and biologic decay and is always less than the physical half life of the radionuclide being used. It is obtained from the curve plot of absolute counting rates over a particular part of the body at suitable intervals after the initial count. Effective half life can be calculated from the physical and biologic half lives by the following equation:

$$\text{effective half life} = \frac{\text{physical half life} \times \text{biologic half life}}{\text{physical half life} + \text{biologic half life}}$$

$$T_e = \frac{T_p T_b}{T_p + T_b}$$

Methods of Applying Radionuclides in Medicine

The use of radioactive nuclides in medicine is predicated on the principle that they behave chemically in the body in exactly *the same way as their stable counterparts.* In other words, the body cannot tell the difference between, for example, radioactive iodine and stable iodine. But once a radionuclide has entered the body, it carries a radioactive tag which allows it to be detected, and its course through the body traced, by special instruments; or, alternately, it can be detected in the blood, secretions, or excretions within or outside the body. Furthermore, it can be measured with reasonable precision in most instances. In this chapter, only basic principles will be covered. The use of radionuclides in therapy will also be touched upon; this is based on the fact that sometimes, as with iodine in the thyroid gland, a radionuclide will concentrate selectively in certain organs. Under such circumstances, a sufficiently large dose can be given to irradiate effectively the target organ.

The following outline summarizes the medical uses of radionuclides according to various indications:

1. *Diagnosis*—general procedure
    a. Compound tagged with radionuclide (radioactive isotope of a normal element substituted for it in the compound).
    b. Injected or ingested in tracer amount—small enough to be diagnostic, but not large enough to deliver intolerable radiation exposure.
    c. Body cannot distinguish between radioactively tagged compound and its stable counterpart.
    d. Fate of tagged element determined by counting with special instruments (GM counter; scintillation counter; scintiscanner; gamma camera)
        (1) Externally as with ^{131}I, thyroid gland; ^{99m}Tc or ^{197}Hg, brain; ^{197}Hg, kidney; ^{99m}Tc or ^{198}Au, liver.
        (2) Internally as in probing internal organs such as brain.
        (3) Excretions as in counting urine in Schilling test for pernicious anemia.
        (4) Secretions as in measuring ^{131}I output in saliva.

2. *Therapy*—methods
    a. *Chemical*
        (1) Selective absorption as ^{131}I in thyroid gland: ^{131}I given orally in therapeutic amount; selectively absorbed in target organ—thyroid—which is irradiated by deposited radionuclide, in this instance, 90 per

cent of therapeutic dose from beta particles, 10 per cent from gamma rays.

(2) Differential absorption as ^{32}P in malignant tumors.

(3) Colloidal dispersion as ^{198}Au in the reticuloendothelial system.

b. *Physical*

(1) External as ^{60}Co and ^{137}Cs in teletherapy, shown in Figure 22.20; and ^{90}Sr in contact therapy, shown in Figure 22.21.

(2) Interstitial as ^{60}Co needles or wires; ^{137}Cs needles; ^{192}Ir wires; ^{198}Au seeds.

(3) Intracavitary as ^{60}Co needles or wires; ^{137}Cs needles; ^{32}P-chromic phosphate or ^{198}Au in pleural or peritoneal malignant effusions.

Radionuclide Instrumentation

The basic procedure in any quantitative diagnostic radionuclide study is to count, with an appropriate detecting or counting instrument, the number of photons or ionizing particles it receives and detects from a radioactive source in a specified interval of time and with a defined geometric arrangement. The source may be either an organ in the body or a specimen outside the body, depending on the examination being conducted.

A general formula for radionuclide counting may be expressed as follows, *provided the sample and the standard are counted under identical conditions:*

$$\frac{\text{mCi in sample}}{\text{mCi in standard}} = \frac{\text{net counting rate in sample}}{\text{net counting rate in standard}}$$

The net sample counting rate, R_S, is the difference between the total counting rate, R_T, and the background counting rate, R_B.

$$R_S = R_T - R_B$$

The background counting rate is obtained with the sample and all other radioactive sources removed from the counting range of the detector. *The counting rate is the total count divided by the time during which the count was taken.* In any radionuclide procedure, the *background counting rate should always be subtracted from the total* to obtain the net counting rate, unless the background rate is so small as to be negligible.

Obviously, we must have a means of counting the radiations emitted by a given radioactive source. Special instruments are available for this purpose; these may be classified as the *radiation counter* itself, and the *device for recording the counts.*

Radiation Counters. The two main types of counters, differing completely in principle, include (1) the Geiger-Muller or GM counter and (2) the scintillation counter.

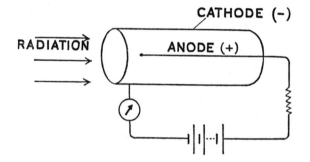

Figure 22.6. Simplified diagram of an end-window GM counter.

1. *GM Counter.* This is an electronic pulse-type device that can count individual ionizing particles or photons. It consists of a central wire anode and a cylindrical cathode enclosed in glass (see Figure 22.6). Within the counter is a suitable gas (often argon and alcohol) under reduced pressure of about ⅛ atmosphere. The GM counter operates within a range of about 1,000 to 1,500 volts. In the *end-window* counter, radiation is admitted at one end; in the *side-window* counter, it is admitted through the side.

The *characteristic curve* of a GM counter is shown in Figure 22.7; the curve is not necessarily the same for all GM counters, although its general shape and significance are common to all of them. Photons or beta particles entering the counter may interact with orbital electrons either in the glass wall or the gas inside. Liberated orbital electrons then ionize other gas atoms within the counter, producing ion pairs. At the *starting potential* of about 800 volts across the counter, as shown in Figure 22.7,

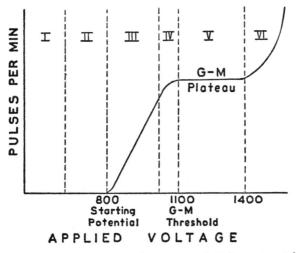

Figure 22.7. Characteristic curve of one type of GM counter. The voltages may be different for other GM counters. The operating voltage should correspond to about the middle of the plateau.

these ions are driven toward the opposite electrodes (that is, positive ions to cathode, negative ions—mainly electrons—to anode); on the way, they gain sufficient speed, due to the applied voltage, to ionize other gas atoms by collision, resulting in a pulse in the circuit. In this region (III) an increase in the applied voltage increases the number of pulses per min, or the counting rate. If an audio circuit is included, each pulse is manifested by a "pop." Another term for *ionization by collision* is *avalanche*. An increase of the applied potential to 1100 volts in this particular GM counter marks the *GM threshold;* the avalanche is no longer limited to a short segment of the wire anode, but rather spreads instantaneously along its full length. From the threshold potential of about 1100 volts up to about 1400 volts the pulses per min or counting rate is less dependent on the applied voltage, this portion of the curve being called the *GM plateau.* Furthermore, in the plateau region the size of the pulse is independent of the energy of the entering particle or photon. The plateau normally has a slight rise which should not exceed about 3 volts per 100 volts. The curve rises steeply at the end of the plateau, representing *continuous discharge;* the GM counter may be damaged if operated in this region.

Before a GM counter is approved, its plateau should be plotted by actual trial and the *operating voltage* selected to correspond to about the midpoint of the plateau, in this case, 1250 volts, to make the counting rate less sensitive to small fluctuations in voltage.

One of the important characteristics of a GM counter is its *dead time,* a brief interval immediately after a pulse during which the counter is unresponsive to radiation. The dead time in most GM counters is about 100 microseconds (that is, 100 millionths or 10^{-4} sec). In counting highly active samples with extremely large counting rates, we must use a correction factor for dead time, although this is seldom necessary in the medical radionuclide department.

Another feature of the GM counter is the possible appearance of fictitious or *spurious counts* representing more than one pulse for a single ionizing event. This is due to ultraviolet light produced when positive ions reaching the cathode are neutralized. Suppression of these spurious counts is called *quenching;* most GM tubes are self quenched by the selection of appropriate gases, such as argon and alcohol vapor. Mainly because these gases are eventually exhausted, a GM tube has a limited, although reasonable, life—about 10^{10} pulses.

The GM tube is efficient in counting beta particles, but is very inefficient in counting gamma rays. In fact, it will count only about 1 to 2 per cent of gamma rays entering the window; its gamma-ray efficiency can be improved by coating the cathode with bismuth.

2. *Scintillation Counter.* Early in the present century, it was already known that ionizing radiation produces flashes of light or scintillations in certain crystals. Because a more efficient counter than the GM was needed for gamma rays, the scintillation counter was developed on the above principle. It is designed to

conduct the light flashes, produced in a crystal by gamma radiation, through a Lucite channel to a photomultiplier tube. This consists of a photosensitive cathode (similar to that in a photocell) and about 11 accelerating anodes or *dynodes* in series, as shown in Figure 22.8. The electrons liberated in the photocathode

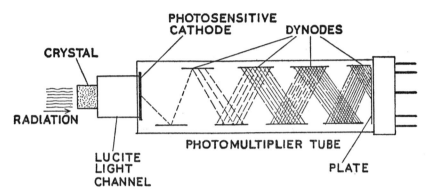

Figure 22.8. Diagram of a scintillation counter. Incoming radiation produces flashes of light (scintillations) in the crystal, one flash per photon or particle. The light flashes are conducted by the Lucite channel to the light-sensitive cathode where photoelectrons are released. These are then accelerated by a series of dynodes (anodes) and are detected as pulses by the counting circuit after they strike the plate.

are accelerated serially along the photomultiplier by the dynodes, eventually giving rise to a pulse in the external circuit. One pulse occurs for each entering photon, provided the latter has a certain minimum energy. Since the scintillation counter has a dead time of only about 5 microsec (5×10^{-6} sec), it is capable of tremendous counting rates. Furthermore, its high degree of sensitivity permits the counting of sources having extremely small activity.

The crystals in scintillation counters are selected according to their intended use. Thallium-activated sodium iodide is preferred in gamma-ray counters, whereas anthracene is used for counting beta particles. The larger the volume of the crystal, the greater is the *sensitivity* or the counting rate for a given activity. Scintillation counters have crystals with diameters varying from a few mm for the probe localization of brain tumors or the measurement of exposure rates about sealed radioactive sources to about 5 in. for rapid scanning of large organs, such as the brain and liver. Even larger crystals (for example, 11 in.) are used in the Anger gamma camera.

The *well counter* is a scintillation counter specially modified for measuring samples of very low activity. A hole has been bored in the crystal of the well counter to accommodate a small bottle containing the weak radioactive sample (see Figure 22.9).

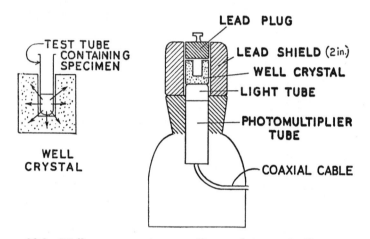

Figure 22.9. Well counter, using a well crystal in a scintillation counter. Thick lead shielding is required to reduce the background counting rate.

Because the crystal surrounds the sample on three sides, the well counter is extremely sensitive; in fact, it can count a source with an activity as low as about 10^{-5} microcurie (μCi). At the same time, the high sensitivity requires heavy lead shielding (usually at least 5 cm of lead) to decrease the background counting rate.

Recording Instruments for Radiation Counters. Special electronic devices receive and register the rapid pulses arising in the circuit of a GM or scintillation counter when it is exposed to ionizing radiation. There are two main types of recording instruments: (1) scaler and (2) rate meter.

1. *Scaler.* This is a recording unit with four main components: an electronic scaling circuit, a mechanical register, a high voltage supply, an interval timer and a "window" or channel selector

that limits the counting to radiation with a particular energy. The *scaling circuit* responds very rapidly to pulses in the GM or scintillation counter (to which it is connected), by lights which flash successively and show accumulated counts. A *mechanical register* is also included to record multiples of 10, 100, or 1,000, according to the scale selected (see Figure 22.10), in the *decade scaler.*

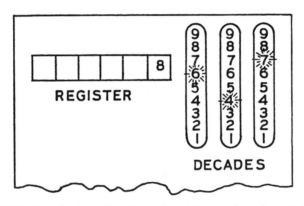

Figure 22.10. Panel of a decade scaler. The reading here totals 8,647 counts.

The scaler includes a *high voltage supply* for the counter. Furthermore, it provides for stabilization of the voltage to prevent fluctuation in the counting rate that might result from instability of the applied voltage.

Two types of *timers* are available to permit calculation of the counting rate which is defined as follows:

$$\text{counting rate} = \frac{\text{number of counts}}{\text{elapsed time}}$$

A *preset timer* is set for a desired time interval, and automatically terminates the counting at the end of the elapsed time. On the other hand, a *preset counting* timer is set for a predetermined number of counts; when this number of counts is recorded, the timer automatically stops the counting process and indicates the time that has elapsed.

2. *Rate Meter.* Operating on a different principle from the

scaler, the rate meter indicates the counting rate in counts per minute by means of a needle moving over a calibrated scale. In a sense, it averages the rate at which the pulses arrive from the GM or scintillation counter.

Due to the random nature of radioactive decay, there may be wide fluctuations in the counting rate (and indicator needle) from instant to instant, especially with sources of low activity. In order to stabilize the indicator needle and improve accuracy, enough time should be allowed for the rate meter to accumulate counts. This is governed by the *time constant* which equals the the product of the resistance and the capacitance of the circuit— *time constant* = *RC.*

The rate meter circuit contains a variable capacitor on which the pulses build up a charge, in parallel with a variable resistor which permits gradual leakage of the charge. As the capacitance and the resistance are increased by means of a control knob, the time constant increases, the charges leak away more slowly, and the needle remains steadier. However, it also takes the needle longer to come to rest (or reach its equilibrium position for a particular counting rate); hence, more time has to be allowed for the needle to come to rest before the reading is taken. In general, to have an accuracy of 2 to 5 per cent, the elapsed time before a reading is made should be *4 times the time constant* (*RC*). For example, if the selected time constant is 10 sec, then one must wait 4 × 10 = 40 sec before reading the counting rate, which will then have an accuracy of 2 to 5 per cent.

The rate meter supplies the high voltage for the operation of a GM or scintillation counter. A separate timer is not required because the counting rate is shown directly in counts per minute.

A rate meter can be used both for routine counting and for radiation surveys in health physics. However, it is especially useful in recording a continuous process such as the progressive loss or uptake of a radionuclide, because it can be connected to a *chart recorder.* This has a continuous roll of paper on which are recorded, in ink, the instantaneous fluctuations of a lever arm activated by the pulses arriving at the rate meter. Multiple channels are available, as in the scaler, to permit selection of the energy "window" of the radiation to be counted.

Sources of Error in Counting

Various errors are inherent in the counting process, the most important ones being (1) statistical errors and (2) recovery time of the counter.

Statistical Errors. Since radioactive decay—a purely random process—obeys the laws of chance, statistical errors affect the accuracy of a given counting rate which, after all, reflects the rate of decay of a given source. As a general rule, it is well to remember that the *reliability* (or reproducibility) *of a given count depends on the total number of counts recorded.* If the same sample is counted repeatedly under identical conditions, the counts will differ from each other because of chance variations (statistical fluctuation). How, then, can we tell which count is correct and how much it differs from the ideal or "true" count? Actually the answer lies in specifying a certain degree of reliability of a given count, based on mathematical laws of statistics. Only the important elementary concepts will be included here.

1. *Standard Deviation, σ* (Greek letter "sigma"), is defined as follows:

$$\sigma = \pm \sqrt{N} \text{ (approx.)} \tag{1}$$

where N is the total number of counts recorded in a given run. The standard deviation means that if we take the range $N \pm \sqrt{N}$ (that is, the range of values from $N - \sigma$ to $N + \sigma$) there is about a 2 to 1 chance that the true count would lie somewhere within this range. For example, if the total count is 6400,

$$\sigma = \pm \sqrt{6400} = \pm 80$$

There is about a 2 to 1 chance that the true count falls between $6400 - 80$ and $6400 + 80$, or between 6320 and 6480.

2. *Per Cent Standard Deviation* (%σ). It is often more convenient to express σ as a percentage of the total count:

$$\%\sigma = \frac{\pm \sqrt{N}}{N} \times 100$$

$$\%\sigma = \pm \frac{\boldsymbol{100}}{\sqrt{N}} \tag{2}$$

$$= \pm \frac{100}{\sigma}$$

In the preceding numerical example,

$$\%\sigma = \pm \frac{100}{\sqrt{6400}} = \pm \frac{100}{80} = \pm 1.3\%$$

The final count may then be stated as $6400 \pm 1.3\%$; there is, again, a 2 to 1 chance that the true count lies between these limits. Figure 22.11 is a curve plot of the per cent standard deviation

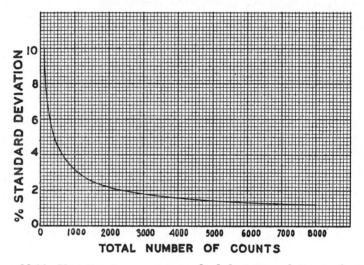

Figure 22.11. Variation in per cent standard deviation relative to the total number of counts.

as a function of the total number of counts, to obviate square root calculation.

We can easily show that the per cent standard deviation becomes smaller (that is, reliability becomes greater) as the total number of recorded counts is increased. Suppose we count a given sample long enough to accumulate 2500 counts. Then,

$$\%\sigma = \pm \frac{100}{\sqrt{2500}} = \pm \frac{100}{50} = \pm 2\%$$

If the same sample is counted long enough to accumulate only 100 counts, then,

$$\%\sigma = \pm \frac{100}{\sqrt{100}} = \pm \frac{100}{10} = \pm 10\%$$

The per cent standard deviation for a given total count also applies to the counting rate.

In actual practice, with preset timing, a sample is counted for a predetermined interval of time; and the background is also counted, but not necessarily for the same interval. The counting rate is then calculated by simply dividing each count by its corresponding time. The true sample counting rate, R (without background), is equal to the gross sample counting rate (including background), R_T, minus the background counting rate, R_B,

$$R = R_T - R_B \qquad (3)$$

As a rule, to keep the per cent standard deviation as small as possible, the ratio R_T/R_B should be as large as possible, preferably at least 9.

In counting a radioactive sample we must know how long the sample and the background must be counted in order to achieve a desired per cent standard deviation in the "true" or net sample counting rate. This information can be obtained from the following equation:

$$\%\sigma_S = \pm \frac{100\sqrt{R_T/t_T + R_B/t_B}}{R_T - R_B} \qquad (4)$$

where $\%\sigma_S$ = desired per cent standard deviation of the net counting rate of the sample
      R_T = total counting rate, including background.
      R_B = background counting rate
      t_T = time required to count sample, including background
      t_B = time required to count background

But, another relationship is needed—the ratio of the time needed to count the sample, to the time needed to count the background, for maximum reliability. This is covered by:

$$t_T/t_B = \sqrt{R_T/R_B} \qquad (5)$$

For example, if the counting rate of the sample is 900 cpm (counts per min) and that of the background is 100 cpm, $\sqrt{900/100} =$ $\sqrt{9} = 3$, so the sample must be counted three times as long as the background.

Let us now apply these equations to an actual problem. Suppose the counting rate of a sample to be 900 cpm and that of the background 100 cpm. How long should the sample and background be counted to provide a net counting rate with a standard deviation of $\pm$ 2%? First we apply equation (5),

$$t_T/t_B = \sqrt{900/100} = 3$$
$$t_B = t_T/3$$

Next we make use of equation (4), substituting 2 for $\%\sigma_S$,

$$2 = \frac{100\sqrt{900/t_T + 100/t_B}}{900 - 100}$$

Now substituting $t_T/3$ for its equal, t_B (found above),

$$2 = \frac{100\sqrt{900/t_T + 100 \times 3/t_T}}{800} = \frac{\sqrt{1200/t_T}}{8}$$

Transposing and then squaring both sides,

$$(16)^2 = 1200/t_T$$
$$t_T = 4.7 \text{ min} \approx 5 \text{ min}$$
$$t_B = t_T/3 = 1.5 \text{ min} \approx 2 \text{ min}$$

The total number of counts accumulated in these time intervals are:

total sample count $= N_T = 900 \times 5 = 4500$ counts
background count $= N_B = 100 \times 2 = 200$ counts

The technologist should always be aware of the standard deviation of any counting procedure, so as to have some idea of its reliability.

Frequently, the problem is the reverse of that described above; we may need to know the per cent standard deviation of a net counting rate after the gross and the background counts have been measured. This can be readily calculated from equation (4).

Recovery Time of Counter. As already mentioned in the discussion of GM and scintillation counters, there is an error at high counting rates due to the inability of the counter to detect radiation during a tiny interval of time following a pulse. This interval is called the *dead time* or *recovery time* of the counter. The higher the counting rate, the greater is the likelihood that two ionizing events may occur so close in time that only one is detected. A simple formula gives the approximate percentage correction for the dead time at extremely high counting rates:

$$\%C = 100 \; \frac{R\tau}{1 - R\tau} \tag{6}$$

where $\%C$ is the percentage correction that must be added to the observed counting rate R, to obtain the corrected counting rate, and τ (Greek letter "tau") is the dead time of the counter in question. For example, with a GM counter having a τ of $2(10)^{-4}$ sec, an observed counting rate of 200 cps should be corrected as follows:

$$\%C = 100 \; \frac{200 \times 2 \times 10^{-4}}{1 - 200 \times 2 \times 10^{-4}}$$

Performing the indicated multiplication first,

$$\%C = \frac{4}{1 - 0.04} = \frac{4}{0.96} = 4\%$$

The corrected counting rate is then

$$200 + 200(0.04) = 208 \text{ cps}$$

With a scintillation counter having a dead time of $(10)^{-6}$ sec, the correction factor is only about 1 per cent even with an extremely high counting rate of 500 cps (the corresponding correction with a GM counter would be approximately 11 per cent).

Efficiency and Sensitivity of Counters

The ability of a particular counter to detect radiation is generally known as its *efficiency*. Yet, this is not a practicable concept as applied in the Nuclear Medicine Department. Radioactive disintegration is a random process accompanied by the emission of radiation. In order to detect all of the radiation, the source would have to be placed within the counter; such counters are, indeed, available, but have only limited use.

Sensitivity of Counters. In practice, a counter is placed at some distance from the source when counting is to be done over an organ or over samples with sufficiently high activity. Because the radiation is emitted in all directions, the counter cannot receive it all, but can "see" only the radiation coming toward it, as shown in Figure 22.12. Furthermore, of the radiation actually entering

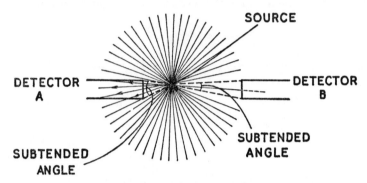

Figure 22.12. Angle of acceptance. The closer the counter is to the source, the greater is the number of particles entering it. Thus, detector *A* would record a higher counting rate than detector *B* with the same radioactive sample.

the counter, only a fraction produces ionizing events and is detected. For example, an ordinary GM counter will detect about 1 to 2 per cent of the entering gamma rays, although this can be improved by coating the cathode with bismuth. These two negative aspects of efficiency—the inability of a counter to see all the radiation, and to detect all the radiation that enters it—are embodied in *counter sensitivity* which is defined by the following equation:

$$f = \frac{\text{counting rate}}{\text{activity}} \qquad (7)$$

where *f* is the sensitivity. Bear in mind that sensitivity is related to a given geometric arrangement of the source and the counter. The units in which *f* are expressed are the same as those selected for activity. For example, *f* may be expressed as counts per sec per μCi, or counts per min per mCi, or any other convenient combination of units. Sensitivity can be used to convert counting rate to activity of a sample, by rearrangement of equation (7),

$$\text{activity} = \frac{\text{counting rate}}{f} \tag{8}$$

provided the same geometric setup, counter, and radionuclides are used as in the initial determination of sensitivity. It also permits the calculation of the sample activity needed to obtain a desired counting rate.

The preceding discussion applies primarily to the determination of the sensitivity of a particular counting system. In actual practice, there are two additional factors that influence the counting rate when counting is done over the body. One is attenuation of the radiation as it passes through tissues or other matter, on its way to the detecting instrument. The other is the scattering of radiation from nearby objects, with entrance of this scattered radiation into the detector. These factors are minimized by using radionuclides whose radiation is not appreciably attenuated, and by keeping the counting system isolated from scattering material. The influence of scattering can be further controlled by narrowing the window of the detecting system to the energy of the primary radiation.

Minimum Detectable Activity. We sometimes need to know the smallest quantity of a radionuclide that a counter can detect to a reasonable degree of reliability, with a given geometric arrangement and type of counter. According to the ICRU, the minimum detectable activity is that activity which, in a given counting time, records a number of counts on the instrument equal to three times the standard deviation of the background count in the same time. The equation specifying minimum detectable activity, A_{min}, is

$$A_{min} = \frac{3}{f}\sqrt{R_B/t_B} \tag{9}$$

where $\sqrt{R_B/t_B}$ is the standard deviation of the background counting rate, and f is the sensitivity of the counter. Since $R_B = N_B/t_B$, we can arrange as $t_B = N_B/R_B$. Substituting this value of t_B in equation (9),

$$A_{min} = \frac{3R_B}{f\sqrt{N_B}} \tag{10}$$

N_B is the number of background counts recorded in a preselected interval of time, usually about ten minutes. For example, if, with a particular geometric arrangement, f is 2000 cps per μCi, R_B is 4 cps, and N_B is 2500 counts in 10 min, then

$$A_{min} = \frac{3 \times 4}{2000 \times \sqrt{2500}} = \frac{3}{500 \times 50} = 0.0001 \ \mu\text{Ci}$$

The importance of the minimum detectable activity, then, is that it indicates the smallest possible sample that can be counted with reasonable accuracy, with a counting setup.

It cannot be overemphasized that the sensitivity and minimum detectable activity both depend on the geometric arrangement of the source and counter, the type of counter, and the radionuclide being considered. Hence, calibration of the equipment should be carried out separately for each counting setup and each radionuclide in use in the department.

Geometric Factors in Counting

Whenever a radionuclide sample is counted, we must take into account (1) the *distance between the source and the counter,* (2) the *presence of scattering material* in the vicinity of the source, (3) *the collimation of the counter,* and (4) *the size of the source.* In Figure 22.12 we show how a source radiates in all directions. *The inverse square law* of radiation applies (approximately, because not a point source), so that the nearer the counter is to the source, the more radiation it "sees." The presence of *scattering material* nearby may introduce an error because of the resulting fictitious increase in counting rate. *Collimation* definitely affects the fraction of the emitted radiation reaching the counter. Figure 22.13 explains the behavior of the same counter with two different degrees of collimation. Note that as collimation is narrowed, the counting rate decreases, but the area of maximum sensitivity depends only on the diameter of the collimator hole. On the basis of this principle, a counter can be so collimated that it can detect, in the thyroid gland, small nodules whose activity is different from that of the surrounding gland. But collimation must not be so severe that the counter fails to see the entire specimen when this is necessary, as in measuring radioiodine uptake by the thyroid gland (see Figure 22.14).

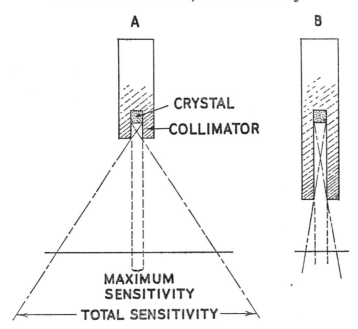

Figure 22.13. Effect of collimation on sensitivity of scintillation counter. In A, with less collimation, the counting rate is higher because the angle of view is large and the crystal "sees" a larger total area than in B where collimation is more severe. However, the area of maximum sensitivity depends on the size of the collimator opening (and not on the depth of the crystal within the collimator); this is the same in A and B. *Modified from Quimby, E. H., Feitelberg, S., and Silver, S. (see Bibliography).*

Because of the fact that sensitivity depends so intimately on the geometric arrangement of the source and counter, equipment should be standardized and used under an identical geometric setup in order to minimize error.

Methods of Counting

Two methods of counting, whether this is to be done on specimens outside the body, or over certain areas of the body, include (1) absolute counting and (2) comparative counting, both of which we shall now describe.

1. **Absolute Counting.** In principle, this is similar to the method of determining the sensitivity, f, of a counter. First the

A   B

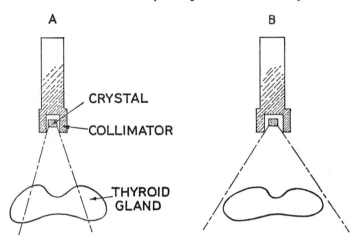

Figure 22.14. Effect of collimation on angle of view of a counter. In *A*, the collimation is so restrictive that the counter does not "see" the entire gland and the counting rate will therefore be too low. In *B*, with less severe collimation, the angle of view covers the entire gland and a more accurate counting rate will be obtained. *Modified from Quimby, E. H., Feitelberg, S., and Silver, S. (see Bibliography).*

sensitivity is measured with a standard known radioactive source (similar to the one to be counted in the patient or specimen) in the same geometric setup as will be used with the clinical material. Suppose the standard source under these geometric conditions has an activity of 1 μCi and a net counting rate of 3000 cps. Then $f = 3000$ cps per μCi. If, now, the specimen is counted under identical conditions (same radionuclide and geometric setup), and the net counting rate is 1000 cps, then according to equation (8),

$$\text{activity of sample} = \frac{\text{counting rate}}{f} \ \mu\text{Ci}$$

$$= \frac{1000}{3000} = 0.33 \ \mu\text{Ci}$$

2. **Comparative Counting.** In clinical radionuclide studies, especially where the tracer method is employed, it is more common to count the sample and standard separately under identical conditions. These net counting rates are then compared directly,

the specimen counting rate being expressed as a percentage of the standard. Comparative counting may be exemplified by tracer studies of the radioiodine uptake by the thyroid gland. A predetermined dose of radioiodine is given to the patient and, after the appropriate time has elapsed, the counting rate over the thyroid gland is measured at a fixed distance (for example, 20 cm) with the "window" set for the proper energy range. The same dose as that given the patient is now measured separately in a neck "phantom" at the same distance. Both counting rates are corrected for background, and

$$\% \text{ thyroid uptake} = \frac{\text{net thyroid cpm}}{\text{net standard cpm}} \times 100 \qquad (11)$$

This procedure is used for 2, 6, or 24 hour uptakes, but more elaborate studies are usually indicated and will be described below.

Important Medical Radionuclides

We shall now summarize the properties of some of the more important radionuclides used in medicine.

Radioactive Iodine. Iodine 131 (^{131}I) is the most commonly used radioisotope of iodine, having a physical half life of 8.1 days and emitting 364-keV gamma rays, and 183-keV (av.) beta particles. Iodine 131, obtained mainly by separation from the fission products of the nuclear reactor, is used as sodium iodide in the diagnosis and treatment of diseases of the thyroid gland, and as a tag for human serum albumin (IHSA or RISA) in lung scanning, blood volume determination, placenta localization, and diagnosis of pericardial effusion. The therapeutic effect of ^{131}I in hyperthyroidism and certain thyroid cancers depends mainly on the *beta particles,* whereas tracer and scanning studies depend on the *gamma rays* which are much more penetrating and can be detected outside the body with a GM or scintillation counter.

Recently, another isotope of iodine—^{125}I—has been under investigation for thyroid scanning. Its longer half life of 60 days permits longer storage and, at the same time, its lower energy— 28 keV—allows better collimation and possibly easier detection of thyroid nodules (although it may hamper detection of retrosternal extension). A 100-μCi scanning dose of ^{125}I delivers about 80 rads to the thyroid gland, as contrasted with about 130 rads from the same dose of ^{131}I, really an insignificant difference.

Radioactive Phosphorus. Phosphorus 32 has a half life of 14.3 days and emits 700-keV (av.) beta particles. It is prepared mainly by neutron irradiation of stable sulfur in the nuclear reactor (see page 414). After administration in suitable form orally or intravenously, ^{32}P tends to concentrate in the bone marrow, spleen, liver, and lymph nodes. It can therefore be used, to a limited extent, in the treatment of certain blood disorders. Its greatest use is in the treatment of polycythemia rubra vera, a condition in which the bone marrow produces an excess of red blood cells. Phosphorus 32 suppresses the overproduction of these cells. In diagnosis, ^{32}P has been used successfully in the location of certain intraocular tumors. An intravenous dose of ^{32}P as phosphate delivers a whole body dose of about 5.4 rads per mCi.

Technetium 99m. The radionuclide ^{99m}Tc has a short though convenient half life of 6 hours. Its gamma radiation, having a relatively low energy—140 keV—is easily collimated and at the same time has adequate penetrating ability to pass through the body and reach the detector. Technetium 99m is at present used mainly as the pertechnetate salt in aqueous solution. Because of the short half life and the absence of beta radiation, large doses can be given with relatively little exposure to the patient; for example, with an intravenous brain scanning dose of 10 mCi, the patient receives a whole body dose of about 0.12 rad. As a result we can achieve high counting rates and good resolution of the scan images. By far the greatest application of ^{99m}Tc has been in brain scanning, although it is coming into wider use in other areas such as lung and liver scanning in the form of ^{99m}Tc-tagged sulfur colloid. We should mention the *technetium "cow"*, a molybdenum (^{99}Mo) generator which decays to ^{99m}Tc. Each day 0.9 per cent sodium chloride solution is passed through the generator, dissolving out the ^{99m}Tc; this is called "milking" the "cow." The resulting ^{99m}Tc is filtered during the milking process and, after calibration, is ready for use.

Radioactive Cobalt. The most widely used radioisotope of cobalt—^{60}Co—is produced in the nuclear reactor. It has a half life of 5.2 years and emits gamma rays of two energies—1.17 and 1.33 MeV (average 1.25 MeV)—and a low energy beta particle (av. 96 keV). Cobalt 60 is available in a variety of forms for short distance therapy—needles, beads, plaques, and wires—in place of radium. More recently, cesium 137 has been found to be an even better substitute for radium. Large ^{60}Co teletherapy units are designed for use in external irradiation.

Vitamin B_{12} or cyanocobalamin contains cobalt as a part of its molecular structure; when this compound is tagged with ^{57}Co or ^{60}Co, it is symbolized by ^{57}CoB$_{12}$ or ^{60}CoB$_{12}$, respectively, and is extensively used in the Schilling test for pernicious anemia (see page 445).

Radioactive Mercury. Two radioisotopes of mercury—^{197}Hg and ^{203}Hg—are important in nuclear medicine. Mercury 197 has a half life of 65 hours and emits gamma rays with principal energies of 67 to 78 keV. Mercury 203 has a half life of 47 days and emits gamma rays with a single energy of 280 keV (and beta particles with average energy of 58 keV, not important in diagnosis). These radionuclides, incorporated in the compound chlomerodrin, are used in brain and kidney scanning. In comparing the radiation dose received by the patient, we find that in brain scanning, a dose of 700 μCi of ^{197}Hg delivers a whole-body dose of 0.012 rad and a kidney dose of 3 rads; whereas, with 700 μCi of ^{203}Hg the whole-body dose is 0.3 rad and the kidney dose is 40 rads. These doses are reduced proportionately for the smaller kidney scanning dose of 150 μCi. Another use of radioactive mercury is in dynamic studies of kidney function called *radioactive renography.*

Radioactive Strontium. Two radioactive isotopes of strontium are now available for scanning—^{85}Sr and ^{87m}Sr. (Strontium 90, used in beta particle therapy of superficial lesions of the eye, will be discussed later.) Strontium 85 has a physical half life of 65 days and emits a 510-keV gamma ray. It is used intravenously as the nitrate salt for bone scanning to detect the presence of metastatic cancer. The whole-body dose is small, approximately 0.3 rad for the usual scanning dose; this is insignificant, especially since its use is limited to cancer patients. Another strontium radionuclide, ^{87m}Sr, has a very short half life of 2.8 hours and emits a 388-keV gamma ray. Because it delivers a much smaller whole-body dose, [87m]Sr is used in children to investigate disorders of bone growth, but requires the use of an yttrium 87 cow and is therefore an expensive procedure.

Radioactive Gold. The radionuclide, gold 198 ([198]Au) in colloidal form, is used extensively in liver scanning. It has a half life of 2.7 days and emits a 412-keV gamma ray and 316-keV (av.) beta particles. As usual with scanning radionuclides, the gamma rays are the ones that are utilized. Virtually all of the energy is deposited in the liver which receives a dose of about 7 rads from an administered dose of 150 μCi of [198]Au.

Radioactive Selenium. When the amino acid methionine is tagged with selenium 75 (^{75}Se), it is called selenomethionine and can be used for scanning the pancreas because it reaches a high concentration in that organ. However, it also appears in the liver in about $\frac{1}{10}$ the concentration of that in the pancreas. Selenomethionine can also be used in scanning the parathyroid glands. Selenium 75 has a physical half life of 127 days and emits gamma rays of several different energies, 95 per cent comprising 270 and 140 keV. The whole body dose is not a limiting factor, and the dose to the pancreas is less than 0.1 rad.

Radioactive Iron. Of the two radioisotopes of iron, ^{59}Fe and [55]Fe, the former is generally more useful in diagnosis. Iron 59, produced by neutron irradiation of stable iron in the nuclear reactor, has a physical half life of 45 days and emits gamma rays of three different energies, 98 per cent of which are 1.29 and 1.10 MeV. It also emits beta particles with an average energy of 118 keV. Various compounds containing ^{59}Fe are used as tracers to study the absorption and storage of iron, and its rate of appearance in the red blood cells, especially in anemia. A diagnostic intravenous dose of 5 μCi of ^{59}Fe delivers a small total-body dose—about 0.16 rad. The corresponding value for oral administration is much less—0.014 rad in normals and 0.1 rad in iron-deficient individuals. The largest dose as far as individual organs are concerned is about 1 to 1.7 rads to the testes.

Radioactive Chromium. The radionuclide chromium 51 (^{51}Cr) is used in the form of chromate to determine red cell volume in polycythemia and in patients undergoing extensive surgery and to determine red cell survival in hemolytic anemias. Chromium 51 has a physical half life of 27.8 days, and emits a 320-keV gamma ray and 242-keV (av.) beta rays. In the chromate form, this nuclide tags red cell, whereas in the chromic form it labels blood plasma.

Examples of the Use of Radionuclides in Medical Diagnosis

It is beyond the scope of this book to give a detailed account of the numerous diagnostic procedures in the Nuclear Medicine Department. Instead, we shall include examples of the main types of procedures based on various principles; these will include (1) *uptake by an organ*, (2) *excretion*, and (3) *dilution*.

1. **Uptake or Tracer Studies.** As mentioned earlier, the thyroid gland has a strong affinity for iodine. If radioactive iodine is taken orally, it will be concentrated in the thyroid gland over a period of twenty-four hours, the concentration depending on the functional capacity of the gland. Obviously, before such a test can be run, the patient must avoid the intake of ordinary stable iodine (^{127}I), as well as drugs that are known to suppress the activity of the thyroid gland. Such restriction may require weeks or months, depending on the type of drug; radiopaque media containing iodine may interfere with further iodine uptake for several months.

Depending on the type of equipment, a tracer dose of 2 μCi to 50 μCi of ^{131}I is taken orally in the morning by the fasting patient. A phantom simulating the neck should be available, made of Lucite or hard paraffin according to ICRU specifications (see Figure 22.15). A plastic test tube containing a standard of the same dose as that given the patient is placed in the hole in the phantom. At prescribed intervals, preferably 2, 6, and 24 hours, counting rates are taken at the same fixed distance—20 cm—from the thyroid and from the phantom (comparative counting). The window is set for a range of 354 to 374 keV. Each time, a body background count should also be made over the lower end of the thigh, which resembles the tissues of the neck without the thyroid; the thigh counting rate should be subtracted from the

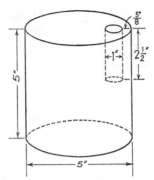

Figure 22.15. Diagram of a standard neck phantom, made of Lucite. The standard radioiodine capsule is placed in the 1-inch well.

thyroid counting rate each time. The per cent uptakes at each counting session (2, 6, 24 hrs) are calculated from the following equation:

$$\% \text{ thyroid uptake} = \frac{100 \ (\text{neck } R_T - \text{thigh } R_T)}{\text{phantom } R_T - \text{room } R_B} \quad (12)$$

where R_T is the total counting rate of each item, and R_B is the background counting rate.

Figure 22.16 shows typical uptake curves obtained in various functional states of the thyroid gland. There is significant over-

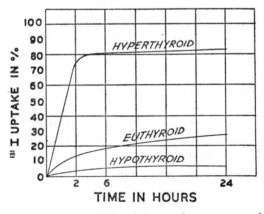

Figure 22.16. Representative 131-iodine uptake curves in the three principle functional states of the thyroid gland.

lap of normal and abnormal values at both ends, this being more serious in the low or hypothyroid region (see Table 22.1). Still, the thyroid uptake of radioiodine is an extremely useful measure of thyroid function, especially when combined with supplementary chemical and radiochemical studies, as well as scanning.

TABLE 22.1

RANGE OF 24-HR UPTAKES OF ^{131}I BY THE THYROID GLAND, IN VARIOUS FUNCTIONAL STATES *

Euthyroid (normal)	8%–33%
Hyperthyroid	more than 33%
Hypothyroid	less than 8%

* These values are typical; actually, there may be much more overlap of the normal and abnormal than that shown here. Other tests are usually necessary to establish the diagnosis.

2. **Excretion of a Radionuclide by the Body.** An example of this is the test for *pernicious anemia,* performed with cyanocobalamin (vitamin B_{12}) labeled with radioactive cobalt, either ^{57}Co or ^{60}Co. Normal persons readily absorb this vitamin from the gastrointestinal tract, storing it in the liver within four to seven days. However, patients with pernicious anemia are unable to absorb it because they lack *intrinsic factor* which is normally secreted by the stomach. The fasting patient is first asked to empty his bladder, the specimen being saved as a pretest control. The patient then swallows a capsule containing a small dose—precalibrated 0.5 μCi of ^{60}CoB$_{12}$ or ^{57}CoB$_{12}$ with ½ glass of water (patient must not have received ordinary B_{12} orally or parenterally during the preceding 2 days). One hour later, a large dose of ordinary, stable B_{12}—1000 μg—is given intramuscularly to flush any absorbed radioactive B_{12} through the kidneys and into the urine which is collected for 24 hours in a 1-gallon plastic bottle with a screw cap. Urine collection must be complete. If the urine volume is less than 1 liter, it is made up to exactly 1 liter; if more than 1 liter, a 1 liter aliquot is taken in another bottle. The same dose of standard as that originally given the patient is made up to 1 liter in the same type of bottle. The standard bottle and the patient's specimen bottle (1 liter) are each

counted in contact with the crystal of a scintillation counter and calculations are made as follows:

if original volume of urine was more than 1000 ml,

$$\% \text{ absorption} = \frac{\text{urine } R_T - \text{background } R_B}{\text{standard } R_T - \text{background } R_B} \times 100$$

if original volume of urine was more than 1000 ml,

$$\% \text{ absorption} = \frac{\text{urine volume (urine } R_T - \text{background } R_B)}{1000 \text{ (standard } R_T - \text{background } R_B)} \times 100$$

An absorption of about 5 per cent or more is considered to be normal. If the absorption is less than this, the examination is repeated after four days, intrinsic factor being given at the same time as the ^{57}Co or ^{60}CoB$_{12}$ capsule. In the presence of pernicious anemia, the absorption should increase to a level above 10 per cent.

3. **Dilution Tests with Radionuclides.** When a radioactive nuclide in a given volume of solution is mixed with an unknown larger volume of water, the nuclide is obviously diluted. If the net counting rate, r_s, per ml of the original sample volume, v_s, is known, and the net counting rate, R_s, per ml of the larger volume is measured under the same conditions as the original sample, then the larger volume, V_s, can be computed by means of the following equation:

$$VR_s = vr_s \tag{13}$$

This equation summarizes the *dilution principle,* which is applied in the determination of circulating blood volume. A small dose of 5 μCi of radioactive iodine-tagged human serum albumin (RISA) is injected intravenously, most often into an antecubital vein. After an interval of about 10 min to allow for admixture with the circulating blood, 5 ml of blood is withdrawn from the opposite antecubital vein, oxalated, and counted in a well counter. The same volume, 5 ml, of a dilute standard (5 μCi in 1 liter) is counted in the same way. Both are corrected for background. The circulating blood volume is then obtained from the following equation:

$$\text{blood volume in ml} = \frac{\text{cpm/ml standard} \times 1000}{\text{cpm/ml blood}} \tag{14}$$

Standard tables are available for estimating the relationship between the observed blood volume and the normal value. The plasma volume can be found by separating the plasma from the sample and counting it in the same manner as the whole blood. After appropriate correction, red cell volume is obtained by subtracting plasma volume from whole blood volume.

Radionuclide Scanning

Any procedure which maps the distribution of radiation within the body, from points outside the body, is called *radionuclide scanning* or *scintiscanning*. The efficiency of such automatic mechanical scanners may be high, especially if high counting rates are possible. Scanning is by no means foolproof, since errors may arise in equipment, method, or interpretation. Considerable experience is necessary to qualify one to interpret the results. We must note that scans are not quantitative—they simply display relative counting rates over various parts of an organ or region.

Mechanical scanners are equipped with a scintillation counter provided with a multiple hole focused collimator, shown in Figure 22.17. Such collimators consist of a block of heavy metal such as tungsten in which are bored multiple holes whose directions converge toward a point as some fixed distance from the face of the detector. The advantage of this type of collimator is that it provides *maximum sensitivity for a given degree of collimation.*

The scintillation counter and its collimator move automatically back and forth over the selected region of the body, recording the counting rates of the administered radionuclide on x-ray film as photographic dots, and on special paper as ink or burn dots. The higher the counting rate in a given area, the closer will be the grouping of the dots. In addition, there is a system for adjusting the background counting rate to increase the contrast in the scan and, within limits, make it easier to detect small abnormalities.

Figure 22.18 shows a block diagram of a mechanical scintillation scanner.

One of the main disadvantages of the mechanical scanner is that it requires a significant time interval for completion, so that the end of the scan occurs significantly later than the be-

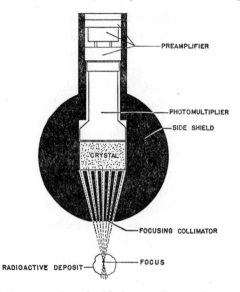

Figure 22.17.   Diagram of a shielded scintillation detector and focusing
collimator used in scanning.

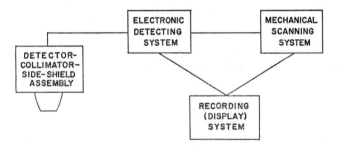

Figure 22.18.   Block diagram of a mechanical scintillation scanner.

ginning. This is especially undesirable in dynamic studies wherein
we wish to follow the deposition or disappearance pattern of a
radionuclide in an organ over a period of time. Several devices
are now available for this purpose, as well as to shorten the time
required for ordinary studies. The *Anger gamma camera,* using
a large detector crystal which measures about 11 in. in diameter
(ordinary scintillation crystals are 3 or 5 in.), can detect and
record simultaneously the radioactivity over a fairly large area,
without mechanical scanning. Another such device is the *auto-*

fluoroscope, which consists of a bank of many detector crystals to accomplish the same end. A Polaroid camera as well as 70 mm and 35 mm roll film may be used to display the scintigram produced by both types of cameras.

At present, scanning is used mainly in the following studies, although new applications are continually being investigated:

1. **Thyroid Gland**—^{131}I or ^{125}I (see page 439) in the form of sodium iodide—for detection of functioning and nonfunctioning nodules, the former being "hot nodules" because they appear dark on the scan; and the latter being termed "cold nodules" because they appear as blank areas in the scan. About 10 to 15 per cent of cold nodules turn out to be malignant tumors. Scanning also shows the size, shape and location of the thyroid gland.

2. **Brain**—technetium 99m (^{99m}Tc) as the pertechnetate (see page 441), and mercury 197 (^{197}Hg) and mercury 203 (^{203}Hg) as chlormerodin (see page 441) are available. Technetium 99m is preferred because it provides a high counting rate, good differential uptake between most lesions and the normal brain, and relatively small irradiation of the patient. The mercurial agents also produce satisfactory scans, but give considerably smaller counting rates which require longer scanning times; furthermore, the patient receives a considerably larger exposure than with ^{99m}Tc. Brain scans are used in the diagnosis of brain tumors (primary and metastatic), subdural hematomas, vascular abnormalities, granulomas, infarcts, and abscesses. In general, brain lesions take up *more* radionuclide than the adjacent normal tissue, so that they appear as darker areas in the scan—so-called "hot" areas.

3. **Liver**—several scanning agents are available, including ^{99m}Tc as a sulfur colloid; gold 198 (^{198}Au) as a colloid; and a dye, rose bengal, tagged with ^{131}I. Liver scans are particularly useful in finding not only metastatic and primary neoplasms but also certain degenerative diseases such as cirrhosis. Lesions replacing normal liver tissue usually take up *less* radionuclide and appear lighter than the normal tissue—so-called "cold" areas. Liver scans also aid in the study of liver function, and in determining the size, shape, and position of this organ.

4. **Kidney**—the mercurials, ^{197}Hg and ^{203}Hg as chloromedrin (see page 441), are used in kidney scanning to aid in the diagnosis of tumors, cysts, granulomas, abscesses, atrophy, malformations, nephritis, and functional impairment.

5. **Lung**—macroaggregated serum albumin tagged with ^{131}I is used to obtain indirect evidence of pulmonary arterial emboli or thrombosis. The albumin particles are large enough to lodge temporarily in the normal pulmonary capillary bed, thereby indicating that blood is reaching the capillaries. In arterial embolism or thrombosis, capillary perfusion is blocked and the albumin particles do not reach the capillaries, a condition manifested in the scan by a "cold" area. However, differentiation from air cysts as in emphysema, from pleural effusions, and from tumor masses is difficult or impossible. Technetium 99m-tagged colloids are also being used in lung scanning.

6. **Bone**—strontium 85 (^{85}Sr) nitrate is most often used in bone scanning, although 99^{m}Tc polyphosphate is also becoming available. Bone scanning has its greatest application in locating metastatic bone cancer in patients with a known primary cancer, often before there is x-ray evidence of such metastases. Strontium concentrates in zones of actively proliferating bone in much the same manner as calcium. Therefore ^{85}Sr can be used in scanning to detect malignant tumor deposits in bone whenever it reacts proliferatively to such tumor invasion, appearing in the scan as "hot" areas. They are not specific for malignant tumor, since some benign tumors, infections, and Paget's disease cause sufficient bone reaction to produce a positive scan.

7. **Miscellaneous**—still experimental, to some degree, are scanning of the heart for myocardial infarction, and the parathyroid glands for hypertrophy. Pericardial effusions can be revealed by scanning. In patients with thyroid cancer the entire skeleton should be scanned after an appropriate dose of radioiodine in searching for possible metastases.

Radionuclides in Therapy

The use of radionuclides in therapy comprises the *internal* and *external* modes of delivery. Only a few examples will be given in order to show the potentialities of this important form of treatment.

Internal Therapy with Radionuclides. Radioactive iodine (^{131}I) is probably the most frequently used radionuclide for internal therapy. Its application is confined to the treatment of hyperthyroidism and certain types of cancer of the thyroid gland.

1. *Hyperthyroidism.* When ^{131}I tracer studies show that the thyroid gland is overactive—hyperthyroid—the medical decision is made as to the therapeutic approach; that is, medical, surgical, or radionuclide. If the latter is selected, a suitable dose of ^{131}I is given orally for internal irradiation and partial destruction of functioning thyroid tissue. The dose is much larger than that used in diagnosis, usually amounting to about 4 to 6 mCi or more, depending on the type of gland, its size, and the uptake percentage. About 90 per cent of the therapeutic effect of ^{131}I is due to its beta particles (maximum energy 608 keV) and about 10 per cent is due to its gamma rays (mainly 364 keV).

2. *Thyroid Cancer.* Approximately 10 per cent of thyroid cancers have the structure and function that make them amenable to treatment with radioiodine. To stimulate metastatic lesions from thyroid cancer to take up ^{131}I, the gland itself should first be removed surgically; or destroyed by a large dose of ^{131}I, provided it can take up enough of the administered dose. This is usually at least 100 mCi, and it may often have to be repeated one or more times. Because of the magnitude of the dose and the amount of radionuclide excreted in the urine, stringent radiation safety precautions have to be observed in ^{131}I therapy of thyroid cancer.

External Therapy with Radionuclides. There are two main types: *teletherapy* and *contact therapy*. The former has, by far, the wider application.

1. *Teletherapy.* In this form of therapy, the radioactive source is placed at a distance of about 50 to 100 cm from the patient. The most widely used source material is ^{60}Co which has a half life of 5.2 years and gives off gamma rays of high energy (1.17 and 1.33 MeV, equivalent to about 2 MV x rays). It also emits low energy beta particles which are removed by suitable filters. The introduction of ^{60}Co therapy marked an important advance in the treatment of deep-seated cancer because the highly pene-

trating gamma rays (HVL = 11 mm Pb) can provide a large depth dose percentage with relatively little skin reaction. Figure 22.19 shows the comparative isodose curves obtained with a ⁶⁰Co teletherapy unit and a 200-kV x-ray machine; it is to be noted that the central axis 10-cm depth dose with ⁶⁰Co is about 55 per cent, as contrasted with about 35 per cent with 200-kV x rays. The comparison is deliberately made at different treatment distances because they are most often used with these types of teletherapy. The treatment time with 200-kV x rays at distances longer than 50 cm is unduly prolonged. Beyond the improved depth dose percentage and the skin-sparing effect of ⁶⁰Co, a third advantage is equal absorption of the high energy radiation in equal masses of bone and soft tissue, thereby reducing the hazard of bone damage. (Review pages 195-196.)

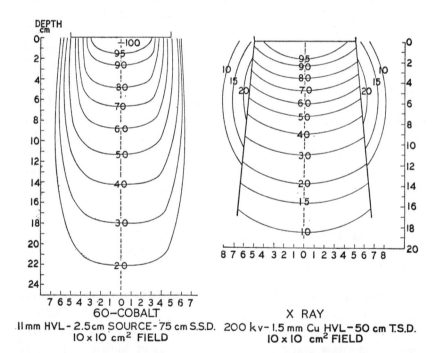

Figure 22.19. Comparison of isodose curves of cobalt 60 and 200-kV x rays. Note the larger central axis depth doses with cobalt 60 at a given depth. There is also less radiation lateral to the primary beam with cobalt 60 because of less lateral scatter.

The ^{60}Co *teletherapy irradiator* consists of pellets or slugs of ^{60}Co doubly encased in stainless steel and mounted in a heavy sphere of lead, tungsten and uranium alloy to reduce the radiation exposure rate outside the sphere to the permissible value. A window in one side of the sphere, which can be opened or closed by remote control, serves as the port through which the gamma rays can be directed at the patient. Figure 22.20 shows a

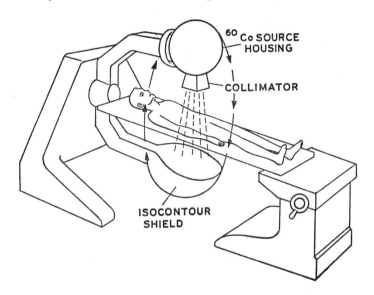

Figure 22.20. Rotational cobalt 60 teletherapy unit with an isocontour shield. The latter always remains opposite the source during rotation, reducing the amount of protective barrier needed in the walls of the therapy room.

typical ^{60}Co teletherapy unit. The output of such a unit may be very large; for example, a 4000 rhm source delivers an exposure rate of 4000 R/hr at 1 meter. Such a source consists of about 4000 Ci of ^{60}Co. Since the gamma rays from this source are nearly monoenergetic, filtration is not used except for removal of the beta particles, this being provided by the steel capsule.

Less widely used at present is the ^{137}Cs (radiocesium) irradiator. Since the gamma rays of ^{137}Cs have an energy of 0.66 MeV, as contrasted with an energy of about 1.25 MeV for ^{60}Co, the depth dose is less with the cesium unit. For example, at a depth of 10 cm in tissue, the depth dose percentage with ^{137}Cs is about 25 per cent less than with ^{60}Co. Furthermore, the specific activity of ^{137}Cs is comparatively small, relative to ^{60}Co, so that shorter treatment distances have to be used in order to keep the treatment time reasonably short; this further decreases the depth dose percentage. Despite these shortcomings, the ^{137}Cs irradiator is used, especially in the larger radiotherapy centers, for therapy at intermediate depths.

2. *Contact Therapy.* Among radionuclides, ^{90}Sr (radiostrontium) is the only one that is used to any significant extent in contact therapy; that is, where the applicator is placed in direct contact with a superficial lesion. Figure 22.21 shows diagrammati-

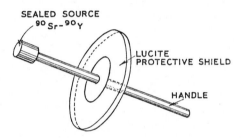

Figure 22.21. Contact 90-strontium applicator (Tracerlab®).

cally the construction of an eye irradiator. The radiation emitted by an ^{90}Sr applicator consists of *beta particles only,* with such poor penetrating ability (av. energy 930 keV) that superficial lesions of the front of the eye can be treated with a relatively small exposure of the highly sensitive lens. At a depth of 9 mm, the exposure rate becomes almost insignificant. Still, the dose to the lens of the eye may be sufficient to induce cataract formation. The half life of ^{90}Sr is 28 years. The source itself measures about 5 mm in diameter and is doubly sealed in a stainless steel container, filtered by thin layers of stainless steel and aluminum.

Table 22.2 summarizes the radionuclides that are most frequently used in medicine. Many others are available for specific purposes, and others may be found, in the future, to be superior to those presently being used.

TABLE 22.2

SOME RADIONUCLIDES OF MEDICAL INTEREST

Nuclide	Symbol	Physical Half Life	Main Radiation		Use in Medicine
			E_γ	$\bar{E}_\beta$	
			MeV	MeV	
Cesium 137	^{137}Cs	30 yr	0.662	0.188	Teletherapy. Needles.
Chromium 51	51Cr	27.8 days	0.320	0.242	Red cell volume and survival.
Cobalt 57	^{57}Co	270 days	0.122	—	Tagged vitamin B$_{12}$ (^{57}CoB$_{12}$) in diagnosis of pernicious anemia.
Cobalt 60	^{60}Co	5.2 yr	1.17 1.33	0.096	Teletherapy. As radium substitute in needles, beads, and wires.
Gold 198	^{198}Au	2.7 days	0.412	0.316	Liver scanning.
Iodine 125	^{125}I	60 days	0.028	—	Diagnosis of thyroid disorders.
Iodine 131	^{131}I	8.1 days	0.364	0.183	Diagnosis and treatment of thyroid disorders. Localization of thyroid cancer metastases. Liver and lung scanning. Diagnosis of pericardial effusion.
Iron 59	^{59}Fe	45 days	1.10 1.29	0.118	Iron absorption and utilization. Iron appearance time in red cells in study of anemia.
Mercury 197	^{197}Hg	2.7 days	0.077	—	Brain and kidney scanning. Radioactive remography.
Mercury 203	^{203}Hg	47 days	0.280	0.058	Same as mercury 197, but less often used because of larger patient exposure.
Phosphorus 32	^{32}P	14.3 days	—	0.700	Treatment of polycythemia rubra vera. Diagnosis of intraocular tumors. Treatment of certain metastatic bone tumors (esp. breast and prostate).
Selenium 75	^{75}Se	127 days	0.270 0.140	0.011	Pancreas and parathyroid scanning.
Strontium 85	^{85}Sr	65 days	0.510	—	Bone scanning for metastases.
Strontium 90	^{90}Sr	28 yr	—	0.196 0.937*	Contact therapy of superficial lesions of eye
Technetium 99m	^{99m}Tc	6 hr	0.140	0.084	Scanning of brain, thyroid, liver, lung, and placenta.

E_γ is energy of principal gamma rays. $\bar{E}_\beta$ is average energy of beta rays.
* Refers to yttrium 90, always associated with strontium 90.

QUESTIONS AND PROBLEMS

1. Define isotope; radioisotope; nuclide; radionuclide; nucleon.
2. What is radioactivity? Artificial radioactivity?
3. Describe the nuclear reactor, including fission, chain reaction, and control.
4. Discuss the two main methods of producing radioisotopes in the nuclear reactor, with equations showing how ^{32}P and ^{60}Co are obtained. What is the principle source of ^{131}I?

5. Discuss the three types of half life. What is the simple formula for the fraction of a radionuclide remaining after a particular number of half lives have elapsed?

6. Explain the construction, operation, and use of GM and scintillation counters. What is meant by dead time?

7. Contrast the rate meter and scaler. What is meant by preset timing? Preset counting?

8. Discuss the sources of error in counting.

9. Suppose the gross sample counting rate (including background) to be about 200 cpm, and the background counting rate to be about 50 cpm. How many sample and background counts should be recorded, respectively, to bring about a 5 per cent standard deviation in the net counting rate? How long would the sample and the background have to be counted?

10. The counting rate of a sample is 400 cps. If one were to use a GM counter with a recovery time of $2(10)^{-4}$ sec, what would be the counting rate, corrected for dead time? What would this be with a scintillation counter having a recovery time of $(10)^{-6}$ sec?

11. Discuss the efficiency of counters, including sensitivity and minimum detectable activity.

12. What are the geometric factors in counting?

13. Name and explain the two principal methods of counting.

14. State the three basic procedures in medical radionuclide diagnosis, and cite examples of each.

15. Discuss the internal and external application of radionuclides in therapy.

16. State important facts about the radionuclides that are useful in medical diagnosis; therapy.

CHAPTER *23* PROTECTION IN RADIOLOGY— HEALTH PHYSICS

THE HARMFUL EFFECTS of exposure to ionizing radiation were largely unsuspected at the time of the discovery of x rays and radium. However, the pioneer workers in this field soon recognized the injurious potentiality of such radiation. It is therefore surprising that so many of the early radiologists and technicians carelessly exposed themselves to x rays and radium, incurring serious local and general radiation injuries which all too often resulted in death.

In the intervening years we have learned a great deal about radiation hazards and their prevention. As a result, protective measures have become ever more stringent, especially in view of the growing use of radionuclides in medicine and industry and the application of atomic energy in weapons testing.

Every radiologic technologist should own and read the following *National Council on Radiation Protection and Measurements (NCRP) Reports,* obtainable from NCRP, P.O. Box 4867, Washington, D.C. 20008:

No.

24 *Protection Against Radiations from Sealed Gamma Sources (1960).*

30 *Safe Handling of Radioactive Materials (1964).*

33 *Medical X-ray and Gamma-ray Protection for Energies up to 10 MeV—Equipment Design and Use (1968).*

34 *Medical X-ray and Gamma-ray Protection for Energies up to 10 MeV—Structural Shielding Design and Evaluation (1970.)*

457

37   *Precautions in the Management of Patients Who Have Received Therapeutic Amounts of Radionuclides (1970).*
39   *Basic Radiation Protection Criteria (1971).*

Additional excellent booklets that are obtainable from the United States Atomic Energy Commission (USAEC), P.O. Box 62, Oak Ridge, Tenn. 37830, include:
Your Body and Radiation.
The Genetic Effects of Radiation.

What is the "safe" limit of exposure to radiation? An attempt will be made to answer this question, as well as to point out steps that can be taken to eliminate the risk of overexposure. The subject of protection or *health physics* will be taken up principally as it affects the *radiologic technologist* and the *patient.* A brief summary of electrical hazards in the x-ray department will be included. Attention will also be directed to the use of radium and artificial radionuclides.

BACKGROUND RADIATION

A sensitive radiation detector, such as a GM counter, will indicate the presence of ionizing radiation even if all known sources, such as radium, x rays, and radioactive nuclides, have been removed from its vicinity. Man has virtually no control over this environmental radiation which is termed *natural background radiation.* Living things obviously experience the same background exposure as the detector. Natural background includes not only external sources of radiation, but also radioactive materials located within the body and the detector itself.

Let us survey the sources of natural background radiation, which may be classified as *external* and *internal.*

1. *External sources* include, first, *cosmic rays* which are of two types. The *primary cosmic rays,* arising in the sun and other stars, consist mainly of high energy protons (more than 2.5 billion electron volts), but also include alpha particles, atomic nuclei, and high energy electrons and photons. *Secondary cosmic rays* are produced by interaction of primary cosmic rays with nuclei in the earth's atmosphere and consist mainly of mesons (a

type of nuclear particle), electrons, and gamma rays. Most cosmic rays observed in the laboratory are of the secondary type. They are so penetrating that they can pass through lead many feet thick.

A second external source of background radiation includes the *naturally radioactive minerals* in the earth itself. These occur in minute amounts almost everywhere in the earth's crust, including building materials; and in larger amounts where the minerals of uranium, thorium, and actinium have been deposited. It must be emphasized that terrestrial background radiation varies from place to place; for example, in Kerala, India, it is about ten times greater than that in most parts of the United States.

2. *Internal sources* include naturally radioactive nuclides incorporated in the tissues of the body, and in the materials of which the detector itself is constructed. The main naturally radioactive nuclides are potassium 40 and carbon 14.

To the natural background radiation must be added the *artificial background radiation* arising from the following sources: medical and dental x rays, occupation (for example, x-ray, radium, and radionuclide technology), environs of atomic energy plants, and atomic bomb fallout. In contrast to natural background, man can exercise some measure of control over artificial background.

Table 23.1 summarizes the average background exposure of the general population of the United States on an annual basis, and on the basis of the accumulated dose, per individual, from conception to age 30 years. This age span is selected because it is estimated that 80 per cent of children are born by the time the parents have reached age thirty. Obviously, the concern with these minute amounts of radiation is for the *genetic effects* (ie, the deleterious effects on future generations, resulting from damage to the reproductive cells), rather than for any possible damage to the individual himself, although even this may not be negligible. Since the values shown in the table are approximate, they should not be applied, with any semblance of precision, to the individual.

TABLE 23.1

AVERAGE RADIATION EXPOSURE PER PERSON DUE TO NATURAL
AND ARTIFICIAL BACKGROUND*
(NOT TO BE APPLIED CRITICALLY TO THE INDIVIDUAL)

Source	*Whole Body Dose*	
	rems/year/person (based on entire population)	*rems conception to age 30* (in procreative segment of population only)
Natural—total	*0.1*	*3.0*
Cosmic	0.03	0.9
Earth and Housing	0.05	1.5
Internal	0.025	0.75
Artificial—total	*0.17 – 0.33*	*1.5 – 4.8*
Medical and Dental	0.15 – 0.3	1.0 – 4.0
Occupational	0.005	0.15
Plant Environs	0.005	0.15
Fallout	0.01 – 0.015	0.3 – 0.45
TOTAL—all sources	*0.3 – 0.4*	*4.5 – 7.8*

* Data from American College of Radiology.

THE MAXIMUM PERMISSIBLE DOSE EQUIVALENT

It is obvious that some sort of safety standard must be established for individuals exposed to penetrating radiations. This standard, the *maximum permissible dose equivalent*, is defined *as the maximum dose of ionizing radiation which an individual may receive without suffering appreciable bodily injury during his lifetime.*

At present the maximum permissible dose equivalent (MPD) is based on the accumulated dose over the entire lifetime of the individual. The *accumulated occupational MPD from x, gamma, or beta radiation to the whole body and to certain radiosensitive organs*—gonads, blood forming organs, lens of the eye, head, and trunk—shall not exceed

$$MPD = 5(N - 18)\,rads \qquad (1)$$

where N is the age in years and is greater than 18. For example, at age 32 the MPD is $5(32 - 18) = 70$ rads. Thus, the average annual MPD is 5 rads; and weekly, 0.1 rad or 100 millirads. An accumulated dose of 3 rads is allowed in any 13-week period. The hands and feet may receive as much as 75 rads per year, but not over 25 rads in any 13-week period (see Figure 23.1). The above requirements apply *only to radiation workers;* mainly because of genetic effects, the **MPD for the general population is 1/10 of the above limits**—0.5 rad per year. The radiation exposure of x-ray personnel for their own diagnosis and therapy is not included in computing the MPD.

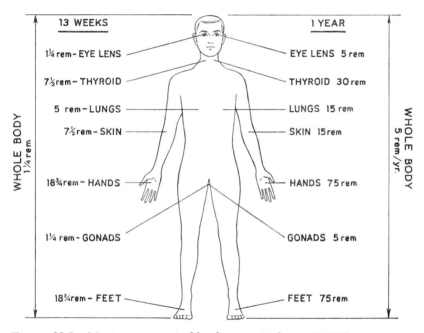

Figure 23.1. Maximum permissible dose equivalents (MPD) to various organs and to the whole body of radiation workers, for 13 weeks (left column) and for 1 year (right column).

The MPD is stated in rads for x, beta, and gamma radiation because equal absorbed doses of these three types of radiation produce virtually the same effects in the tissues. However, to produce a given effect in tissues, a smaller absorbed dose (rads) is required with heavy particles, such as alpha particles and neutrons, than with x rays. Thus, *one rad of alpha particles or fast neutrons produces a greater tissue effect than one rad of x rays,* and it would therefore be incorrect to add the absorbed doses of such different kinds of radiation in arriving at the total MPD. In order to be able to express on a common scale, for protection purposes only, the irradiation incurred by exposed persons for all types of radiation, the quantity *dose equivalent (DE)* has been introduced by the ICRU. By definition, the *DE* is the product of the absorbed dose and appropriate modifying factors which depend on the particular radiation hazard involved. For our purposes, it is sufficient to mention the *quality factor (QF)* only. Thus,

$$DE = \text{absorbed dose} \times 1$$
$$= 1 \text{ rad} \times 1 = 1 \text{ rem}$$
or, 1 rad = 1 rem *for x, beta, and gamma rays*

The unit of *DE* is the *rem,* and the value of *QF* varies with the type of radiation. For x rays of moderate energy (200 kV), beta rays, and gamma rays, the value of *QF* has been arbitrarily chosen to be 1. Therefore,

$$DE = \text{absorbed dose} \times 1$$
$$= 1 \text{ rad} \times 1 = 1 \text{ rem}$$

or, 1 rad = 1 rem *for x, beta, and gamma rays*

On the other hand, the *QF* for fast neutrons is 10 (eg, 1 rad of fast neutrons is 10 times as effective as 1 rad of x rays in producing cataract); hence,

$$DE = \text{absorbed dose} \times 10$$
$$= 1 \text{ rad} \times 10 = 10 \text{ rems}$$
or, 1 rad = 10 rems *for fast neutrons*

This means that with fast neutrons, 1 rad gives a *DE* of 10 rems insofar as cataract induction is concerned. The *QF* for alpha particles is 20, so that 1 rad of this radiation gives a *DE* of 20 rems. It must be emphasized that the rems of different kinds of radiation are equivalent (*DE*) and can be added. For example, if an individual were exposed to a mixture of radiations and received 1 rad from x rays, 1 rad from fast neutrons, and 1 rad from alpha particles, the total absorbed dose would be 3 rads. However, this would seriously underestimate the radiation hazard. By converting these doses to *dose equivalents*, we would obtain:

x rays	1 rad × 1	= 1 rem
fast neutrons	1 rad × 10	= 10 rems
alpha particles	1 rad × 20	= 20 rems

total 31 rems

Thus, the dose equivalent, in this case 31 rems, provides a more realistic measure of the radiation hazard when the quality factor is different from 1. However, in the average hospital Radiology Department exposure is usually limited to radiation with a quality factor of 1 so that the MPD may be stated in rads.

PERSONNEL PROTECTION FROM EXPOSURE TO X RAYS

Let us now turn to the problems associated with the *occupational* exposure of x-ray personnel, under three main headings:

1. Determination of radiation safety in the Radiology Department, called *radiation monitoring.*
2. Deleterious effects to the individual resulting from excessive exposure.
3. Protective shielding of radiographic and radiotherapeutic installations.

Radiation Monitoring. In the Radiology Department we can monitor whole-body exposure without special equipment or training (a qualified health physicist can do this more accurately, but not ordinarily required). Several methods are available. The most precise is that employing a *pocket dosimeter* which outwardly resembles a fountain pen, but which contains a thimble ionization chamber at one end. The exposure may be read by means of a separate electrometer; or, in the self-reading type, by means of a built-in electrometer. A dosimeter of this type, while sensitive to exposures up to 0.2 R (200 milliroentgens), is unreliable in inexperienced hands. Besides, it does not provide a permanent record, although it is useful in monitoring procedures with radioactive materials.

The *film badge* offers the most convenient method of personnel monitoring under average conditions. Commercial laboratories specialize in furnishing and servicing the film badges. The x-ray film badge consists of a dental film, covered by a copper stepped wedge filter to show the quality of the radiation, and backed by a sheet of lead foil to absorb scattered radiation from behind the badge. It is usually worn on the clothing over the front of the hip or chest. Whether the badge should be worn inside or outside the lead apron is still debatable; it is probably good practice to wear it in the front of the neck of the apron to monitor exposure of the thyroid gland and eyes. A fresh badge is furnished regularly by the laboratory, to be used the following week. At the end of this time, it is returned to the laboratory where the film is processed under standard conditions and compared densitometrically with standard films exposed to known amounts of radiation. Having been obtained in this manner, the exposure in roentgens received by the wearer's film badge is reported to the Radiology Department for permanent filing. Badges should never be exchanged among personnel and should be clearly marked for identification. When exposure levels are well below the weekly MPD (0.1 R), the film badge should be worn and returned *monthly,* especially if the work load is reasonably constant. The one-month period is preferred because of convenience and greater accuracy of calibration.

A new kind of personnel monitor, still in the developmental stages, is the *thermoluminescent dosimeter* (TLD). This is based on the principle that certain crystalline materials are capable of storing energy when they are exposed to radiation, because of the trapping of valence electrons in crystal lattice defects. Upon being heated under strictly controlled conditions, the electrons return to their normal state, the stored energy being released in the form of light. Measurement of the light by a photomultiplier device gives a measure of the initial radiation exposure, since the two are very nearly proportional. The dosimeters can be returned to a commercial laboratory for readout if desired. Lithium fluoride (LiF) is at present the most promising material for TLD. There are many advantages of TLD over film badge monitoring: (1) detectors are cheap and small—for example, 1 mm × 6 mm—

and can be sealed in Teflon to avoid mechanical damage, (2) direct reading is available at any time, (3) response to radiation is proportional up to about 1000 R, (4) response is almost independent of radiation energy (film badge is not), (5) response is very similar to that of human tissues, (6) wide exposure range detectable—10 mR to 100,000 R, (7) accuracy may be about ±5 per cent (film badge varies from ±10 to 50 per cent), (8) it is possible to incorporate detector in jewelry to make it unobtrusive. For these reasons, it is very likely that TLD will eventually become the method of choice in personnel monitoring.

Harmful Effects of Radiation. Why has health physics become so important in radiology? The answer lies in the fact that radiation is always harmful to tissues, the amount and nature of any damage being dependent on the type of radiation (LET), the absorbed dose, the size and location of the treated region, and the fractionation or number of days over which the treatment is administered. It is the differential damage to a given type of tumor as compared with that to normal tissues that underlies successful irradiation therapy. The term *therapeutic ratio* has been applied to the ratio of the normal tissue tolerance dose to the tumor lethal dose. Remember that the therapeutic exposure of patients is a matter quite different from the small exposure that technologists may receive over a prolonged period of time. The latter problem concerns us here. *The objective of a program in health physics is to keep the exposure of personnel as far as possible below the maximum permissible level to avoid the slightest radiation injury.* But suppose overexposure is incurred; what sort of harmful effects may be anticipated? These are three in number: *local, general, and genetic.*

1. *Local Effects.* Small occupational exposures should not cause local changes in the skin. Several hundred rads within a few days are required to produce erythema (skin reddening), and such a dose would be unlikely except in therapy. However, even with very small exposures repeated over a period of years, late irradiation effects may occur in the form of dry skin which is subject to cracking, ulceration, and eventual cancer production. Another harmful effect is the formation of cataract (opacity) in

the lens of the eye; relatively small exposures are required, on the order of a few hundred rads, because of the marked radio-sensitivity of the lens.

2. *General Effects*. When the whole body is exposed to x or gamma radiation, there is always the possibility of harmful effects. The *blood-forming organs* (bone marrow and lymphatic tissues) are especially susceptible. No certainty exists as to the presence of a threshold—that is, a minimum level of exposure below which there is no damage, and above which damage abruptly appears. Such *a threshold or lower limit has not been demonstrated.* We know that an absorbed dose exceeding 25 rads to the whole body in one exposure can produce discernible, though temporary depression of the manufacture of white blood cells. Higher doses cause more severe and prolonged depression of the blood count, especially the granulocytes and platelets. Chronic exposure, even to moderate doses of about 10 rads per year, may cause leukemia. Changes in the red blood cell count indicate prolonged and heavy exposure.

Evidence in experimental animals indicates that chronic exposure to radiation causes *shortening of life span,* increasing in severity as the exposure is increased. Statistical studies indicate a similar effect on humans subjected to exposures significantly greater than the MPD.

Exposure during *pregnancy* is especially hazardous to the *fetus.* Maximum sensitivity of the fetus occurs during the first three months or so of pregnancy, wherein absorbed doses of the order of 40 or 50 rads can produce significant congenital anomalies. Sensitivity gradually decreases as pregnancy continues beyond this period. Therefore, radiologic procedures should be kept at a minimum during pregnancy, especially in the early months.

Not likely to be encountered in the Radiology Department are the serious damaging effects produced by *acute exposure to large amounts of radiation*—in the range of 100 to 1000 rems or more to the whole body. Exposures of this magnitude are more apt to occur accidentally in atomic energy installations, and give rise to the *acute radiation syndrome.* The severity of this condition depends on the amount of exposure. A *supralethal dose* of 600 rems or more to the whole body causes the fulminating form of the acute radiation syndrome, consisting of nausea, vomiting, and prostration within one or two hours, followed by step-like rising fever. There is associated loss of appetite, diarrhea, dehydration, and profound drop in the white blood count. Hemorrhages occur in the skin and intestinal tract, terminating in death from sepsis in one to two weeks. With a *midlethal dose* of 300 to 500 rems the effects are somewhat less pronounced, about one-half the exposed individuals dying in one to two months. With a *sublethal dose* of 100 to 300 rems there may be mild nausea, vomiting, and malaise lasting about 12 to 24 hours. After a latent period of about 3 to 5 weeks, loss of hair occurs, although it usually regrows. There is a drop in all of the blood cell counts, and many individuals develop cataracts in 2 to 6 years. Most recover, but some enter a chronic stage, remaining feeble and anemic for months. Statistically, there is an increased incidence of leukemia in the survivors—about nine times that in the unexposed population, as found in the survivors of the atomic bombing of Hiroshima and Nagasaki in World War II.

3. *Genetic Effects.* A serious problem associated with exposure to radiation is the *genetic hazard*; that is, the danger of harmful effects on future generations. This is due to the extreme vulnerability of the genetic material (particularly the chromosomes in the sperm and ova) to radiation damage. There is no known lower limit of dosage that is safe, and the damage is approximately proportional to the absorbed dose. Because the genetic material is transmitted to offspring during reproduction, it is obvious that any radiation-damaged genetic material will be similarly handed down. This becomes especially serious when large populations are exposed and hence it is important to protect the gonads (ovaries and testicles) during diagnostic procedures whenever possible. Chromosomal damage is manifested by *mutations* or freaks; these tend to be hidden by pairing with normal chromosomes during reproduction, only to become manifest in subsequent generations. Many such mutations die very early in pregnancy and therefore never appear. Insofar as technologists are concerned, their contribution of damaged genetic material to future generations is nil, especially if whole body exposure is well below the maximum permissible.

With careful monitoring, the likelihood of incurring radiation injury is virtually nonexistent. When, under *unusual* circumstances, there is evidence of damage to the blood-forming organs, corrective steps should be taken immediately. First, the affected individual should be removed until he has recovered completely. Second, a trained health physicist should survey the Radiology Department and recommend the best means of eliminating the hazardous conditions. However, it must be emphasized here that even if conditions are absolutely safe, each technologist must observe certain rules of conduct. In other words *even in the best protected department, the technologist is subect to overexposure if he does not make full use of the available protective facilities, without exception.*

PROTECTIVE SHIELDING IN RADIOGRAPHIC AND RADIOTHERAPEUTIC INSTALLATIONS

Radiation Sources. All x-ray tubes must be of rayproof type; that is, the manufacturer must enclose the tubes in a metal housing that reduces leakage radiation to a prescribed safe level. (Leakage radiation is that radiation which penetrates the protective housing.) The specifications for such protective enclosures are given in *NCRP Report #33*. Leakage radiation for the average *radiographic tube* shall not exceed an average value of 100 mR/hr at 1 meter, the required shielding being about 1.5 mm (about $\frac{1}{16}$ in.) lead. With *orthovoltage x-ray therapy tubes* (150 to 500 kV), the exposure rate from leakage radiation must not exceed 1 R/hr at 1 meter. In addition to protective housing, specifications are prescribed in the same report governing collimation, exposure timers, filter systems, shielding of the patient, and other factors.

Radioactive sources used in *teletherapy*—cobalt 60 and cesium 137—must be doubly sealed in stainless steel and tested for leakage by the manufacturer who must provide a certificate attesting to the absence of leakage. In addition, leakage tests must be conducted every six months by the user (this also includes semiannual leakage tests for sealed radium sources such as needles and tubes, and for strontium 90 eye applicators). A leakage test for radioactive sources is performed by carefully wiping the collimator leaves and edges of the final collimator aperture with a long-handled cotton applicator moistened with dilute detergent solution. The swab can either be sent in a special container to a commercial laboratory or it can be tested by the method of Rose—the cotton is held near the unshielded window of a GM survey meter (30 mg/cm² window). Any increase over background counting rate with earphones indicates a minimum leak of 10^{-9} Ci. A leak only ten times as great (that is, 10^{-8} Ci) will give thousands of counts per min and kick the indicator needle off the scale. When this occurs, the equipment must be promptly checked by a health physicist or other qualified person. Radium needles and tubes as well as strontium 90 applicators can be tested similarly. In addition, radium needles and tubes should be placed in a stoppered test tube together with activated charcoal for 24 hours, after which time the charcoal is tested as above by Rose's method. Leaking needles must be returned promptly to the supplier for repair.

Wall Protection. This requires the use of built-in *protective barriers* which are shields of radiation-absorbing material(s) used to reduce radiation exposure. Wall protection should be planned in advance by a radiation physicist qualified in this field, to avoid expensive alteration after the building has been completed. On the other hand, poor planning may result in excessive wall protection which becomes unnecessarily expensive. An important basic principle in permanent barriers is that joints and holes must be occluded by the same or equivalent protective barrier as the wall. Proper wall protection varies with the *energy of the radiation* modified by certain factors described in *NCRP Report #33.*

Terms that often apply to wall protection are "controlled" and "uncontrolled" areas. A *controlled area* is one that is under the supervision of the Radiation Safety Officer; an *uncontrolled area* is not.

We may conveniently discuss wall protection in (1) radiography, (2) orthovoltage therapy, and (3) megavoltage therapy. Exact wall barrier thicknesses may be obtained for any particular installation from appropriate publications (*NCRP Report #34*). However, we shall indicate the type of barrier ordinarily needed in each general area. In this discussion, the *useful beam* refers to the beam that passes through the window and filter of the x-ray tube, otherwise known as the primary beam.

1. *Radiography.* As a general rule, the *primary protective barrier* (that is, where the useful beam can strike the wall *directly*) in radiography up to 140 kV should consist of $\frac{1}{16}$ in. lead (Pb) extending $7\frac{1}{3}$ ft up from the floor when the x-ray tube is 5 to 7 ft from the wall. This barrier also takes care of leakage radiation; that is, radiation coming through the tube housing. A *secondary protective barrier* (that is, where only scattered and leakage radiation will strike the wall) requires $\frac{1}{32}$ in. Pb under the same operating conditions. This secondary barrier extends from the top of the primary barrier to the ceiling, but these barriers should overlap about $\frac{1}{2}$ in. Ordinary plaster often suffices as a secondary barrier without added lead in the radiographic range, especially in large radiographic rooms. Control booths should have the same protection as walls, and radiation should scatter at least *twice* before reaching the opening of the booth.

2. *Orthovoltage X-ray Therapy*—about 150 to 500 kV. Here the primary protective barrier is about $\frac{3}{8}$ in. Pb extending $7\frac{1}{3}$ ft up from the floor. The secondary barrier of $\frac{1}{8}$ in. Pb should overlap the primary barrier about $\frac{1}{2}$ in. and extend to the ceiling.

Protection of the floor and ceiling in the above two types of installations depends strongly on beam energy, and other factors, and must be determined in advance by a qualified expert.

The leaded glass observation port should have the same lead equivalent as the adjacent wall and should be overlapped about $\frac{1}{2}$ in. by the lead in the wall. Leaded glass ordinarily requires about four times the thickness of sheet lead for equivalent protection; for example, 1 in. leaded glass is equivalent to $\frac{1}{4}$ in. sheet lead.

In wall construction economy may be achieved by using laminated lead building blocks. Another, possibly less expensive, method is to nail sheet lead with lead-headed nails between double studs placed flatwise. If sheet lead is used and a thickness of $\frac{1}{8}$ in. or more is required, it is much easier and cheaper to apply it in multiple layers of $\frac{1}{16}$ in. because this thickness can be cut with suitable shears and can be lifted more readily. One-eighth inch lead must be chiseled and is difficult to lift because of its weight. Furthermore, it is much easier to overlap $\frac{1}{16}$-in. Pb at seams and joints.

3. *Megavoltage Therapy*—0.5 to 10 million volt range (cobalt 60 and megavoltage x ray in same energy region). Building concrete of density 2.35 g/cc (147 lb/cu ft) is most economical for wall material here because it provides both adequate protection and structural strength. Only approximate values of thickness will be given—accurate advance planning by a qualified expert is imperative. In the average cobalt 60 installation about 3 ft or more of concrete of the above density is required for the primary barrier, and about 2 ft for the secondary barrier. This will vary, of course, with the energy of the beam, size of the source, and other protective factors. *Leakage radiation* is an important item, since the housing for cobalt 60 provides only relative protection and the radioactive source radiates continuously. The exposure rate from leakage radiation must be added to that from scattered radiation (the two together comprise *stray radiation*) in calculating wall thickness for the secondary barrier. Radiation that has been scattered twice is usually reduced to a safe level by a barrier of $\frac{1}{4}$ in. Pb, but this must be verified. An observation port of leaded glass must be about 12 in. thick for the average megavoltage therapy room. A less expensive and more satisfactory method uses closed circuit television to monitor the patient, with an auxiliary system of mirrors for viewing the patient, in the event of failure of the television system.

WORKING CONDITIONS

Certain general rules have been laid down for personnel safety. These should be impressed repeatedly on all Radiology Department personnel, since their cooperation is mandatory.

1. Never expose a human for demonstration purposes alone.

2. Never remain in a radiographic or radiotherapy room while an exposure is in progress. In general, approximately 0.1 per cent of the useful beam is scattered perpendicular to the beam at a distance of 1 meter from the patient.

3. Never hold a patient for therapy, since it takes but one

200-R treatment with a 250-kV beam to exceed the weekly MPD for the technologist. Patients should *rarely* be held for radiography, since many efficient devices are available, in addition to the technologist's ingenuity. However, if a patient must be held for radiography, whenever possible this should be done by a person not habitually exposed to ionizing radiation and he should be protected by a lead apron and lead gloves.

4. Give yourself the same protection as a loaded cassette!

5. Wear a lead-lined apron having 0.5 mm Pb equivalent in fluoroscopy and, when not actually assisting the radiologist, stand either in the control booth or behind the radiologist.

6. Check lead-lined gloves periodically for cracks by a radiographic test using intensifying screens and exposure factors of 100 kV and 10 mAs at a 40-in. focus-film distance.

PROTECTION SURVEYS

Protection surveys include evaluation of potential exposure incidental to the use of medical x-ray or gamma-beam equipment. A complete survey must be made *initially* by a *qualified expert,* then repeated after any change in conditions that might affect exposure of personnel. If additional shielding is required, the area should be resurveyed after the added barrier has been installed. This situation might arise, for example, in changing from a smaller to a larger cobalt 60 source.

If, in the opinion of the qualified expert, there is reasonable probability that a person in a controlled or uncontrolled area may receive more than 10 mR (0.01 R) in one week, then one or more of the following steps should be taken:

1. Determine the cumulative dose in the area in question.

2. Use personnel monitoring in the area in question.

3. Add barrier material to comply with authoritative recommendations (*NCRP Report #34*).

4. Impose restrictions on the use of the equipment, or on the direction of the beam.

5. Impose restrictions on the occupancy of the area if this is controlled.

A tissue-equivalent phantom such as piled sheets of Masonite must be used to check the safety of the secondary barrier for scattered radiation.

All "on-off" control mechanisms (control panel, entrance door, emergency cutoff switch) should be checked semiannually and repaired if necessary.

Warning signs must be posted in a prominent place in the appropriate areas as follows:

a. **RADIATION AREA** sign in area where prevailing exposure rate is more than 5 mR/hr but less than 100 mR/hr.

b. **HIGH RADIATION AREA** sign at entrance to any area in which the exposure rate is 100 mR/hr.

Whenever any restriction has been placed on the equipment relative to the use factor of the useful beam, such as limitation of beam angulation to prevent penetration of an inadequate barrier, such restriction must be enforced.

All reports of calibrations and surveys should be made in writing and signed by the qualified expert and filed *permanently*. The report should indicate whether a resurvey is needed, and when.

Electrical power to megavoltage equipment must be *locked in the "off" position* when equipment is not in use.

Instructions must be posted conspicuously in the control area describing the steps to be taken in the event equipment cannot be turned off.

The name, address, and phone number of the Radiation Safety Officer should be posted in a conspicuous place in the control area so that he may be reached in the event of an emergency. A substitute should be available, on call, if the Radiation Safety Officer is unavailable.

PROTECTION OF THE PATIENT FROM X RADIATION

We must take every possible precaution to avoid unnecessary exposure of the patient during diagnostic studies. The development of a skin reaction as a result of *fluoroscopy* or *radiography* is inexcusable, and in many states, has been ruled in the courts as evidence of malpractice.

Patient Protection in Fluoroscopy

Fluoroscopy is the source of greatest hazard, and before a patient is subjected to this type of examination, a careful history should be taken to ascertain the recency and number of previous fluoroscopic studies. X-ray exposures are cumulative over many years and further exposures must be governed by what has gone before.

Several precautions should be observed to minimize the danger of excessive radiation in fluoroscopy:

1. **Dark Adaptation.** The radiologist, before starting conventional fluoroscopy (in a room which should be as light-tight as the darkroom) should sit in a dark room or wear special dark adaptor goggles for 20 to 30 minutes. This permits the retina of the eye to attain maximum sensitivity to the relatively dim fluoroscopic image. Such careful dark adaptation is necessary to make a more *rapid* and more *accurate examination.* Dark adaptation is unnecessary with x-ray image intensification.

2. **Intermittent Fluoroscopy.** It is good practice to use intermittent activation of the fluoroscopic tube. During the brief dark intervals, the eyes have an opportunity to rest and re-adapt. Furthermore, this decreases patient exposure with either dark or bright (image intensified) fluoroscopy.

3. **Restriction of Field Size.** The size of the fluoroscopic field must be limited by suitably collimating lead shutters placed between the tube and the patient. As we have seen, the skin and depth doses both decrease as the area of the field is decreased. Hence the fluoroscopic beam should be restricted to the smallest area that contains the part under observation, to reduce the exposure of both the patient and personnel.

4. **Correct Operating Factors.** It is well known that an increase in kV produces an increased brightness of the fluoroscopic image. Dr. Russell H. Morgan has shown experimentally that if, with increasing kV, screen brightness is kept constant by correspondingly decreasing mA and increasing filtration, the patient's exposure in R per min will actually be less than that at the lower kV and higher mA. In other words, for a *given image brightness,* fluoroscopic exposure of the patient decreases as kV is increased

and mA decreased. Since improvement in this regard becomes insignificant above 100 kV, a range of 85 to 100 kV, 4 mA, and 3 mm Al filter are considered to be optimum factors in *dark* fluoroscopy. The mA should be reduced to 1.5 to 2 for bright fluoroscopy (image intensification). The *focus-table* distance must be at least 12 in., but preferably 15 in. At distances greater than this, further improvement is so small as to be of doubtful value.

5. **Filtration.** We can increase the hardness of an x-ray beam by using a suitable filter. This removes relatively more soft than hard rays, thereby decreasing the amount of radiation absorbed by the skin. In fluoroscopy, when filtration is increased from 1 mm Al to *3 mm Al,* the exposure rate to the patient is reduced to about *1/4;* yet, there is only slight loss of screen brightness which can be restored by a small increase in kV or mA.

6. **Maximum Permissible Exposure.** In recent years, concern for overexposure of patients in diagnostic radiology has shifted from possible erythema, to the hazards of cumulative small doses to the patient during his lifetime. However, there is as yet no official limit to *diagnostic* x-ray exposure of patients. Obviously, this should be reduced to a minimum consistent with medical necessity. The dose rate at the table top in fluoroscopy should be *less than 10 R per min.* (*NCRP Report #33*). Optimum factors have been discussed: 85 to 100 kV, 4 mA, 15 in. focus-table distance, 3 mm Al filter, small aperture. Even so, the patient may receive as much as 50 R during dark fluoroscopy of the gastrointestinal tract.

Patient Protection in Radiography

Every effort should be made to reduce general exposure in radiography, especially to the *gonads,* because of the possibility of genetic damage (see Table 23.2). Protection of the gonads can be aided by lead rubber shielding and by severe collimation of the beam, unless this interferes with the examination.

The roentgen exposure in various radiographic examinations has been determined by a number of workers, with fair agreement. Table 23.2 shows approximate skin and gonadal exposures based on published data (see Bibliography). Exposures have been reduced significantly by the use of adequate *filtration;* thus,

TABLE 23.2

RADIATION EXPOSURE OF PATIENTS DURING VARIOUS
RADIOGRAPHIC PROCEDURES USING BEST AVAILABLE
PROTECTIVE MEASURES
(HARWELL, GREAT BRITIAN)*

Examination	kV	mAs	FFD	Added Filtration	Exposure per Film With Backscatter		
					Skin	Testes	Ovaries
			in.		mR	mR	mR
Sinuses	80	40	27	3 mm Al	1,040	0.1	0.05
Chest, PA	90	3	72	3 mm Al	8	0.01	0.02
Thoracic Spine, AP	75	80B	43	3 mm Al	480	1.0	1.2
Lumbar Spine, AP	75	80B	43	3 mm Al	480	0.5**	95.0
Lumbar Spine, Lateral	85	300B	43	3 mm Al	2,000	2.25	270.0
Lumbosacral, Lateral	90	400B	43	3 mm Al	3,000	2.0	350.0
Abdomen, AP	75	60B	43	3 mm Al	360	0.5**	75.0
Abdomen, Prone (barium)	90	20B	43	3 mm Al	130	1.5	20.0
Pelvis, AP	75	80B	43	3 mm Al	480	20.0**	80.0
IVP, Renal, AP	75	80B	43	3 mm Al	480	0.5**	95.0
IVP, Bladder, AP	75	80B	43	3 mm Al	480	10.0**	80.0
Chest Fluoro (Image Intensifier)	75	90G 3 min	18	5 mm Al	900	3.0	3.0
Barium Meal Fluoro (Image Intensifier)	75	150G 5 min.	18	5 mm Al	1,500	5.0	5.0

* Data of Ardran, G. M., and Crooks, H. E. (Courtesy of authors and British Journal of Radiology.)
B = Moving grid.
G = Stationary grid.
** Lead Rubber Protection.

a 3 mm Al filter reduces the skin exposure to about ¼ that without a filter, with only a small change in radiographic density and contrast. Patient exposure can also be reduced by an *increase in kV*, radiographic density being maintained constant by a decrease in mAs. Table 23.3 shows the reduction in male gonadal exposure with increased filtration and lead shielding.

The important factors in reducing radiographic exposure of the

TABLE 23.3

REDUCTION IN EXPOSURE OF MALE GONADS (TESTES) WHEN
3 mm Al FILTER IS ADDED*

Examination	Added Filtration	kV	Testes Uncovered	Testes Covered 1 mm Sheet Lead
			mR	mR
Pelvis, AP 15 x 12 film	0	65	2,000	42
	3 mm Al	65	670	24
Lumbar Spine, AP 10 x 12 film	0	68	24	5
	3 mm Al	68	6	2

* Data of Ardran, G. M., and Crooks, H. E. (Courtesy of authors and British Journal of Radiology.)

patient—so-called *minimum exposure radiography*—may be summarized as follows:

1. Beam *filtration* by 3 mm Al equivalent.
2. Careful beam *collimation.*
3. *Lead shielding* of gonads, when possible.
4. Use of *films and intensifying screens having as high a speed as is consistent with good quality radiography.*
5. *Optimum film processing.*
6. Use of highest practicable *kilovoltage.*
7. Limitation of cassette-front filtration to *0.5 mm Al equivalent.*
8. *Careful technic* to avoid unnecessary repeat examinations.

In any radiographic procedure, the patient should be questioned about previous x-ray exposures, both diagnostic and therapeutic. The total exposure for medical diagnosis should be kept at the lowest level consistent with medical needs. It is essential that a careful permanent record be made of each exposure, including the area exposed, kV, mAs, filtration, and focus-film or focus-skin distance. Radiodiagnosis during pregnancy should be strictly limited both as to use and to number of exposures to protect the sensitive fetus.

Patient Protection in Therapy

The problem of protection of the patient in *therapy* is entirely different from that existing in diagnosis. Depending on the lesion being treated, the radiologist varies radiation dosage to obtain the desired effect. In some instances, the required exposure is so small that no visible skin reaction is anticipated. At the other extreme lies the treatment of malignant tumors necessitating heavy exposure which may produce a skin reaction.

TABLE 23.4

THICKNESS OF LEAD REQUIRED TO REDUCE THE INCIDENT
RADIATION TO 0.5 PER CENT OF THE INITIAL EXPOSURE
RATE, WITH BEAMS OF VARIOUS DEGREES OF HARDNESS

kV	Half Value Layer	Thickness of Lead for 0.5% Transmission		Usual Thickness of Lead Available
		mm	in.	
60	0.5 mm Al	0.21	0.008	1/100 in.
100	1.0 mm Al	0.68	0.026	3/100 in.
140	0.5 mm Cu	1.52	0.06	1/16 in.
200	1.0 mm Cu	1.97	0.08	5/64 in.
220	2.0 mm Cu	2.77	0.11	1/8 in.
250	3.0 mm Cu	3.45	0.14	5/32 in.

Various size cones or collimators are used to restrict the area of the field being treated with orthovoltage radiation, since one should avoid producing skin reaction beyond the actual treatment field. Frequently, the treatment field is delimited by lead shields which are arranged to protect the tissues lying directly beneath the shield. The thickness of such lead barriers depends on the kV and filtration of the x-ray beam. Data in the article by Trout and Gager show the thickness of lead needed, with beams of various half value layers, to reduce the exposure rate to 0.5 per cent of its initial value. Table 23.4 is based on this article and gives the necessary information for the more commonly used treatment factors. The last column in the table indicates the nearest lead thicknesses that are ordinarily available and may safely be used with the beam of a given half value layer.

In irradiation therapy in the 1 to 2 million volt range a lead block 4 in. thick is required to shape the beam. The block should be at least 15 in. from the skin to avoid contamination of the beam with secondary electrons which intensify the skin reaction.

ELECTRICAL PROTECTION

In recent years, as most radiology departments have been equipped with shockproof equipment for therapy and radiography, nonshockproof equipment has become obsolete. With modern equipment, the shock hazard has been eliminated, provided the ground connections are intact and the cables are in satisfactory condition. These should be inspected periodically.

Oddly enough, one of the commonest sources of electrical danger exists in the darkroom. All exposed electric outlets should be *grounded*. The technologist should always observe the *one-hand rule:* never reach for an electric fixture while the other hand is submerged in the processing tanks or is touching a conductor.

High-voltage, low-amperage shock tends to "throw" the victim, whereas low-voltage, high-amperage shock tends to "hold him." In the latter instance, do not grasp the victim directly, but first either open the main switch or remove him by means of a dry board, a dry wad of newspaper, or a dry rope. Then, if shock is severe, seek medical aid, in the meanwhile applying closed cardiac massage and mouth-to-mouth resuscitation.

Closely allied to the electric hazard is the *danger of explosion.* When certain gas anesthetics reach a sufficiently high concentration in the atmosphere, they may ignite explosively if a spark occurs either in the electrical contacts of the equipment or from a static electrical discharge. Modern x-ray equipment is shockproof, but it is *not necessarily sparkproof.* To minimize the danger, all equipment should be carefully grounded. Sparkproof electric outlets and sparkproof x-ray illuminators should be used in all operating rooms. If such precautions cannot be taken, then it is essential that non-explosive anesthetics be used. The explosive anesthetics include ether-nitrous oxide mixture, ether, ethylene, and cyclopropane.

PERSONNEL PROTECTION FROM RADIUM EXPOSURE

Just as with x rays, inadvertent exposure of personnel to radium involves *general body radiation* and *local radiation.* These two aspects of radium protection will be considered separately. They also apply to other sealed radionuclide gamma-ray sources. Sealed sources must be tested periodically for leakage.

Protection from Whole-body Exposure to Radium

Whole-body exposure to radium may occur during:

1. Storage
2. Manipulation
3. Transportation

The *storage* of radium can be made relatively safe by providing containers of the proper lead equivalent, according to the tables in *NCRP Report #24*. **Lead** and **distance** are the two best means of protection that we have. However, large thicknesses of lead are required because of the great penetrating power of gamma rays (see pages 484-486 for method of calculating safe distances and barriers).

Table 23.5 shows the thickness of lead shielding required to reduce to the MPD, the exposure from various quantities of ra-

TABLE 23.5

THICKNESS OF LEAD STORAGE CONTAINER NEEDED
FOR PROTECTION FROM RADIUM GAMMA RAYS,
BASED ON 40-HOUR WORK WEEK*

| Radium mg | Distance | |
| | 1 Meter | 2 Meters |
	cm Lead	cm Lead
50	5.7	3.4
100	7.0	4.6
200	8.3	5.8

* Based on *NCRP Report #24.*

dium stored at different distances. In this table, note that 100 mg radium should be stored in a container with a wall thickness of 7 cm lead at 1 meter (3 ft). At 2 meters (6 ft) the thickness is reduced to 4.6 cm lead. With these factors, the MPD (0.1 R per week) will not be exceeded during a 40-hour workweek. In general, more lead is required for larger quantities of radium, but this is not proportional, as shown in the table; the half value layer of radium gamma rays is 1.3 cm lead (Pb). Note that as the amount of stored radium is doubled, one additional HVL (1.3 cm Pb) is needed for protection. As the distance is halved, 2 HVL (2.6 cm Pb) must be added. Why?

During the manipulation of radium, as in loading needles, capsules, or other applicators, a large whole body exposure could be accumulated. This can be reduced to a safe limit if a wide, L-shaped lead block *at least 2 inches thick* is placed between the radium and the operator. It should have, at the top, an inclined lead glass visor, equivalent to at least 3 mm lead (see Figure 23.2).

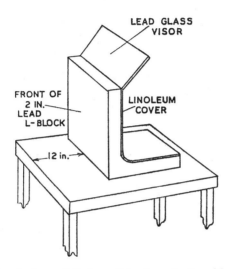

Figure 23.2. Front view of L-shaped lead protective block for preparing radium applicators. The operator stands in front of the block so that the upright lead section protects his body and the inclined lead glass visor protects his face. Lead bricks can be placed around the L-block for added protection.

Protection from Local Exposure to Radium

For protection of the hands from *local injury,* the above general measures should first be carefully observed. Then, the additional precaution of handling radium only with long forceps should be strictly adhered to. *Speed of handling* is as important as distance, since both reduce the exposure of the skin. *Never, under any circumstances, should radium be picked up directly with the fingers.* Long forceps or some similar instrument should be used. Lead rubber gloves offer no protection from the gamma rays of radium, because it is impracticable to line gloves with sufficient lead thickness; and secondary radiation induced in lead of ordinary thickness would actually increase exposure. Local injury by radium may not become visible for a number of years, depending, of course, on the exposure. It consists, in the early stages, of reddening and dryness of the skin. Later, there is brittleness and splitting of the nails, followed by keratoses, atrophy, and ulceration of the skin. Finally, cancer may develop in the ulcerated areas.

Transportation of radium requires long-handled containers lined with sufficient lead. For example, 100 mg radium should be carried in a container lined with 2 cm lead suspended by a handle 45 cm long; or, the container may have a 1 cm lead thickness and a handle 60 cm long. Transportation should be as fast as possible, and should preferably be done by someone who does not regularly handle radium if it is to be carried about frequently.

PROTECTION IN NUCLEAR MEDICINE

Precautions are necessary in the handling of radionuclides because of the harmful effects of chronic exposure to radiation, as already described. But the problems associated with radionuclides differ in many respects from those prevailing in the diagnostic x-ray department; hence, the need of a separate section on health physics related to them.

Types of Radiation. Before discussing the harmful effects of radiation emitted by radionuclides and the methods of protection, let us review the types of radiation involved. In general, radioactive substances give off three types of radiation: alpha particles, beta particles, and gamma rays. Figure 23.3 shows their respective half value layers as an indication of their penetrating abilities. Since medical radionuclides emit only gamma or beta rays, our discussion will be limited to them. Not only are gamma rays much more penetrating than the beta particles, but they are also different in character, being electromagnetic waves (photons), whereas beta radiation consists of high speed electrons. Consequently, there are differences in the details of protection from these two types of radiation.

Harmful Effects. These have already been described in detail on pages 465-468. The same effects are produced by beta and gamma radiation as by x radiation, for equal absorbed doses.

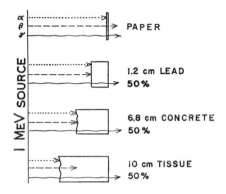

Figure 23.3. Comparative absorption of alpha, beta, and gamma rays in various materials. Also shown are the half value layers (HVL) of 1 MeV gamma rays in lead, concrete, and body tissues.

However, beta radiation is less penetrating than the others, its effects being limited to the skin with external exposure, or to its immediate vicinity with internally deposited beta-emitting sources. On the other hand, the effects of gamma rays are manifested deep within the body even with external exposure.

Maximum Permissible Dose Equivalent. Since the radiations from artificial radionuclides used in medicine include only beta and gamma rays, the maximum permissible dose can be stated in rads, as already described on pages 460-461. The same formula applies:

$$MPD = 5(N - 18) \text{ rads}$$

where N is the age. It must be emphasized again that this formula applies to *chronic whole body exposure of professional radiation workers,* including technologists. The same limit applies to exposure of the gonads and the eyes. The MPD to the hands, forearms, feet, and ankles is 15 times that to the whole body. For more specific data on MPD, see Figure 23.1 and *NCRP Report #39.*

Principles of Radiation Protection. There are basically four factors, suitable combinations of which are used to reduce or eliminate radionuclide hazards. These include: (1) *distance,* (2) *shielding,* (3) *time of exposure,* and (4) *limitation of activity* of source to smallest required amount. In this section, we shall assume that the radioactive sources are sealed in the form of capsules or needles. Liquid preparations of radionuclides, except for the hazard of spill, ingestion, or inhalation, may also be regarded as sealed sources insofar as external protection is concerned.

1. *Distance.* This is the easiest and least expensive method of reducing the exposure rate to the permissible level. In the case of gamma rays, the inverse square law applies with reasonable accuracy for the purpose of protection in storage and handling. Long forceps for picking up capsules or needles (^{131}I or ^{60}Co), and for removing ^{131}I solutions and other radioistopes from shipping containers, will reduce significantly the exposure of the hands and forearms. With highly active sources, remote handling equipment is necessary to keep the exposure down to the permissible level.

The *storage* of radioactive materials that emit *gamma rays* should be as remote as possible from areas habitually occupied by personnel and patients. The following formula can be used to calculate the *safe distance* or *danger range* for *gamma-ray emitters;* that is, the distance at which the exposure rate approximates the MPD in a 40-hour work week:

$$D = 22\sqrt{A\Gamma}\ cm \qquad (3)$$

where D is the safe distance, A is the activity of the radionuclide in mCi, and Γ (Greek capital *gamma*) is the specific gamma-ray constant (defined as the exposure rate in R/hr at a distance of 1 cm from a point source of a radionuclide having an activity of 1 mCi). For example, what is the safe distance from a capsule of ^{131}I containing 10 mCi? The value of Γ for ^{131}I is 2.2 R/hr at 1 cm. Therefore,

$$D = 22\sqrt{10 \times 2.23} = 22\sqrt{22.3}$$

$$= 22 \times 4.7 = 103\ cm\ or\ 3\tfrac{1}{3}\ feet$$

The values of Γ for several of the more commonly used radionuclides are shown in Table 23.6. In actual practice, distance alone is usually impracticable, so shielding is needed to provide adequate protection (see below).

2. *Shielding.* Protective materials placed between a radioactive source and its surroundings to reduce the exposure rate are

TABLE 23.6

SPECIFIC GAMMA-RAY CONSTANTS (Γ) OF SOME RADIONUCLIDES*

Nuclide	Γ
	R/mCi-hr at 1 cm
^{198}Au	2.32
^{60}Co	13.0
^{131}I	2.23
^{137}Cs	3.2
^{226}Ra	8.25 (0.5 mm Pt capsule)

* Based on *ICRU Handbook 86* (experimental values).

called *shields* or *barriers.* Proper shielding contributes materially to the reduction of *gamma-ray exposure rate*, especially when combined with distance. The shield may be incorporated in the container itself, or it may be placed as a barrier around the source in the storage or working area. *Lead* is the most satisfactory shielding material under ordinary conditions. Shielding is effective because of the absorption of radiation in matter. The fraction of radiation (of given quality) absorbed by a barrier depends on the *atomic number* and *thickness* of the barrier material. It is convenient, in health physics, to specify barrier thickness in HVL because one HVL of any particular material is equivalent in shielding ability to one HVL of any other material. The HVL, it may be recalled, is that thickness of any material that reduces the exposure rate of radiation by one-half. The number of HVL's, *n*, required to decrease the exposure rate from a gamma-ray source with an activity A mCi to the permissible level at a given distance *d* cm from the source is given by the following equation:

$$2^n = \frac{480 A \Gamma}{d^2} \qquad (4)$$

For example, what is the thickness of lead barrier in HVL needed to reduce the gamma radiation from 100 mCi of ^{131}I to the permissible level at a distance of 2 meters (200 cm)? Substituting in equation (4),

$$2^n = \frac{480 \times 100 \times 2.23}{200 \times 200}$$

$$2^n = 2.67$$

$$n \log 2 = \log 2.67$$

$$n = \frac{\log 2.67}{\log 2} = \frac{0.43}{0.30} = 1\tfrac{1}{3} \text{ HVL}$$

Since the HVL of the ^{131}I gamma radiation is 0.3 cm lead,

$$1\tfrac{1}{3} \text{ HVL} = 1\tfrac{1}{3} \times 0.3 = 0.4 \text{ cm lead}$$

In actual practice in the average Nuclear Medicine Department, a gamma-ray emitting nuclide can be stored temporarily in its shipping container. For more permanent storage, it should be shielded with lead or steel bricks, 2 inches in thickness, on all sides that might constitute radiation hazard. A survey meter should be used routinely to check the adequacy of the barrier.

Shielding of *beta radiation* differs from that of gamma rays. Since the range of beta particles in soft tissues is usually limited to several mm, the external hazard can be readily controlled. With 'beta radiation having a HVL of less than 1 mm of skin, the weekly MPD is 1.5 rads. For higher energy than this, the MPD is less, 0.5 rad per week. In either case, however, the lens of the eye must absorb no more than 5 rads per year. It is well known that the range of a beta particle depends on its energy. Weak or low energy beta particles (that is, less than 0.3 MeV) are absorbed by the container, by the cornified layer of the skin, or by several feet of air. On the other hand, the medical radionuclides ^{32}P (maximum beta energy 1.7 MeV) and ^{90}Sr-^{90}Y (maximum beta energy 2.18 MeV) require special shielding to protect against brems radiation produced when the beta particles interact with atoms. Plastic materials such as Lucite or polystyrene are preferred because they are light-transparent and because their low atomic number makes them efficient absorbers of brems radiation. About ¼ in. plastic suffices for ^{32}P, and about ⅜ in. for ^{90}Sr. Since air contributes significantly to the protective barrier, the handling of these beta emitters, especially when sealed, poses no difficult problem if a sufficient thickness and area of plastic shielding are used.

3. *Time of Exposure.* Since the total exposure is equal to the exposure rate times the time, it is obvious that the faster a given

procedure is carried out, the smaller is the exposure incurred by the operator. Each different procedure should first be practiced with blank (nonradioactive) materials until adequate speed has been attained. Only then should the procedure be conducted with radioactive material.

4. *Limitation of Activity of Stored Radionuclides.* The exposure rate is directly proportional to the activity of the source. Furthermore, the exposure rates from all sources must be added to obtain the total exposure rate. Hence, the quantities stored should be kept at a minimum consistent with the requirements of the department.

Radiation Monitoring. Besides requiring that all personnel wear film badges (see page 463), every Nuclear Medicine Department should conduct periodic surveys to determine the exposure rates from stored radionuclides (radium as well as artificial radionuclides); from possibly contaminated work areas and waste receptacles, especially when liquid preparations are used; and from possibly contaminated clothing, shoes, instruments, and hands. Such surveys are advisable at least weekly, and even more often when deemed necessary. Whenever patients receive radionuclides in quantities other than tracer doses, clothing, sputum cups, bedpans, urinals, and other utensils that may have become contaminated must be surveyed before they can be declared "safe." All contaminated articles must be stored in special bins at a safe distance until their activity has decreased to a value a few times that of background.

Various types of survey instruments are available. The two most popular ones are the *survey meter* (a portable battery-operated rate meter with GM counter) and the *cutie pie* (see Figures 23.4 and 23.5). These instruments read directly in mR/hr (milliroentgens per hour). Besides, the survey meter indicates counts per minute and is usually provided with earphones for audible detection of counts. A survey meter operating continuously as a GM counter should be placed at the exit of any laboratory where high level activity prevails, so that a constant check can be maintained on personnel leaving the laboratory.

Special Precautions—Nursing Procedures. Complete details for every possible radionuclide procedure are beyond the intent

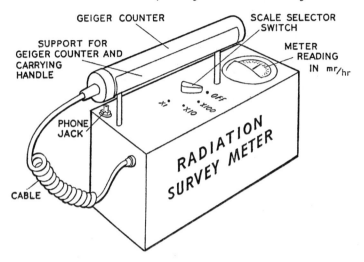

Figure 23.4. Battery-operated portable survey meter with GM counter. A phone jack is included for audible detection. (Based on Nuclear-Chicago Model.)

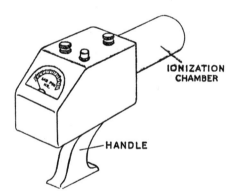

Figure 23.5. Battery-operated radiation meter known popularly as the "cutie pie." Ionization chamber type. (Based on Tracerlab®, Inc., Model.)

of this book. Reference should be made to publications of the Atomic Energy Commission, or the appropriate State Agency or local Health Department. However, certain principles are exemplified by the more commonly used radionuclides. In any event, children and pregnant women should be kept out of the room of any patient receiving internal gamma-ray therapy.

1. *Waste Disposal.* All radioactive wastes must be disposed of according to the regulations of the State Health Department, especially where high activity is concerned. With ^{131}I in quantities used for tracer studies and in treatment of hyperthyrodism, and with ^{32}P used in diagnosis and treatment, no special precautions are necessary for waste disposal. But with ^{131}I used in the treatment of thyroid cancer and heart disease there is such high activity in the urine that contaminated linens must be monitored by the radiation safety officer. However, patients receiving radionuclides internally may freely use the regular toilet facilities regardless of the dose. Hospitalization is required only if the dose of ^{131}I exceeds 30 mCi.

2. *Nursing Procedures.* Hazard to nursing personnel can be minimized (see NCRP Report No. 37 for details):

a. *^{131}I in Hyperthyroidism.* During the first three days, rubber gloves should be worn in handling emesis basins, bedpans, and urinals; linens should be handled in the same way if the patient is incontinent. The gloves should be washed for two minutes with soap and running water before being removed from the hands, and then set aside for surveying. If the patient is ambulatory, he may use the bathroom.

b. *^{131}I in Thyroid Cancer or Heart Disease.* There is very high activity in the urine and sweat, but the external hazard is small (actually, much less than in radium therapy). The nurse should wear rubber gloves as above. No baths are to be given during the first forty-eight hours. Linens should be surveyed. The patient should be surveyed and safe distances and exposure times posted. Children and pregnant women must not be allowed to visit. The patient may have bathroom privileges, but contaminated linens require surveying by the radiation safety officer, and disposal according to regulations.

c. *^{32}P in Therapy.* No precautions are necessary other than rubber gloves if the patient is incontinent or vomits within the first 24 hours. A responsible member of the Nuclear Medicine Department should be notified of such occurrence.

d. *^{198}Au, Ra, and ^{60}Co (internal) Therapy.* Radioactive gold in pleural or peritoneal effusions requires monitoring the patient

TABLE 23.7

SAFE WORKING DISTANCES AND DAILY EXPOSURE TIMES
PERMISSIBLE FOR PERSONNEL ATTENDING PATIENTS
RECEIVING VARIOUS DOSES OF RADIONUCLIDES.
BASED ON MPD OF 0.1 RAD PER 40-HR. WEEK*

^{131}I or ^{198}Au (*first 2 days*)	*Ra or Rn*	*Maximum Safe Daily Exposure Times in min at Indicated Distances*		
		1 ft	*2 ft*	*6 ft*
mCi	mCi			
25	—	20	90	800
50	—	10	45	400
150	50	3	15	140
200	—	2½	12	100
300	100	1½	7½	70

* Adapted from *NCRP Report No. 37*. Maximum exposure times with ^{60}Co (internal therapy, such as needles or wires) are about 3/5 of those for Ra or Rn.

and posting safe distances and exposure times. Rubber gloves should be used during the first 48 hours. Dressings from contaminated wounds must be monitored.

The safe distances and daily exposure times for personnel attending patients with nuclides of various activities are shown in Table 23.7. These distances and times should be posted in a conspicuous place. Children and pregnant women are not allowed to visit patients undergoing internal therapy with gamma-emitting radionuclides. Radium precautions are stricter because of the higher energy of the gamma rays of radium than of ^{131}I or ^{198}Au.

The following list of unsafe practices has been compiled by the Atomic Energy Commission. It provides a useful check list of the safety of one's own nuclear medicine department.

1. Inadequate planning—excessive time required for carrying out procedures.

2. Improper monitoring.

3. Inadequate shielding.

4. Inadequate use of trays and paper covering of work areas when radioactive liquids are used.

5. Pipetting solutions by mouth instead of remote apparatus.

6. Poor work habits, such as smoking or eating in the laboratory.

7. Inadequate use of proper outer garments, such as rubber gloves and laboratory coats, when handling high level radioactive liquids.

8. Improper waste disposal.

9. Improperly designed fume hoods in high-level laboratories.

10. Changing levels of activity without modifying procedures to maintain low hazard.

11. Failure to keep adequate radiation safety records of radionuclides received, waste disposal, and personnel exposure.

12. Failure to post special signs to mark storage and disposal areas.

RADIATION HYGIENE

Assuming that all of the above safety factors have been followed and the technologist has been most conscientious in observing the rules, can we be certain that hazards have been completely removed? As far as we know, there have been no harmful effects from exposures approximating the MPD. Periodic blood counts, although of value in detecting early radiation injury, are a poor substitute for preventive monitoring of radiation by means of film badges or other devices. Film badges are readily available through commercial radiation laboratories. If the permissible dose is exceeded, the situation must be corrected without delay.

Upon accepting a new job, radiologic technologists should have a complete blood count and physical examination. In addition, radium workers should have their hands inspected by a radiologist or dermatologist at frequent intervals.

Certain hygenic policies should be established. The technologist should have at least two, and if possible, four weeks' vacation each year, with at least two weeks consecutive. Vacation should preferably be just that, and away from x and gamma rays. The working schedule should conform to an average 8-hour day with at least one full day off each week. Radiologic rooms should be well-lighted and well-ventilated.

It is reasonably certain that if adequate protective measures are instituted, if the technologist is always careful to make full use of the available protective equipment and takes proper personal precautions against careless exposure, and if proper general hygiene is practiced, the danger from these radiations is almost nonexistent. This is attested to by the fact that it is the rare exception for a careful technologist to incur radiation injury.

QUESTIONS AND PROBLEMS

1. What is background radiation due to? What is the average individual background exposure per year from natural and from artificial background?
2. What is meant by maximum permissible dose equivalent? What is the whole-body MPD for an 8-hour day? For one week? For one year?
3. Describe two methods by which one may determine the amount of whole-body radiation he is receiving in his radiology department.
4. Discuss the effects of excessive whole-body exposure to penetrating radiations. What are the locally damaging effects of overexposure to ionizing radiation?
5. What materials are widely used in protective barriers? What material (and thickness) is required for the walls of ordinary radiographic rooms? For 250-kV x rays? For cobalt 60 gamma rays?
6. How can one avoid the local deleterious effects of x rays? Of radium and artificial radionuclides.
7. Define dose equivalent; quality factor. What is the unit of *DE*? Of what importance are these concepts?
8. Discuss minimum exposure radiography in (1) fluoroscopy and (2) radiography, including the important factors in each.
9. Assume that the conditions of your x-ray department are safe, from the standpoint of penetrating radiations. State three rules of conduct on your part that will help eliminate the hazard of exceeding the permissible dose.

10. What are the best means of protection from radium gamma rays?
11. Name and describe the four main protective measures with radionuclides.
12. Find the gamma-ray danger range from 100 mCi of ^{131}I. Suppose this source is to be stored at a distance of 1 meter from an occupied area; what thickness of lead would be required to reduce the exposure to the MPD?
13. Why should x-ray equipment be grounded? Does an ordinary steam-heating pipe afford a satisfactory ground connection? A gas pipe? A water pipe?
14. What is the one-hand rule? Why should it always be observed?
15. Discuss the basic precautions of nursing personnel in attending patients undergoing internal radium and artificial radionuclide therapy. Why should children and pregnant women not be permitted to visit patients undergoing internal therapy with gamma-emitting radionuclides?

APPENDIX 1   *ANSWERS TO PROBLEMS*

Chapter 1.

  1. (a) $\frac{1}{2}$ (b) $\frac{3}{4}$ (c) $\frac{2}{3}$ (d) $\frac{4}{5}$

  2. (a) $1\frac{4}{5}$ (b) $1^{71}\!/_{84}$ (c) $1^{29}\!/_{30}$

  3. 25 bu

  4. (a) $1\frac{1}{3}$ (b) $1\frac{1}{63}$ (c) $1^{13}\!/_{27}$

  5. b $=$ ac   c $=$ b/a

  6. (a) 8   (b) -3 (c) $2\frac{1}{3}$ (d) $4\frac{4}{9}$ (e) 44

  7. (a) $\frac{2}{3}$ (b) 5   (c) $10\frac{1}{2}$ (d) 15

  8. 20 cm

  9. tripled

10. 12.56 in

11. 600 sq ft

12. about $3\frac{3}{4}$ min

13. 3,600,000

14. 4.24×10^5

15. 1.6×10^6

Chapter 2.

11. 20 C

12. 104 F

Chapter 4.

  7. 12

  8. At. no. 6

      Mass no. 8

Chapter 6.

  8. 2 amps

9. 150 volts
11. 1.8 Ω; 4, 2, and ⅔ amps in branches; 6⅔ amps in main lines.
13. 2200 watts = 2.2 kw

Chapter 10.

3. 110,000 volts = 110 kV
10. 100/230 or 10/23

Chapter 13.

9. ⅒ sec

Chapter 16.

9. 10 gal.

Chapter 18.

3. Image diameter 10 in. Percentage magnification 11%. Magnification factor 1.11
10. 80 R/min
11. 11 mAs
12. 180 mA

Chapter 19.

14. 16 in.
16. 8 in. × 10 in.

Chapter 22.

9. Total count 1060, background count 130. Total counting time 5.3 min; background counting time 2.6 min.
10. 435 cps, corrected, with GM counter. No correction needed for scintillation counter.

Chapter 23.

12. 10.8 ft; 1.0 cm lead.

BIBLIOGRAPHY

American College of Radiology: A practical manual on the medical and dental use of x rays with control of radiation hazards.

Ardran, G. M., and Crooks, H. E.: Gonad radiation dose from diagnostic procedures. *Brit. J. Radiol.*, 30:295-297, 1957.

Batson, Oscar V., and Carpentier, Virgina E.: Stereoscopic depth perception. *Am. J. Roentgenol.*, 51:202-204, 1944.

Bloom, W. L., Jr., Hollenbach, J. L., and Morgan, J. A.: *Medical Radiographic Technic.* 3rd ed. Springfield, Thomas, 1969.

Chamberlain, W. E.: Fluoroscopes and fluoroscopy. *Radiol.*, 38:383-425, 1942.

Corrigan, K. E., and Hayden, H. S.: Diagnostic studies with radioactive isotope tracers. *Radiology*, 59:1, 1952.

Crabtree, J. I., and Henn, R. W.: Developer solutions for x-ray films. *Med. Radiog. & Photog.*, 23:2-12, 1947. (Published by Eastman Kodak Co., Rochester, N.Y.)

Eastman Kodak Company, Rochester, N.Y.:
   The Fundamentals of Radiography.
   Radiography in Modern Industry.
   Sensitometric Properties of X-ray Films.

Egan, R. L.: *Technologist Guide to Mammography.* Baltimore, Williams & Wilkins, 1968.

Field, T., and Seed, L.: *Clinical Use of Radioisotopes.* Chicago, The Year Book Publishers, Inc., 1961.

Fuchs, W. C.: *Principles of Radiographic Exposure and Processing.* Springfield, Thomas, 1969.

Handbooks, U.S. Department of Commerce, National Bureau of Standards, Superintendent of Documents, Washington, D.C.
   84. *Radiation Quantities and Units. International Commission on Radiologic Units and Measurements (ICRU Report 10a).*
   85. *Physical Aspects of Radiation (ICRU Report 10b).*
   86. *Radioactivity (ICRU Report 10c).*
   87. *Clinical Dosimetry (ICRU Report 10d).*

88. *Radiobiologic Dosimetry (ICRU Report 10e).*

89. *Methods of Evaluating Radiological Equipment and Materials (ICRU Report 10f).*

Hendee, W. R.: *Medical Radiation Physics.* Chicago, Year Book, 1970.

Humphries, R. E.: Personal communication.

Johns, H. E., and Cunningham, J. R.: *The Physics of Radiology,* 3rd ed. Springfield, Thomas, 1969.

Kieffer, J.: Analysis of laminographic motions and their values. *Radiology,* 33:560-585, 1939.

Littleton, J. T., Rumbaugh, C. L., and Winter, F. S.: Polydirectional body section roentgenography: A new diagnostic method. *Am. J. Roentgenol.,* 89:1179-1193, 1963.

McGann, M. J.: Plesiosectional tomography of the temporal bone. *Am. J. Roentgenol.,* 88:1183-1186, 1962.

National Council on Radiation Protection and Measurements (NCRP) Reports: NCRP, P.O. Box 4867, Washington, D.C., 20008.

24. *Protection Against Radiations from Sealed Gamma Sources (1960).*

30. *Safe Handling of Radioactive Materials (1964).*

33. *Medical X-ray and Gamma-ray Protection for Energies up to 10 MeV—Equipment Design and Use (1968).*

34. *Medical X-ray and Gamma-ray Protection for Energies up to 10 MeV—Structural Shielding Design and Evaluation (1970).*

37. *Precautions in the Management of Patients Who Have Received Therapeutic Amounts of Radionuclides (1970).*

39. *Basic Radiation Protection Criteria (1971).*

O'Mara, R. E., Ruzicka, F. F., Jr., Osborne, A., and Connell, J., Jr.: Xeromammography and film mammography: completion of a comparative study. *Radiology,* 88:1121-1126, 1967.

Paul, J. E., Razzak, M. A., and Sodee, D. B.: *Textbook of Nuclear Medicine Technology.* St. Louis, Mosby, 1969.

Quimby, E. H., Feitelberg, S., and Silver, S.: *Radioactive Isotopes in Clinical Practice.* Philadelphia, Lea & Febiger, 1968.

Ridgway, A., and Thumm, W.: *The Physics of Medical Radiography.* Reading, Mass., Addison Wesley, 1968.

Schulz, R. J.: *Primer of Radiation Protection.* New York, GAF Corporation, 1969.

Selman, J.: *The Basic Physics of Radiation Therapy.* Springfield, Thomas, 1960.

Ter-Pogossian, M.: *The Physical Aspects of Diagnostic Radiology.* New York, Hoeber, 1967.

Trout, E. D., and Gager, R. M.: Protective materials for field definition in radiation therapy. *Am. J. Roentgenol.*, 63:396-408, 1950.
Trout, E. D., Kelley, J. P., and Cathey, G. A.: The use of filters to control radiation exposure to the patient in diagnostic roentgenology. *Am. J. Roentgenol.*, 67:946-963, 1952.
United States Atomic Energy Commission (USAEC), P.O. Box 62, Oak Ridge, Tenn. 37830.
    Your Body and Radiation.
    The Genetic Effects of Radiation.
Wolfe, J. N.: Xerography of the breast. *Radiology*, 91:231-240, 1968.
————— Xerography of the bones, joints, and soft tissues. *Radiology*, 93:583-587, 1969.

INDEX

A

Absorbed dose, 183-184
  bone *vs.* soft tissue, 184
  depth dose, 190-192
  radium gamma rays, 404
Absorption, radiation
  alpha particles, 398, 405
  beta particles, 399, 411, 416
  bone, 184
  Compton, 177-179, 187
  filters, 192-195
  gamma rays, 405, 482-483
  pair production, 179-180, 197
  photoelectric, 175-177, 187
  scattering, 177-179
  true, 175-177
  x rays, 170-181
Accelerator, 282-283
Acceptor atom, 221
Acetic acid, 286
Actinium, 398
Active layer, intensifying screen, 260, 263-264, 267
Activity (radionuclide)
  minimum detectable, 435-436
  specific, 416
  units, 403, 416
Acutance, (radiographic sharpness) 299
Air core transformer, 123
Algebra, 8-13
Alnico, 84
Alpha particles, 398, 415
Alternating current (ac), 105-108
  advantages, 111-112
  curve, 106-107
  cycle, 107
  effective value, 108
  frequency, 107-108
  generator, 103-106

  measurement, 118-119
  peak value, 108
  power loss, 111-113
  production, 103-108
  properties, 108-110
  rectification, 136-147
  RMS value, 108
Alum (hardener), 286
Aluminum filter, 169, 184
  in radiographic protection, 476-477
Ammeter, 71-72, 116, 118-119
  ballistic mAs meter, 237-238
  filament, 236-237, 241
  milliammeter, 225, 237, 240, 241
Ammonium thiosulfate (fixer), 286
Ampere (unit), 68
Anesthetic explosion hazard, 479
Ångstrom (unit), 28
Angular distribution (heel effect), 354-356
Angulation, anode face
  radiographic tube, 204-205
  therapy tube, 214
Anion, 52
Annihilation reaction, 180
Anode, x-ray tube
  angulation, 203-205, 214
  atomic number, target, 202
  cooling, 213, 214
  double focus, 200-202
  field emission, 252-254
  focus (focal spot), 200, 202
  heat dissipation, 211-212
  heat production, 210-212
  heat storage, 210-211
  heat units, 210-211
  line focus principle, 203, 204
  molybdenum
    focusing cup, 199-200
    target, 204, 319